INTRODUCTION TO
SEMIMICRO QUALITATIVE ANALYSIS

Eighth Edition

J. J. Lagowski

C. H. Sorum

PEARSON

Prentice
Hall

Upper Saddle River, NJ 07458

Library of Congress Cataloging-in-Publication Data

Lagowski, J. J.
  Introduction to semimicro qualitative analysis.—8th ed. / J. J. Lagowski, C.H. Sorum.
    p. cm.
  Includes indexes.
  ISBN 0-13-046216-0
  1. Chemistry, Analytic—Qualitative. 2. Microchemistry. I. Sorum, C. H. (Clarence Harvey). II. Title.

QD98.M5S67 2004
543'.22—dc22

2004015738

Project Manager: Kristen Kaiser
Editor-in-Chief, Science: John Challice
Executive Marketing Manager: Steve Sartori
Vice President and Director of Production and Manufacturing, ESM: David W. Riccardi
Production Editor: Robert Saley
Manufacturing Manager: Trudy Pisciotti
Manufacturing Buyer: Lynda Castillo
Managing Editor Audio/Visual Assets: Patty Burns
AV Editor: Abby Bass
Art Director: Jayne Conte
Cover Designer: Bruce Kenselaar
Cover Image: Spike Walker / Stone / Getty Images, Inc.

© 2005, 1991, 1983, 1977, 1967, 1960, 1953, 1949 by Pearson Education, Inc.
Pearson Prentice Hall
Pearson Education, Inc.
Upper Saddle River, New Jersey 07458

Pearson Prentice Hall ® is a trademark of Pearson Education Inc.

Printed in the United States of America.

ISBN 0-13-046216-0

Pearson Education Ltd., *London*
Pearson Education Australia Pty., Limited, *Sydney*
Pearson Education *Singapore*, pte. Ltd
Pearson Education North Asia Ltd., *Hong Kong*
Pearson Education Canada, Ltd., *Toronto*
Pearson Educatión de Mexico, S.A. de C.V.
Pearson Education–Japan, *Tokyo*
Pearson Education Malaysia, Pte. Ltd.

Contents

Preface

The subject of qualitative analysis has its roots in the works of Boyle (1627–1691) who, some argue, laid the foundations for the subject by using flame colors, spot tests, "fumes," precipitates, specific gravity, and solvent action as analytical tools. The modern expression of qualitative analysis can be traced directly to Fresenius (1818–1897) whose thesis, published in 1842, was the basis for a successful student manual on qualitative analysis. This manual was revised repeatedly to keep the subject up to date as chemistry passed through a phase of uncertain notation and nomenclature and experienced the reforms instigated by Cannizzaro's use of Avogadro's hypothesis to clarify a number of apparently troublesome facts, and which clarified the "nature of the molecule." An American edition, translated by S. W. Johnson, appeared in 1869.

The current interest in the need for students to "learn some descriptive chemistry" can be satisfied by a course in qualitative analysis. Although "descriptive chemistry" may be difficult to define precisely, most chemists would agree that a major component of this subject is the chemistry of the more common elements and their compounds. From this point of view, the chemistry of elements and their compounds is probably best expressed as the chemical processes that occur in an aqueous environment—processes that involve considerations of metathetical, redox, and complexation reactions, namely, the directions in which these processes occur (thermodynamics) as well as how rapidly they occur (kinetics).

This kind of practical descriptive chemistry can be, and has been, discussed on the printed page, but we believe it is best learned by the student in a laboratory setting. We choose to use, as the basis of descriptive chemistry, the chemistry of the compounds of common metal cations formed with a relatively small number of anions. The focus of this kind of chemistry is qualitative analysis, a format that most students find engaging and informative. The basis for our analysis scheme is the relative solubilities of the chlorides, sulfides, and hydroxides of a representative selection of cations. Thus, we present a process for learning descriptive chemistry in the format of a scheme of analysis where students are challenged to bring to bear their manipulative and observational skills to provide the basis for identifying a substance or a mixture of substances. In our experience, most students enjoy the "hunt" for the unknown. We recognize the fact that wet methods of qualitative analysis are not usually used in modern chemical laboratories; the scheme of analysis presented here is designed as a method of learning descriptive chemistry, not necessarily of providing a practical modern scheme of analysis.

The interest in exposing our students early to the principles of "green chemistry"—using less chemicals and smaller volumes of solution in a benign solvent—is also met by providing them with an experience in the semimicro methods of qualitative analysis, which is the focus of the experimental processes described here.

The experimental procedures outlined here are in consonance with the current "less-than-macro" trends in laboratory instruction. All the benefits of this strategy—small quantities of chemicals involved, simpler equipment, fewer hazards—accrue from the use of the semimicro methods of qualitative analysis described.

The chapters in this edition are distributed into three parts. In Part I, we describe the strategy of qualitative analysis. We believe it is pedagogically beneficial for students to have a review of the principles readily available when they are engaged in the details of laboratory work. Accordingly, in Part II, we present the concepts involved in qualitative analysis, systematically dealing with the nature of chemical compounds, including ionic, covalent, and coordination compounds, and the chemical reactions they undergo, stressing solution processes and equilibria. The discussion in Part II is intended to support the practical aspects of qualitative analysis that appear in Part III. The well-tested analytical procedures of Part III from previous editions remain, for the most part, intact. Within each chapter of Part III is a general survey of the chemistry of the species in a qualitative analysis group of ions, followed by the experimental details for the ions in that group.

We have attempted to direct the reader's attention to certain important sections of Part III by using the following pictorial devices.

This symbol indicates a cautionary section, that is, a place in the directions where a safety problem could arise if the procedure is not followed carefully.

This symbol indicates a key idea, a place in the directions where a principle described in Part II is applied. The reader would benefit from recognizing the relationship of the principle to the action being described at the key symbol.

We express our appreciation to users of previous editions, colleagues, and students for suggestions, comments, and criticisms, which were helpful in preparing this revision. We would also like to thank the following reviewers for their suggestions: Gerhard Buchbauer of the University of Vienna, Philip W. Crawford of Southeast Missouri State University, Lisa C. Price of Bennett College, and Duanne E. Weisshaar of Augustana College. For invaluable assistance in the preparation of the manuscript for this edition, we express great appreciation to Rita D. Wilkinson. The editorial staff at Prentice Hall are gratefully acknowledged for their extensive contributions to the production of this volume.

J. J. L.
Austin, Texas

Note to Instructors

After a short overview of the strategy of qualitative analysis in Part I, we intend Part II, the theory of qualitative analysis, to be a review of the important principles of chemistry that pertain to the laboratory-oriented phenomena that are normally associated with qualitative analysis as described in Part III. Many of the questions at the end of the chapters in Part III serve to illustrate the principles developed in Part II. Conversely, the problems at the end of the chapters in Part II anticipate some of the more practical problems listed in Part III.

A list of the required equipment appears in Appendix III. We find it useful to most students to be introduced to each piece of equipment and special technique at the time it is first used in the plan of analysis described here. The equipment is of the standard semi-micro type available at most supply houses. Only a few items need to be prepared, but these can be made by the student. If H_2S gas is used as a precipitating reagent, the student will need to prepare several hydrogen sulfide bubbling tubes from 6-mm glass tubing, as shown in Fig. 10.2 and described in the accompanying text. If each student is to prepare H_2S by heating a mixture of sulfur, paraffin, and asbestos, simple generators of the type illustrated in Fig. 10.2 must be set up. Stirring rods will need to be cut from 3-mm glass rods and fire-polished.

A list of all reagents used with directions for preparing all solutions is also given in Appendix III.

The experimental procedures and specific tests have been checked numerous times; if the directions given are followed with care, good results will be obtained. We would appreciate being informed of errors, inconsistencies, or ambiguities in the procedures.

In general, net equations are used to describe reactions that occur in solution in conformity with the principle that equations should indicate the predominant species in the system. No effort has been made to present the detailed mechanisms of the more complex reactions.

PART

I

GENERAL
STRATEGY

The Strategy of Qualitative Analysis

All the world is made of chemical substances, most usually occurring as mixtures. Chemists make substances, either new—never been made—or known. Substances exhibit chemical and physical properties, and the *characterization* of substances involves establishing their properties, the grouping of which is unique to each substance. No two substances exhibit *all* of the same properties.

One of the most important tools in the characterization of substances is *analysis*. Qualitative analysis answers questions concerning the identity of the species that make up a substance. *Quantitative* analysis is concerned with the relative amounts of these species. So, in a sense, the results of *qualitative* analysis are necessary before *quantitative* analysis techniques can be applied. Two general methods of analysis exist.

1. *Instrumental methods* generally exploit a difference in physical properties of the species of interest; they are rapid, but often require specialized equipment.
2. *Chemical methods*, on the other hand, are generally applicable in ordinary laboratory settings, often using only test tubes and common reagents. Most importantly, chemical methods provide a vehicle for understanding the details of some common types of chemical reactions as well as exposing the student to some interesting and useful descriptive chemistry.

In general, it would be an immensely difficult task to attempt to identify a single substance in a mixture by its chemistry if there were no restrictions on the nature of the substances that constitute the mixture. Fortunately, compounds are divided naturally into two general types: molecular compounds and ionic compounds. The majority of the molecular compounds are carbon-containing and are the purview of organic chemistry, while the ionic compounds tend to be the derivatives of the metallic elements and fall into the realm of inorganic chemistry. This division provides a convenient basis for our current discussion of qualitative analysis. For the most part, the chemistry of inorganic compounds is the chemistry of ions, especially if the inorganic compounds are dissolved in aqueous solutions, with the important exceptions in the field known as *organometallic chemistry*. Metals tend to form cations, and nonmetals tend to form anions. Our studies here are focused on inorganic qualitative analysis, which is concerned with determining which cations and anions are present in pure substances and in mixtures of substances.

A complementary system of qualitative analysis, which is often the subject of a separate course of instruction, is available for organic compounds.

A complete system of qualitative inorganic analysis would include methods for detecting *all* cations and anions of all of the elements. Such procedures would have to include cations not only of the common metals—such as zinc, copper, and tin—but also of the less common metals—such as platinum, gold, rhenium, and einsteinium. Correspondingly, the less common anions—such as vanadate, tellurate, and molybdate—would have to be included with the more common chloride, sulfate, and nitrate.

Here we limit the scope of the qualitative analysis of inorganic compounds by focusing on *common* metal species. By "common" we mean the species that would normally be found in the typical laboratory or in most of the world in which we live. By limiting our study of qualitative analysis to inorganic compounds, we have decreased the scope of our problem, but it is still complex and challenging. Since we are primarily interested in presenting an understanding of the *methods* of qualitative analysis and the *principles* that underlie these methods, we concentrate our attention on only a few common and representative cations and anions. The ions we consider are those formed by the common metals, except for the ammonium ion (NH_4^+), which has properties similar to those of the alkali metal cations (see Table 1.1). The anions to which we limit our attention are, for the most part, those derived from common acids and/or the simple oxy anions.

The basis for identification of substances lies in their chemical and physical properties. The simplest properties used for identification are those that can be directly observed by the experimenter. Thus, the color of a substance and the results of its reaction with various reagents are commonly used for identification purposes. If we limit ourselves to reagents that produce *observable* effects, we must then form colored substances (in solution or as precipitates), form gaseous products that can be visually seen to leave the solution, produce a characteristic odor, or cause a previously insoluble substance to dissolve. In other words, a chemical change involves a change in the number of phases present or a change in color. Many other important reactions, such as the neutralization of an acid with a base, do not usually produce visual indications of occurring and are not useful in qualitative analysis.

TABLE 1.1 IONIC SPECIES OF THE QUALITATIVE ANALYSIS SCHEME

Silver Group	Copper-Arsenic Group	Aluminum-Nickel Group	Barium-Magnesium Group
Cations			
Ag^+	Hg^{2+}	Al^{3+}	Ba^{2+}
Pb^{2+}	Pb^{2+}	Cr^3	Ca^{2+}
Hg_2^{2+}	Bi^{3+}	Fe^{3+}	Mg^{2+}
	Cu^{2+}	Mn^{2+}	Na^+
	Cd^{2+}	Co^{2+}	K^+
	As^{3+}	Ni^{2+}	NH_4^+
	Sn^{4+}		
Anions			
Cl^-	S^{2-}	CrO_4^{2-} (chromate)	
Br^-	SO_4^{2-} (sulfate)	PO_4^{3-} (phosphate)	
I^-	SO_3^{2-} (sulfite)	AsO_4^{3-} (arsenate)	
NO_3^-	CO_3^{2-} (carbonate)		
$C_2H_3O_2^-$ (acetate)	BO_3^{2-} (borate)		

A single common example will suffice to illustrate these points. A solution can be identified as containing Cl^- ions by treating it with a solution of $AgNO_3$, with which it produces a white precipitate of silver chloride (AgCl).

$$Ag^+ + Cl^- \rightarrow AgCl(s) \text{ [white solid]} \tag{1.1}$$

In this case, the sample, which was suspected to contain Cl^-, was caused to react with a reagent that produces a visual effect (a white precipitate). The supposition that AgCl and not some other unknown, insoluble substance was formed can be verified by considering the properties of AgCl; this substance is insoluble in aqueous HNO_3, which is a factual piece of chemical information. Thus, if we isolate the white solid and treat it with aqueous HNO_3 and it fails to dissolve, we can conclude that the original solution contained Cl^-, a conclusion based on two chemical observations: first, that a white precipitate was formed and, second, that this precipitate is insoluble in nitric acid.

From the preceding example, it might appear that the process of qualitative analysis is simply one of observing and matching properties. The example, however, does not represent the usual situation, which invariably involves a mixture of several substances rather than just one pure substance. Under these circumstances, reagents used to identify a substance should possess, ideally, a specific reaction with each compound of the mixture. Even a cursory knowledge of chemistry suggests it is highly improbable that a *single reagent* exists that would, for example, distinguish components in a mixture of four substances. The reagent might react with one to provide a distinguishing product, but not with the other three. How are we to establish the presence of any one of the other three, which, remember, are now mixed with the excess unreacted reagent? In other words, we now have a new mixture that possibly contains four substances: three of the original substances and the known reagent. Unless something else is done, we could continue forever adding other substances (as possible reagents to effect identification) to the original unknown mixture, forming increasingly complex mixtures. The most efficient way out of this dilemma is to arrange the chemistry so that the original unknown mixture is separated into several parts, each containing one or more of the unknown substances, and, then, to proceed to the analysis of the separated parts. In general, qualitative analysis involves a series of separations and identifications.

To illustrate the logic and the methods of qualitative analysis, let us consider a solution containing some of the cations and anions listed in Table 1.1 as an example. If we do not know how many cations and anions are present, which is often the case, the only safe position is to assume that the solution contains *all* of these ions. With this assumption, we are forced to perform experiments and make observations that will help us decide on the presence or absence of *every* ion. In the real world of qualitative analysis, we hardly ever have an unknown about which we know nothing. Often the unknown sample is derived from a system that provides some clues concerning the possible species present. For example, the analysis of a sample of natural waters—a lake or stream—is not likely to contain cations that form insoluble sulfides or hydroxides. Thus, if the water comes from a "sulfur spring," it probably contains sulfide anions, and will probably not contain cations that form *insoluble* sulfides. Thus, knowledge of the solubilities of metal sulfides will help eliminate certain cations from our consideration of the problem.

One approach would be to search for a reagent that would give a characteristic reaction with each ion of the unknown. Remember that we can identify an ion because it forms a product distinguishable from another ion, and that the distinction is usually a difference in color. Even the most optimistic student cannot hope to find a reagent that would give different colored products with the 24 cations and/or 13 anions in Table 1.1. Consider what would happen if such a reagent were available. It would be useful if the unknown contained only one cation: a mixture of cations would give a confusing mixture of colored products.

A second possible strategy would be to dissolve the unknown mixture in water and to find reagents that are specific for (i.e., react with) only one cation or anion. If such a reagent existed and the reaction yielded a precipitate, we could *filter* the reaction mixture and then test the filtrate with another reagent. Thus, from a series of precipitations and separations, we could obtain the necessary information to determine the presence or absence of each ion. Unfortunately, such an ideal set of reagents does not exist for the collection of ions we have chosen as our example in Table 1.1. With only a few exceptions, a reagent that gives a characteristic reaction with one ion gives a characteristic reaction with other ions in the group as well, or else its characteristic reaction with one ion is interfered with by other ions. That is, a given ion must be generally free of the presence of other (interfering) ions before it may be identified.

At this point, it should be obvious that the identification of the ions in our mixture requires that a *separation* be made. It is possible to develop a system in which the ions could be separated and identified one at a time. The more practical method, employed in all systems of qualitative analysis, involves the separation of cations into relatively small groups, which are then separated into smaller and smaller groups until only one cation remains in solution. The separated cations can then be easily identified by their characteristic chemical reactions.

Let us take a simple example to illustrate the kind of chemistry and logic involved in our discussion thus far. Imagine a colorless aqueous solution that might contain Ag^+ and Cd^{2+}. The analytical question is, which cations are present? There are four obvious answers to this question:

1. Neither Ag^+ nor Cd^{2+} is present.
2. Only Ag^+ is present.
3. Only Cd^{2+} is present.
4. Both Ag^+ and Cd^{2+} are present.

The safest strategy to follow is to assume the worst case, that the solution contains both cations. By making this assumption, you are forced to address the presence or absence of both species, a situation that includes the other possibilities.

Before you can attack the problem, you need to know some of the chemistry of the species involved. Table 1.2 is a summary of the aqueous solution chemistry of these two cations. The

TABLE 1.2 THE PROPERTIES OF SOME COMPOUNDS OF Ag^+ AND Cd^{2+}

Cation \ Anion	Cl^-	S^{2-}	OH^-
Ag^+	White ppt ($AgCl$)	Black ppt (Ag_2S)	Brown ppt (Ag_2O)
Cd^{2+}	Soluble ($CdCl_2$)	Yellow ppt (CdS)	Sparingly soluble White ppt ($Cd(OH)_2$)

table lists the properties of the species formed when aqueous solutions of the cations are mixed with the anions indicated. Thus, mixing a solution of a soluble sulfide (S^{2-}) with a solution containing Ag^+ produces a black precipitate of silver sulfide (Ag_2S).

$$2Ag^+ + S^{2-} \rightarrow Ag_2S(s) \text{ [black]} \tag{1.2}$$

A yellow precipitate, CdS, forms when a solution of a soluble sulfide is mixed with a solution containing cadmium ion.

$$Cd^{2+} + S^{2-} \rightarrow CdS(s) \text{ [yellow]} \tag{1.3}$$

The data in Table 1.2 suggest at least one way to answer the original question: Which cations are present in the solution? A key to the solution of the problem lies in the observations that the chloride ion does *not* form a precipitate with Cd^{2+} and that the other two reagents (S^{2-} and OH^-) form insoluble substances with both relevant cations. This difference in the properties forms the basis for the identification of both cations.

Figure 1.1 is a flow diagram of the sequence of reactions and the logic involved in one possible solution to the problem. This solution involves the use of Cl^- and then S^{2-}. Reversing the order will give no useful information because, if the unknown contains both Cd^{2+} and Ag^+, the addition of S^{2-} will precipitate both Ag_2S and CdS. Since Ag_2S is black, its presence will tend to mask any yellow CdS that may also be present. The order of addition indicated in Figure 1.1 permits the separation of Ag^+ (as AgCl) from Cd^{2+} as well as the identification of Ag^+ (white AgCl). Notice that the flow diagram indicates that the *solution* from the precipitation reaction—not the reaction mixture or the precipitate—is then tested with S^{2-}. Since the solution contains no Ag^+, black Ag_2S will not form. If a yellow precipitate forms, Cd^{2+} must have been present. A similar strategy using Cl^- and OH^- as reagents will also answer the question posed—Which species are present, Ag^+ or Cd^{2+}?

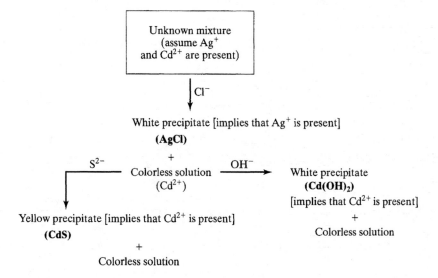

FIGURE 1.1   A flow diagram describing the analysis of a solution containing a mixture of Ag^+ and Cd^{2+}.

The example given in Figure 1.1 illustrates a number of important ideas in qualitative analysis.

- Colors of precipitates are important for the identification of species.
- The formation of precipitates also provides the basis for separating species.
- The order of use of reagents is often critical in obtaining reliable observations on which sound deductions can be based.

If we concentrate only on the cations listed in Table 1.1, a method of separation can be readily described on the basis of the known chemistry of these ions. The cations are separated into four *groups* by the addition of three specific *group reagents*. The unknown mixture is dissolved in water so that we work with an aqueous solution. The addition of the group reagent causes the cations of that group to precipitate; the remaining ions remain in solution, so that a separation of that group can be made. The outline of the group precipitation scheme is shown in Figure 1.2 in the form of a flow diagram. You should study this flow diagram carefully because it presents the essence of the whole plan of group separation of the cations.

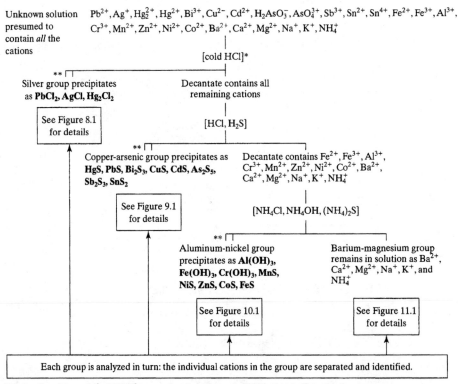

Unknown solution presumed to contain *all* the cations $Pb^{2+}, Ag^+, Hg_2^{2+}, Hg^{2+}, Bi^{3+}, Cu^{2-}, Cd^{2+}, H_2AsO_3^-, AsO_4^{3+}, Sb^{3+}, Sn^{2+}, Sn^{4+}, Fe^{2+}, Fe^{3+}, Al^{3+}, Cr^{3+}, Mn^{2+}, Zn^{2+}, Ni^{2+}, Co^{2+}, Ba^{2+}, Ca^{2+}, Mg^{2+}, Na^+, K^+, NH_4^+$

[cold HCl]*

Silver group precipitates as **$PbCl_2$, AgCl, Hg_2Cl_2**

See Figure 8.1 for details

Decantate contains all remaining cations

[HCl, H_2S]

Copper-arsenic group precipitates as **HgS, PbS, Bi_2S_3, CuS, CdS, As_2S_5, Sb_2S_3, SnS_2**

See Figure 9.1 for details

Decantate contains $Fe^{2+}, Fe^{3+}, Al^{3+}, Cr^{3+}, Mn^{2+}, Zn^{2+}, Ni^{2+}, Co^{2+}, Ba^{2+}, Ca^{2+}, Mg^{2+}, Na^+, K^+, NH_4^+$

[NH_4Cl, NH_4OH, $(NH_4)_2S$]

Aluminum-nickel group precipitates as **$Al(OH)_3$, $Fe(OH)_3$, $Cr(OH)_3$, MnS, NiS, ZnS, CoS, FeS**

See Figure 10.1 for details

Barium-magnesium group remains in solution as Ba^{2+}, $Ca^{2+}, Mg^{2+}, Na^+, K^+$, and NH_4^+

See Figure 11.1 for details

Each group is analyzed in turn: the individual cations in the group are separated and identified.

* A bracketed formula [cold HCl] means that the substance (HCl in this case) is added as a reagent.
** The notation ⌐⌐ means separation of the precipitate from the solution by centrifuging and decanting. The double vertical parallel lines at the left refer to the precipitate and one vertical line at the right refers to the decantate (the remaining solution).

FIGURE 1.2 The group separation scheme.

Note that the use of the term "group" has nothing to do with the location of the species in the periodic table.

- *The Silver Group.* Dilute hydrochloric acid is the group reagent (precipitant) for the first group (called "the silver group") that is separated. Separation of the cations of the silver group from all the other cations depends on the fact that the chlorides of silver (AgCl), lead (PbCl$_2$), and mercury (I) (Hg$_2$Cl$_2$) are insoluble in dilute HCl, whereas the chlorides of all the other cations in the scheme are soluble.
- *The Copper-Arsenic Group.* The group reagent for the second group (called "the copper-arsenic group") is H$_2$S. Separation of the cation of the copper-arsenic group depends on the fact that their sulfides are insoluble in dilute HCl, whereas the sulfides of the remaining cations are soluble.
- *The Aluminum-Nickel Group.* The group reagent for the third group (called "the aluminum-nickel group") is a mixture of NH$_4$OH and (NH$_4$)$_2$S, which precipitates a mixture of hydroxides and sulfides that are insoluble in alkaline solution.
- *The Barium-Magnesium Group.* The ions of the fourth group (called "the barium-magnesium group") form soluble chlorides, sulfides, and hydroxides and remain in solution after all the ions of the first three groups have been precipitated. There is no group reagent for the barium-magnesium group.

The identification of the ions present in each group involves a consideration of the chemistry of the ions in the group. These details are discussed in the appropriate parts of Chapters 8–14 and are summarized in flow diagrams in Figures 8.1 (Silver Group), 9.1 (Copper-Arsenic Group), 10.1 (Aluminum-Nickel Group), and 11.1 (Barium-Magnesium Group).

SUMMARY

It should be apparent that the separation and identification of the cations follows an orderly plan that, as illustrated by the flow diagrams in Figures 8.1, 9.1, 10.1, and 11.1, will require careful and *systematic* laboratory work. An understanding of the details of the plan requires knowledge of a large number of principles and facts. Rather than having you attempt to memorize the entire scheme, the following program of work is suggested to ensure that these facts and principles are used repeatedly and, thus, become second nature to you. A practice (or "known") solution containing all the cations of a particular group should be analyzed first. Then an unknown solution (or solid) for that group should be analyzed, using the same steps taken with the known. The series of group knowns and unknowns for the four groups may be followed by an unknown that contains a combination of all four cation groups (a general unknown) as a soluble mixture of cations or in the form of an alloy that must be dissolved. By engaging in this overall process, you will progress through all the possible cations (Table 1.1) in solutions of increasing complexity. You will also observe the colors and nature of precipitates—fine, dense, etc.—making them an important part of your knowledge base of descriptive chemistry.

Our study of qualitative analysis involves the separation of ions into groups with common properties and the identification of the ions within a group, all processes occurring in aqueous solutions. Thus, it is useful to connect the descriptive chemistry that will engage you in qualitative analysis to key ideas that you have probably studied in previous chemistry courses.

PART

II

CONCEPTS IN QUALITATIVE ANALYSIS

The Nature of Chemical Reactions

2

Chemical reactions are a part of the core ideas in the study of qualitative analysis. Although a myriad of chemical reactions occur, there are only a relatively few *types* of reactions. Over the years, chemists have sorted all the chemical reactions into the following few types:

1. Combination reactions
2. Decomposition reactions
3. Metathesis reactions
4. Displacement reactions
5. Oxidation–reduction (redox) reactions

The first three kinds of reactions (1, 2, and 3) are fundamentally different from the other two (4 and 5). A short discussion of the basis for this classification scheme now follows.

Combination Reactions. These reactions are characterized by the formation of more complex substances (compounds) from two or more simpler substances, which may be elements or compounds. Examples of combination reactions follow:

$$4Na(s) \ + \ O_2(s) \ \rightarrow \ 2Na_2O(s)$$
$$\text{element} \qquad \text{element} \qquad \text{compound}$$
(2.1)

$$2SnO(s) \ + \ O_2(s) \ \rightarrow \ 2SnO_2(s)$$
$$\text{compound} \qquad \text{element} \qquad \text{compound}$$
(2.2)

$$CaO(s) \ + \ CO_2(s) \ \rightarrow \ CaCO_3(s)$$
$$\text{compound} \qquad \text{compound} \qquad \text{compound}$$
(2.3)

Combination reactions are rarely involved in schemes of qualitative analysis.

Decomposition Reactions. These reactions are characterized by the conversion of more complex species to simpler species, which are often elements, but can also be compounds.

Decomposition reactions often require high-energy conditions. Examples of decomposition reactions follow:

$$2H_2O(l) \rightarrow 2H_2(g) + O_2(g) \tag{2.4}$$
compound   element   element

$$2H_2O_2(l) \rightarrow 2H_2O(l) + O_2(g) \tag{2.5}$$
compound   compound   element

$$Mg(OH)_2(s) \rightarrow MgO(s) + H_2O(s) \tag{2.6}$$
compound   compound   compound

Decomposition reactions are sometimes used in qualitative analysis schemes.

Metathesis Reactions.   These reactions are characterized by a "switching" of partner species, exemplified by the general reaction

$$AB + XY \rightarrow AY + XB \tag{2.7}$$

Specific examples include:

$$AgNO_3(aq) + NaCl(aq) \rightarrow AgCl(s) + NaNO_3(aq) \tag{2.8}$$

$$HCl(aq) + NaOH(aq) \rightarrow HOH(H_2O) + NaCl(aq) \tag{2.9}$$

$$MnS(s) + 2HCl(aq) \rightarrow MnCl_2(aq) + H_2S(g) \tag{2.10}$$

Metathesis reactions are very commonly found in qualitative analysis schemes. These kinds of reactions are characterized by the formation of precipitates (Eq. 2.8) that have characteristic colors and are often useful in identification of species as well as separation processes. Acid–base reactions (Eq. 2.9) are also quite often found in qualitative analysis schemes.

Displacement Reactions.   These reactions are characterized by the replacement of an element in a compound by another element. Examples of displacement reactions include:

$$Zn(s) + CuSO_4(aq) \rightarrow ZnSO_4(aq) + Cu(s) \tag{2.11}$$

$$2K(s) + 2H_2O(l) \rightarrow 2KOH(aq) + H_2(g) \tag{2.12}$$

$$Cl_2(g) + 2NaBr(aq) \rightarrow 2NaCl(aq) + Br_2(l) \tag{2.13}$$

Oxidation–Reduction Reactions.   These reactions, often abbreviated as "redox reactions," are characterized by changes in the oxidation states of chemically bound species (see Appendix I for a brief review of oxidation numbers). Redox chemistry is often found in qualitative analysis schemes. A typical, useful redox reaction is the use of concentrated nitric acid to dissolve insoluble copper sulfide:

$$3CuS(s) + 2NO_3^- + 8H^+ \rightarrow 3Cu^{2+} + 3S(s) + 2NO(g) + 4H_2O \tag{2.14}$$

These classes of reaction are treated in detail in subsequent chapters. Here we are interested in the global principles that govern all the reactions in each class. Accomplishing the practical goal of qualitative analysis requires manipulating the chemical environment of the species of interest to yield a predictable reaction of a known type.

We are fundamentally interested in reactions that allow us to separate and/or identify chemical species. Since we would like to control these reactions, it is reasonable to inquire into the areas that clarify why reactions occur. We would like to know why reactions occur, to what extent they occur with respect to the concentrations of the reactants, and how the extent of reaction can be affected by changes in pressure, temperature, and other factors over which we have control. The basis for predicting the answers to such questions is best understood from the standpoint of kinetics and thermodynamics. *Kinetics* involves the study of the factors that influence the rates of chemical reactions. *Thermodynamics* is concerned with the energy relationships among the reacting species and the products. We present here an overview of the principles that govern both the rate of a reaction (kinetics) and the extent to which it can occur (thermodynamics).

KINETICS

The Rate of a Reaction

At this point it is important to consider the meaning of the rate or velocity of a reaction. The words *rate* and *velocity* have the usual connotation of a process involving time, such as the rate at which an automobile moves. For example, if an automobile moves at a constant velocity of 30 miles/hour, we would expect it to be, after 1 hour, 30 miles away from the place we started to count time. Similar ideas prevail for rates of chemical reactions. Thus, the rate of a chemical reaction is the number of chemical events that occur in a given length of time. These chemical events may be indicated by the disappearance of reactants or the appearance of products. For example, let us consider the reaction of two substances, A and B, to form products, C and D, as shown in Eq. (2.15).

$$A + B \rightarrow C + D \qquad (2.15)$$

We can imagine that at zero time only A and B molecules are present in the system; as time passes, A and B molecules disappear, while C and D molecules appear. If we could analyze the reaction mixture at any time after zero time—the time the reaction was started—we would expect to find the amounts (or concentrations) of A and B to be less than the initial amounts, while the amounts (or concentrations) of C and D would be greater. The results of such an experiment are best summarized graphically, as in Figure 2.1; only the amounts of A and C are plotted on this graph.

As might be expected, the rates of chemical reactions can be expressed quantitatively. For example, at time t_1 we might find that the concentration of A was A_1 in the reaction mixture for Eq. (2.15); at some later time t_2, the concentration of A was found to be A_2. Then the change in concentration of A during the time interval $t_2 - t_1$ is $A_2 - A_1$, and the rate, r, of the reaction can be expressed as

$$r = \frac{A_2 - A_1}{t_2 - t_1} = -\frac{\Delta A}{\Delta t} \qquad (2.16)$$

The symbol Δ is a standard way of expressing an incremental change in a quantity. Since the concentration of A has changed by the increment ΔA during the time interval Δt, the symbol

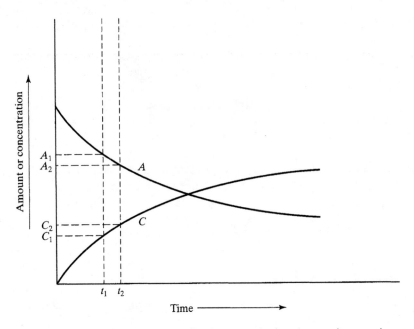

FIGURE 2.1 A plot of the concentration (or amount) of a substance in a reaction at different time intervals. The reaction in question is given by Eq. (2.15), and the line labeled A represents the concentration of the species A in Eq. (2.15). The line labeled C represents the concentration of species C in this process.

$\Delta A/\Delta t$ means *the change in concentration of A that occurs in the time interval* Δt, which is the rate of reaction. In a comparative sense, the larger the ratio $\Delta A/\Delta t$, the greater the rate, which physically means the reaction proceeds more rapidly. It should be apparent that the rate of the process shown in Eq. (2.15) could be determined experimentally by measuring how rapidly A or B is lost or how rapidly C or D is gained. Thus, experimentally, the rate of decrease of A (or B) and the rate of increase of C (or D) must be numerically the same [Eq. (2.17)].

$$-\frac{\Delta A}{\Delta t} = -\frac{\Delta B}{\Delta t} = \frac{\Delta C}{\Delta t} = \frac{\Delta D}{\Delta t} \qquad (2.17)$$

The negative signs in Eq. (2.17) indicate that the concentrations of A and B decrease with time.

Graphically, the rate of the reaction—as expressed by either $\Delta A/\Delta t$ or $\Delta C/\Delta t$ at a given time (for example, at t_1)—corresponds to the slope of the curve shown in Figure 2.1; the slope is the tangent to the curve at the time (t) in question. Thus, if we are able to analyze a reaction at various times after starting it, we can find the rate of the reaction at any time by measuring the slope of the experimental curve at that point. It is the observations of how the rates of chemical reactions change with a variety of factors that provide insight into the detailed processes, or *mechanisms,* of chemical reactions. We will not develop the details here of how the experimental rates lead to an understanding of the mechanism for a chemical reaction because the rates of the reactions used in qualitative analysis schemes are sufficiently rapid so as not to be a practical problem. The reader interested in mechanisms of reactions is referred to any modern general chemistry textbook.

The Initial Rate of Reaction. An inspection of Figure 2.1 indicates that the slope of the curve giving the amount of reactant (or product) as a function of time changes regularly with time; for example, the value of $\Delta C/\Delta t$ starts out to be a large number initially and approaches zero (a horizontal slope) at large values of t. This behavior suggests that the rate of the reaction might be dependent on the concentration of reactants, since their concentrations also decrease with time.

A useful measure of the rate of a reaction is the change in the concentration of the reactants (or products) over a very short period at the very start of the reaction when a negligible fraction of reactants is consumed. We can symbolize the initial rate of r_0, which, according to our previous discussion, is the slope of the concentration *versus* the time curve (Figure 2.1) at zero time.

Factors Affecting the Rate of Reaction

The experimental results of numerous studies on a variety of chemical reactions lead to a simple pattern of behavior that can be summarized succinctly.

Concentration of Reactants. A comparison of the experimentally determined initial rates of reaction for mixtures of various compositions shows that, in general, the rates depend on the concentration of each reactant raised to some power. Thus, for Eq. (2.15), the rate can be defined as

$$r \propto [A]^x[B]^y \tag{2.18}$$

or

$$r = k[A]^x[B]^y \tag{2.19}$$

Equation (2.19) is called the *rate law* for the reaction shown in Eq. (2.15). The proportionality constant k is called the *rate constant* for the reaction and is a measure of the relative velocity with which reactions occur; high values of k mean rapid reactions, and small values indicate relatively slow reactions. The power to which the concentration of the reactant is raised, x for [A] and y for [B] in Eqs. (2.18) and (2.19), is called the *order of the reaction* with respect to that reactant; numerically, the order of the reaction can be zero, an integer, or a fraction. For example, the rate law for Eq. (2.15) might experimentally be found to be

$$r = k[A][B]^0 \tag{2.20}$$

We would say that the reaction is zero order in B and first order in A. Both the rate constant, k, and the order of the reaction are determined experimentally.

Temperature. If reactions are studied at different temperatures, we find that the rates of reactions generally increase with increasing temperature. A very rough rule of thumb has emerged for the effect of temperature on reactions: The rates of most reactions will approximately double for every 10°C rise in temperature at or near 25°C. This observation is reflected in the natural instinct of chemists to heat a reaction mixture to cause the reaction to speed up. A consideration of the rate law for any reaction [such as Eq. (2.19)] indicates that the variation in rate with temperature must arise from a change in the rate constant; after all, the concentrations of the species — the only other factors in Eq. (2.19) — cannot change with temperature.

Catalysts and Inhibitors.   Reaction rates can be affected in some cases by the presence of certain substances that apparently do not enter into the process. For example, the rate of reaction between SO_2 and O_2 to give SO_3 [Eq. (2.21)] is markedly increased by the presence of NO.

$$2SO_2(g) + O_2(g) \rightarrow 2SO_3(g) \tag{2.21}$$

At the completion of the reaction, the NO can be recovered unchanged. Substances that increase the rate of reaction are called *catalysts*. *Inhibitors* are substances that decrease the rate of a given reaction. Catalysts and inhibitors affect the rate of reactions, but appear not to enter into the process in the sense that they are present unchanged after the reaction is completed. Thus, in the conversion of SO_2 to SO_3 described in Eq. (2.21), without NO the reaction occurs at an exceedingly slow rate. However, the addition of a trace of NO speeds up the production of SO_3. When the reaction is completed, the NO is still present in the original trace amount. Obviously the NO must have produced an intermediate species with either the SO_2 or O_2 that reacted rapidly; this more rapid path releases the NO, which can then go back through the cycle. Catalysts provide a different, more efficient pathway to the desired products.

Collision Theory

To understand the details of how experimental factors affect the rate of chemical reactions, we must consider the processes as they occur at the molecular level. It is logical to assume that only the molecules that collide have a chance to react. We can visualize that in a reaction mixture containing A and B molecules, the molecules are moving independently and randomly bumping into each other. Thus, it is not surprising that, for the most part, the rates of chemical reactions depend on the concentrations of the reactants. For example, let us assume that the reaction between A and B [Eq. (2.15)] is first order in both A and B [Eq. (2.22)], which means the rate law has the *form* given by Eq. (2.22).

$$r = k[A][B] \tag{2.22}$$

We would expect that an increase in the concentration of A *or* B would increase the probability of a collision between A and B molecules. If A and B do not collide, we cannot expect a reaction to occur. Note that the concentration of the products C and D does not enter into the consideration of the rate of the reaction as it is written in Eq. (2.15). The rates of heterogeneous reactions, such as the reaction of oxygen gas with solid coal, can be affected by increasing the surface area between the two reacting phases; this is equivalent to changing the concentration of the reactants in a homogeneous (one-phase) reaction. For example, a single lump of coal will burn rather slowly, but the equivalent amount of finely pulverized coal will explode when it reacts with oxygen at the same temperature.

The increase in the rate of a reaction with increasing temperature can be traced to a rise in the number of effective collisions. Kinetic molecular theory suggests that, at a given temperature, the molecules in a sample exhibit a distribution of energies, and an increase in temperature shifts the maximum of the distribution to higher energies (Figure 2.2). Not all molecular collisions lead to a chemical reaction. Rather, a threshold energy E_a, called the *activation energy,* is required to activate the molecules sufficiently to permit a reaction to occur. Unless the molecules involved possess this minimum energy, they will not react even though they collide. Inspection of Figure 2.2 shows why an increase in the temperature leads to an increase in the

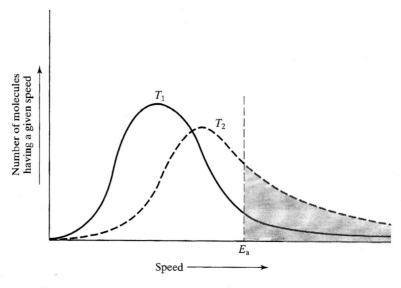

FIGURE 2.2   According to the kinetic molecular theory, the molecules of a sample of matter exhibit a distribution of velocities about some average value, as shown by the curve labeled T_1. At higher temperatures (curve T_2) the average velocity shifts to higher values. E_a represents the minimum energy required for a given chemical reaction.

number of successful collisions. It is reasonable to assume that only molecules that have energies equal to or greater than some minimum energy are activated sufficiently to react when they collide. This situation is illustrated in Figure 2.2, where E_a is the activation energy. At the lower temperature T_1, the area under the curve to the right of E_a is proportional to the number of activated molecules. An increase in temperature to T_2 shifts the distribution of velocities to give a greater area under the curve to the right of E_a, hence, a greater proportion of activated molecules. Thus, an increase in temperature of a reaction increases its rate for two reasons: The probability of a collision increases because the average molecular velocities increase, and a higher proportion of molecules in the mixture have more than the minimum energy necessary for reaction.

The energy change involved in chemical reactions is conveniently displayed on an energy level diagram, as illustrated in Figure 2.3. The horizontal axis—called the *reaction coordinate*—represents the *course of the reaction,* during which we imagine the two molecules approaching each other in a favorable orientation as the reaction proceeds. The reactants (A and B) have a certain energy content with respect to the products (C and D in this case); the overall change in energy for the process is $E_1 - E_2$. Before A and B react, however, they must form an *activated complex* that is of higher energy than the reactants by an amount E_a, the activation energy. The activated complex then yields products, giving up $E_a + \Delta E$ of energy in the process.

Catalysts affect the course of a reaction by decreasing the activation energy required for the reaction; this often involves the establishment of a new pathway of lower energy for the process. The relative energies involved in such a relationship are illustrated in Figure 2.4.

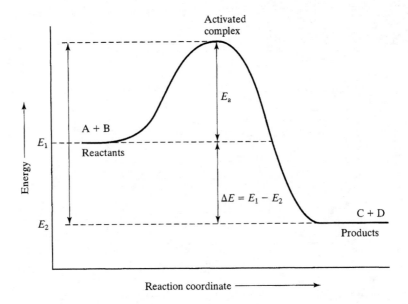

FIGURE 2.3   The relationship among the energies of the reactants, products, and activated complex for the reaction $A + B \rightarrow C + D$.

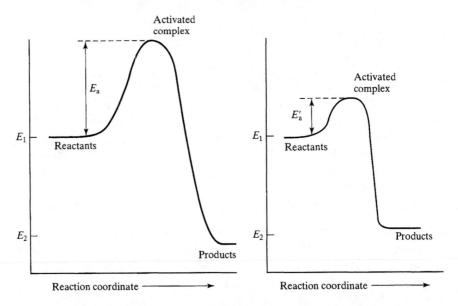

FIGURE 2.4   Adding a catalyst to a chemical reaction has the effect of producing a lower energy pathway, usually by way of a different activated complex, between reactants and products. For example, the left-hand graph represents the energy relationships without a catalyst, whereas the right-hand diagram indicates the energy relationships in the presence of a catalyst.

The relative energies of the products and reactants are still the same, whether a catalyst is present or not. The addition of a catalyst lowers the activation energy, which means that at the same temperature more molecules have energies equal to or greater than the minimum energy required for the reaction to occur. This is readily seen in the graph in Figure 2.5.

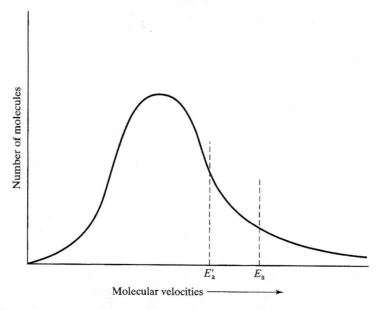

FIGURE 2.5   The presence of a catalyst lowers the activation energy from E_a to E'_a. Thus, at the same temperature, a larger number of molecules, as measured by the area under the curve to the right of the activation energy, have energies in excess of the activation energy, and the reaction proceeds more rapidly.

CHEMICAL EQUILIBRIUM

The curve in Figure 2.1 indicates that the reaction apparently stops after a given point in time; that is, the concentration of the product, C, gradually increases to a limiting value, after which it remains constant. A similar conclusion can be reached from the concentration of the reactant, A; after a time, the concentration of A apparently stops changing.

   The correct interpretation of the static concentrations of the reactants and products after a given time involves the recognition that the products reach a concentration where their rate of reaction to re-form the reactant molecules, the reverse reaction [Eq. (2.23)], occurs as rapidly as the forward reaction [Eq. (2.24)].

$$C + D \rightarrow A + B \tag{2.23}$$

$$A + B \rightarrow C + D \tag{2.24}$$

The net result is that the products of the forward reaction [Eq. (2.24)] are formed at the same rate as they disappear in the reverse reaction [Eq. (2.23)]; Eq. (2.25) expresses this situation mathematically.

$$r_f = r_r \tag{2.25}$$

A chemical reaction that has attained this state is said to be at *equilibrium*. Although the concentrations of the species are constant at equilibrium, the process is dynamic. Numerous analogies are available to enhance understanding of the condition of equilibrium. For example, consider the occupancy of a large public office building at midmorning; we can imagine that the concentration of people in such a structure builds up to a level that is essentially constant for a period of time. Although the number of people may remain constant, their identities change as some people leave and others come into the building; when the rate of loss of people is equal to the rate of gain of people, the total number present at any one time does not change, even though the identities of the individuals can change.

The Equilibrium Constant

When a chemical system is at equilibrium, a balanced equation still describes the mass relationships for both the forward and reverse processes. The equilibrium condition is conventionally designated by two opposing arrows:

$$aA + bB \rightleftharpoons cC + dD \tag{2.26}$$

The coefficients a, b, c, and d are those necessary to give a balanced equation. A chemical system at equilibrium is characterized by an *equilibrium constant,* which is the ratio of the product of the concentrations of the original species to the product of the concentrations of the reactant species, the concentration of all species being raised to the power represented by their coefficients. Thus, for Eq. (2.26) the equilibrium constant expression (K_{eq}) is given by

$$K_{eq} = \frac{[C]^c [D]^d}{[A]^a [B]^b} \tag{2.27}$$

The symbol [C] is read as the *molar concentration of C* (the number of moles of C per liter of solution). Equations (2.28) to (2.30) and the associated equilibrium constants illustrate the relationship between the equilibrium constant and the corresponding balanced equation.

$$2H_2(g) + O_2(g) \rightleftharpoons 2H_2O(g) \qquad K = \frac{[H_2O]^2}{[H_2]^2[O_2]} \tag{2.28}$$

$$PCl_5(g) \rightleftharpoons PCl_3(g) + Cl_2(g) \qquad K = \frac{[PCl_3][Cl_2]}{[PCl_5]} \tag{2.29}$$

$$3H_2(g) + N_2(g) \rightleftharpoons 2NH_3(g) \qquad K = \frac{[NH_3]^2}{[H_2]^3[N_2]} \tag{2.30}$$

Each system at equilibrium at a given temperature exhibits a characteristic value of the equilibrium constant. Large values of K_{eq}, such as 3×10^8, mean that the concentrations of the products are relatively high compared to the concentrations of reactants, which is another way of saying the reaction favors the products. A small value of the equilibrium constant, such as 4×10^{-5}, means that the reactants are favored over the products or that the reaction has not proceeded very far toward the products.

Equilibrium Calculations

Since all reactions have a finite value for K_{eq}, no chemical process goes to completion, although many processes have very large values for the equilibrium constant and may be assumed to go to completion for all practical purposes. If the equilibrium constant for a process is known, we have all the information necessary to establish the concentration of all species present in the system.

As an example of the method of determining the concentration of all species present in an equilibrium system, let us consider the decomposition of PCl_5 in the gaseous phase, which occurs according to Eq. (2.29). At 250°C, the value of the equilibrium constant, an experimentally determined quantity, is 0.0415. What would be the concentration of all species if 1.5 mole of PCl_5 were placed in a 1.0-L flask and allowed to come to equilibrium at 250°C? The answer to this question starts with the realization that the *identity* of the species present is established by the chemical equilibrium expression in (Eq. 2.29), and the quantitative relationship among the concentrations of the species present is given by the corresponding equilibrium constant

$$K = \frac{[PCl_3][Cl_2]}{[PCl_5]} = 0.0415 \tag{2.31}$$

Remember also that the conversion of products to reactants is governed by the coefficients of the balanced equation (2.29).

Imagine first the conditions before equilibrium is established. The molar amounts are expressed in tabular form:

	Substance		
	PCl_5	PCl_3	Cl_2
Initial amounts (moles)	1.5	0	0

Now consider that some unknown amount—call it x—of PCl_5 decomposes to reach equilibrium. The table can be expanded to include this change in the amounts of substances present. We lose x moles of PCl_5, but the balanced equation predicts a gain of x moles of PCl_3 and x moles of Cl_2.

	Substance		
	PCl_5	PCl_3	Cl_2
Initial amounts (moles)	1.5	0	0
Change (moles)	$-x$	$+x$	$+x$

Mass conservation helps us predict the equilibrium amounts of the substances in the flask.

	Substance		
	PCl_5	PCl_3	Cl_2
Initial amounts (moles)	1.5	0	0
Change (moles)	$-x$	$+x$	$+x$
Equilibrium amounts (moles)	$1.5 - x$	x	x

Recall that equilibrium constants are expressed in terms of the molar concentrations of the species involved, so that we must take into account the volume of the container, which is 1.0 L in this case.

	Substance		
	PCl_5	PCl_3	Cl_2
Initial amounts (moles)	1.5	0	0
Change (moles)	$-x$	$+x$	$+x$
Equilibrium amounts (moles)	$1.5 - x$	x	x
Equilibrium concentration (M)	$(1.5 - x)/1$	$x/1$	$x/1$

The concentrations in the table can be substituted into the equilibrium expression, Eq. (2.31), to give

$$K = 0.0415 = \frac{[PCl_3][Cl_2]}{[PCl_5]} = \frac{(x)(x)}{(1.5 - x)} \tag{2.32}$$

Eq. (2.32) can be arranged to give

$$x^2 + 0.0415(x) = 0.06225 \tag{2.33}$$

This quadratic expression can be solved (see Appendix II) for x to yield

$$x = 0.2296 \quad \text{or} \quad x = -0.2711 \tag{2.34}$$

The positive root ($x = 0.2296$) is the only one that makes sense physically; thus, the concentration of all species is given by substitution into the last line of the table.

$$[PCl_5] = 1.5 - 0.2296 = 1.27\ M$$
$$[Cl_2] = 0.23\ M$$
$$[PCl_3] = 0.23\ M$$

The foregoing analysis of the PCl_5 dissociation illustrates that it is always possible to relate the equilibrium constant to an expression of the equilibrium concentration of all species present.

If we know, or can establish, the equilibrium concentration of all species present, it is possible to calculate the numerical value of the equilibrium constant. For example, consider the dissociation of HI at 450°C [Eq. (2.35)].

$$2HI(g) \rightleftharpoons H_2(g) + I_2(g) \tag{2.35}$$

At this temperature, an experiment shows that 1.0 mole of HI in a 1.0-L flask produces 0.125 mole of I_2. This is sufficient information to allow us to calculate the equilibrium constant for Eq. (2.35) using the organizational scheme described previously. Initially, the amounts of reactants and products are given by:

	Substance		
	HI	H_2	I_2
Initial amounts (moles)	1.0	0	0

The initial HI reacts to produce 0.125 mole of I_2. The balanced equation allows us to determine the change in the amount of each species.

	Substance		
	HI	H_2	I_2
Initial amounts (moles)	1.0	0	0
Change (moles)	$-(2)(0.125)$	$+0.125$	$+0.125$

The equilibrium amounts can be established by the principle of mass action to be:

	Substance		
	HI	H_2	I_2
Initial amounts (moles)	1.0	0	0
Change (moles)	$-(2)(0.125)$	$+0.125$	$+0.125$
Equilibrium amounts (moles)	0.250	0.125	0.125

And, of course, the equilibrium concentrations become:

	Substance		
	HI	H_2	I_2
Initial amounts (moles)	1.0	0	0
Change (moles)	$-(2)(0.125)$	$+0.125$	$+0.125$
Equilibrium amounts (moles)	0.250	0.125	0.125
Equilibrium concentration (M)	0.250/1.0	0.125/1.0	0.125/1.0

The equilibrium constant for Eq. (2.35) is

$$K = \frac{[H_2][I_2]}{[HI]^2} \qquad (2.36)$$

Substituting the equilibrium concentrations from the previous table into Eq. (2.36) gives the value of the equilibrium constant.

$$K = \frac{[H_2][I_2]}{[HI]^2} = \frac{(0.125)(0.125)}{(0.250)^2} = 0.0625 \qquad (2.37)$$

The examples given above show the general procedures for establishing the concentration of all the species present in an equilibrium mixture if the equilibrium constant is known and for establishing the value of the equilibrium constant from an analysis of the equilibrium mixture. Although the examples are focused on gas phase equilibria, the process works equally well with solution equilibria, which predominate in the scheme of qualitative analysis presented here.

A good example of the use of kinetic and thermodynamic principles in qualitative analysis involves the generation of H_2S to provide the source of S^{2-} used to precipitate Groups II and III in the scheme used in this book; see Part I, Chapter 1 for an overview of the scheme of analysis. Rather than prepare gaseous H_2S for the analysis scheme, the procedure calls for the generation of H_2S, *in situ*, by the hydrolysis of thioacetamide (Eq. 2.38).

$$CH_3CSNH_2 + 2H_2O \rightarrow CH_3CO_2H + NH_3 + H_2S \tag{2.38}$$

The temperature dependence of this reaction is such that, at room temperature, there is essentially, no reaction, but it proceeds rapidly at about 80°C. A one-molar aqueous solution of CH_3CSNH_2 at 80°C will produce a solution saturated with H_2S very rapidly. In addition, the rate of the reaction described by Eq. (2.38) is markedly increased in the presence of acid. Thus, the procedure for the preparation of an H_2S-containing solution for use in qualitative analysis incorporates a number of important principles. Thus, the rate of hydrolysis of thioacetamide is negligible at room temperature so that a stock aqueous solution of this substance can be prepared and used on demand; the rate of hydrolysis of thioacetamide is increased by increasing the temperature; and the rate of hydrolysis is increased in the presence of an acid.

An example of the manipulation of an equilibrium process to effect a separation of ionic species in qualitative analysis occurs in the separation of the cations of the copper-arsenic group from those of the aluminum-nickel group (Figure 1.2). The cations in these two groups are separated on the basis of the relative solubilities of the corresponding sulfides. The sulfides of the copper-arsenic group are much more insoluble than those of the aluminum-nickel group. This means that only a relatively small concentration of S^{2-} is necessary to precipitate the copper-arsenic group sulfides compared to the sulfide concentration necessary to precipitate the aluminum-nickel group sulfides. We could, of course, attempt to regulate the S^{2-} concentration by adding an appropriate sulfide ion-containing reagent sufficient to cause the metal ions to precipitate. This strategy, although it is the most direct approach, is not very practical; since we do not necessarily know the concentrations of the metal ions in solution, we would have no idea how much S^{2-} to add to effect the separation desired.

The plan of action chosen involves producing the appropriate concentration of S^{2-} from a molecular substance (H_2S) in solution, which, as mentioned earlier, is produced through the hydrolysis of thioacetamide [Eq. (2.38)]. The S^{2-} source is H_2S, which undergoes stepwise ionization in aqueous solution.

$$H_2S(aq) \rightleftharpoons H^+(aq) + HS^-(aq) \tag{2.39}$$

$$HS^-(aq) \rightleftharpoons H^+(aq) + S^{2-}(aq) \tag{2.40}$$

The overall process—the sum of Eqs. (2.39) and (2.40)—is given by Eq. (2.41)

$$H_2S(aq) \rightleftharpoons 2H^+(aq) + S^{2-}(aq) \tag{2.41}$$

and the corresponding equilibrium constant is written as

$$K = \frac{[H^+]^2[S^{2-}]}{[H_2S]} \tag{2.42}$$

The sulfide ion concentration, $[S^{2-}]$, of a solution saturated with H_2S can be obtained by rearranging Eq. (2.42).

$$[S^{2-}] = \frac{K[H_2S]}{[H^+]^2} \tag{2.43}$$

Equation (2.43) clearly shows that the sulfide ion concentration in an aqueous solution of H_3S is a function of $[H^+]$—the acidity of the solution. At high values of $[H^+]$ the sulfide ion concentration is low, and at lower acidity—basic solutions—the sulfide ion concentration is high. From a practical standpoint, if the H_2S concentration is kept high (saturated), the $[S^{2-}]$ in solution can be regulated by adjusting the $[H^+]$. The experimental details are given in Part III.

Le Chatelier's Principle and Changing the Point of Equilibrium

Since a system at equilibrium is a balance of two opposing dynamic processes, it is possible to shift the point of this balance by operating on factors that may affect the rate of the opposing process differently. Take, for example, the equilibrium condition described by Eq. (2.26), which is reproduced here in Eq. (2.44).

$$aA + bB \rightleftharpoons cC + dD \tag{2.44}$$

If we remove some C species from the equilibrium mixture, the rate of the reverse reaction will decrease (because there are now fewer C molecules to react with the same number of D molecules), but the rate of the forward reaction will be unchanged. Under these conditions, more A and B will disappear, and the system will be momentarily unbalanced; however, after a time, the rates of the forward and reverse reactions will again be equal. At the new equilibrium point there will be relatively less reactants (A and B) and more product (D). In other words, the point of the equilibrium has been shifted toward the products. Henri Louis Le Chatelier stated a generalization that describes all equilibrium systems: If a system that is in a dynamic equilibrium is subjected to some stress, the equilibrium will shift in the direction to relieve that stress. Thus, addition of Cl_2 to the system described in Eq. (2.29) will cause the equilibrium to shift to the left (increasing the concentration of $PCl_5(g)$ to relieve the momentary stress of a larger concentration of Cl_2 than permitted at the original point of equilibrium).

$$PCl_5(g) \rightleftharpoons PCl_3(g) + Cl_2(g)$$

The concentration of PCl_3 will decrease compared to the original equilibrium value. An important point that needs to be mentioned here is that the value of the equilibrium constant is the same at the old and the new equilibrium conditions; only the relative concentrations of the species have changed, but the ratio of the product of the concentrations remains the same. The value of the equilibrium constant will change only with temperature.

A change in temperature can also affect the position of a system at equilibrium. All reactions either evolve heat to the surroundings (*exothermic reactions*) or absorb heat (*endothermic reactions*). Two examples are given in Eqs. (2.45) and (2.46).

$$N_2(g) + 3H_2(g) \rightleftharpoons 2NH_3(g) + 92.26 \text{ kJ} \tag{2.45}$$

$$PCl_5(g) + 92.55 \text{ kJ} \rightleftharpoons PCl_3(g) + Cl_2(g) \tag{2.46}$$

If we consider heat (thermal energy) as a product [the exothermic reaction in Eq. (2.45)] or as a reactant [the endothermic reaction in Eq. (2.46)], application of Le Chatelier's principle gives the direction in which the point of equilibrium shifts when the temperature of an equilibrium system is changed. Thus, raising the temperature causes the point of equilibrium in Eq. (2.45) to shift toward the left; a shift in this direction has the effect of absorbing thermal energy, which can be considered as an attempt by the system to compensate for an increase in temperature. Raising the temperature on the equilibrium shown in Eq. (2.46) causes the point of equilibrium to shift to the right. Of course, lowering the temperature of these equilibria will have the opposite effect of that described above. In summary, the point of equilibrium for exothermic reactions is shifted toward reactants when the temperature is raised, whereas the equilibrium point for endothermic reactions is shifted toward products; a decrease in temperature has the opposite effect.

The general scheme of qualitative analysis described in the section of this book on laboratory practice contains numerous examples of the application of Le Chatelier's principle. Most of the examples deal with equilibria established in aqueous solutions, and discussions of these are presented in more appropriate chapters of Part II. One point should be made here, however, concerning the practical application of Le Chatelier's principle: The principle can be used as a guide to the ways in which the concentration of a given species in an equilibrium system can be adjusted (either increased or decreased) to achieve a desired chemical result. The use of the hydrogen ion concentration, $[H^+]$, to affect the concentration of S^{2-} is an example of Le Chatelier's principle in qualitative analysis. Recall from Eq. (2.41) that H_2S is the source of S^{2-} through the ionization process, reproduced here in Eq. (2.47).

$$H_2S(aq) \rightleftharpoons 2H^+(aq) + S^{2-}(aq) \qquad (2.47)$$

Increasing the $[H+]$ in this solution shifts the equilibrium to the left, which has the effect of decreasing S^{2-}. Conversely, decreasing the $[H^+]$ in solution—by making it basic—shifts the equilibrium to the right; in doing so the S^{2-} concentration increases over what it was before.

THERMODYNAMICS

Another way to describe the position of equilibrium in a chemical system is based on considerations of energy. The following mechanical analogy will assist us to understand chemical systems. Consider a ball resting on the side of a hill, position A in Figure 2.6.

The ball possesses potential energy (E_A) by virtue of its position. Under ordinary conditions the ball will roll to the bottom of the hill (position B) and come to rest with an energy of E_B. The ball/hill system is more stable when it has attained the minimum in energy, E_B, which corresponds to the ball at the bottom of the hill. In other words, the ball/hill system spontaneously passes from a less stable arrangement (ball on the side of the hill) to a more stable arrangement (ball at the bottom of the hill) of lower energy.

Chemical systems are similar to the mechanical one described in Figure 2.6. A mixture of molecules has a certain amount of associated energy, consisting of kinetic energy of motion of the molecules (which is dependent on temperature) and the bond energies of the molecules present. If it is possible for this mixture to change to a new mixture of lower energy, it will do so. The free energy is a chemical measure of the energy content of a substance and is symbolized by G. The free energy of a substance can be determined by a variety of experimental methods, which are beyond the scope of our discussions here, and is usually expressed on a molar basis. Thus, for

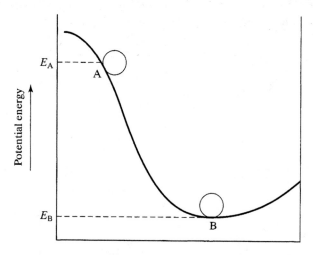

FIGURE 2.6   The relative energy of a physical system can be expressed in an energy level diagram.

any mixture of molecules, such as n_A, moles of A, and n_B, moles of B, the total energy of the system, G_{total}, is given by Eq. (2.48):

$$G_{total} = n_A G_A + n_B G_B + \cdots \qquad (2.48)$$

where G_A, G_B, ... are the free energies of the substances, A, B, ..., respectively.

Free Energy and Equilibrium

Consider the process described in Eq. (2.15) as an example: Let 1.0 mole of each of the reactants be placed in the same 1.0-L flask. The free energy of this system, G, before a reaction occurs is given by Eq. (2.49); there are no products present.

$$G = (1.0)G_A + (1.0)G_B \qquad (2.49)$$

Then assume, for example, the system attains equilibrium [Eq. (2.50)] after 0.4 mole of A and 0.4 mole of B have reacted.

$$A + B \rightleftharpoons C + D \qquad (2.50)$$

At equilibrium, the flask contains 0.4 mole of C, 0.4 mole of D, 0.6 mole of A, and 0.6 mole of B. The free energy of the system at equilibrium is given by Eq. (2.51):

$$G_{eq} = (0.6)G_A + (0.6)G_B + (0.4)G_C + (0.4)G_D \qquad (2.51)$$

It should be apparent from this example that the free energy of the system containing any mixture of reactants and products can be calculated; the results of such calculations for Eq. (2.51) are given in Figure 2.7. Note that only one mixture of products and reactants gives the minimum free energy possible (G_{eq}) for the system; all other possible mixtures lead to systems with higher

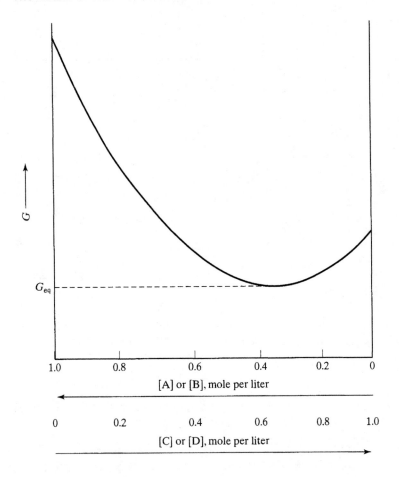

FIGURE 2.7 The free energy (G) of a chemical system depends on the relative proportions of the species present in the system. Here is plotted the free energy of the system $A + B \rightleftharpoons C + D$ as a function of composition. The minimum in free energy, G_{eq}, occurs when [A] and [B] are 0.6 mol/L and [C] and [D] are 0.4 mol/L.

free energies than the minimum. Thus, it is not surprising that all these other mixtures will spontaneously go to (react to form) the equilibrium mixture that has the minimum free energy, a situation similar to the observation that all possible positions of the ball in Figure 2.6 will spontaneously go to that of the ball in the valley.

The relationship between the equilibrium constant for a reaction and the free energy—the source of which will not be discussed here—of a system at equilibrium is given by Eq. (2.52):

$$\Delta G^0 = -RT \ln K \tag{2.52}$$

where ΔG^0 is the difference in standard free energy between products and reactants, R is the ideal gas constant, T is the absolute temperature of the system, and $\ln K$ is the symbol for the natural logarithm (base e) of the equilibrium constant K. Thus, if we know the standard free energy minimum of a system, we can calculate a value of the equilibrium constant for that system from Eq. (2.52).

SUMMARY

We have discussed, in a broad sense, the principles that govern chemical reactions. Chemical reactions occur spontaneously because the energy of the mixture of molecules is greater than some possible minimum value that occurs when the system is at equilibrium (see Figure 2.7). Chemical reactions always attain an equilibrium state, given sufficient time. It is possible to upset the equilibrium point of a reaction by either adding or removing products or reactants; Le Chatelier's principle governs the direction in which an equilibrium will shift when a chemical stress is applied.

Subsequent chapters elaborate on the classes of reactions described in the first paragraph of this chapter.

PROBLEMS

2.1 Consider the hypothetical rate-determining reaction found to be second order in A and first order in B.

$$2A + B \rightarrow C + D$$

a. Write the expression for the rate law for the initial reaction rate, r_0.

b. What would be the effect on r_0 if the concentration of A were doubled and all other factors remained the same?

c. What would be the effect on r_0 if the concentration of B were doubled and all other factors remained the same?

2.2 Consider the following reaction:

$$A + 2B \rightarrow C$$

It was determined experimentally that the rate of formation of the product C is independent of the concentration of A and quadruples when the concentration of B is doubled.

a. Write the expression for the rate law for this reaction.

b. When the initial concentrations of A and B are 0.3 M and 0.2 M, respectively, the initial rate of formation of C is 4×10^{-3} M/s. What is the rate constant for the reaction?

c. Under the same experimental conditions described in part (b), what is the initial rate of reaction when the initial concentrations of A and B are 0.1 M and 0.5 M, respectively?

2.3 Consider the chemical system

$$H_2(g) + Br_2(g) \leftrightarrows 2HBr(g) + 70.3 \text{ kJ}$$

Indicate whether the equilibrium concentration of HBr is increased, decreased, or unaffected when the following changes are made on the system after equilibrium is established.

a. Temperature is increased.

b. A catalyst is added.

c. Br_2 is added to the system.

d. HBr is added to the system.

e. The system is placed in a smaller container.

f. Pressure is applied to the system.

g. Which, if any, of the factors listed in (a) through (f) operate because the value of the equilibrium constant changes?

2.4 Consider the formation of HBr from the elements

$$H_2(g) + Br_2(g) \rightarrow 2HBr(g) + 70.3 \text{ kJ}$$

How would you expect the initial rate of formation of HBr to be affected as the following changes are made? Assume in each instance that all other factors remain unchanged. Give your answer in reference to the rate of reaction that would occur if 1 mole of H_2 and 1 mole of Br_2 were brought together at 1 atm pressure, in a 1-L vessel, maintained at room temperature.

a. The temperature is increased.
b. A catalyst is added.
c. More Br_2 is added initially.
d. The reaction is allowed to occur in a smaller container.
e. The reaction is allowed to occur at a higher pressure.

2.5 Write the equilibrium constant expressions for the following reactions and give the units.

a. $N_2(g) + 3H_2(g) \leftrightharpoons 2NH_3(g)$
b. $2SO_3(g) \leftrightharpoons 2SO_2(g) + O_2(g)$
c. $2H_2(g) + O_2(g) \leftrightharpoons 2H_2O(g)$
d. $3O_2(g) \leftrightharpoons 2O_3(g)$
e. $NO_2(g) + SO_2(g) \leftrightharpoons NO(g) + SO_3(g)$

2.6 PCl_5 dissociates and reaches equilibrium according to the following equation:

$$PCl_5(g) \leftrightharpoons PCl_3(g) + Cl_2(g)$$

Under a certain set of conditions PCl_5 was experimentally determined to be 20 percent disso-ciated. Two moles of PCl_5 were introduced into a 1-L flask under the same conditions and equilibrium was established.

a. How many moles of each species are present in the flask at equilibrium?
b. What is the concentration of each species present in the flask at equilibrium?
c. Calculate the equilibrium constant for the dissociation of PCl_5.

2.7 An equilibrium mixture

$$CO(g) + Cl_2(g) \leftrightharpoons COCl_2(g)$$

contained 1.50 moles CO, 1.00 mole Cl_2, and 4.00 moles $COCl_2$ in a 5-L reaction flask at 100°C. Calculate the value of the equilibrium constant for the system at 100°C.

2.8 One mole of NH_3 was introduced into a 1-L reaction flask at 800°C. When the equilibrium

$$2NH_3(g) \leftrightharpoons N_2(g) + 3H_2(g)$$

was established, 0.6 mole of H_2 was present in the flask. Calculate the value of the equilibrium constant for this reaction at 800°C.

2.9 Consider the equilibrium discussed in Problem 7. Give the concentration of all species present if 49.5 g of $COCl_2$ is allowed to come to equilibrium at 100°C in a 5-L flask.

2.10 What fraction of $COCl_2$ undergoes dissociation in the situation described in Problem 2.9?

2.11 At 25°C the value of the equilibrium constant is 5.65×10^7 for the reaction

$$H_2(g) + C_2H_4(g) \leftrightharpoons C_2H_6(g)$$

Calculate the value of ΔG^0 for this system.

2.12 Calculate the value of the equilibrium constant for the reaction

$$2NO(g) + O_2(g) \leftrightharpoons 2NO_2(g)$$

using the following information. After mixing 1 mole of NO and 1 mole of O_2 in a 2-L flask at 1 atm and 10°C, the system was allowed to attain equilibrium. At equilibrium the concentration of NO was found to be 0.3 M.

2.13 The value of the equilibrium constant for the reaction

$$N_2(g) + 3H_2(g) \leftrightharpoons 2NH_3(g)$$

is 9.5×10^{-4} at 500°C. Calculate the equilibrium concentration of each of the components if the starting amounts are 1.00 mole of N_2 and 1.00 mole of H_2 in a 2-L flask at 500°C.

2.14 The standard free energy change for the reaction

$$2HI(g) \leftrightharpoons H_2(g) + I_2(g)$$

is -2.6 kJ at 200°C. Calculate the equilibrium constant for this reaction.

The Nature of Chemical Compounds

Your understanding of qualitative analysis will be enhanced if you recognize the fundamental nature of certain classes of pure substances, how they interact with water, and the nature of the species present in aqueous solutions. In this chapter we consider the constitution of pure substances and some important properties of aqueous solutions.

PHYSICAL PROPERTIES OF COMPOUNDS

Clues to the nature of chemical compounds come from a study of their physical properties. An intellectual link can be formed between the results obtained from the measurement of a physical property—usually made on a bulk sample—and the behavior of the constituents of the sample on an atomic level, if we understand the physical basis for the property in question. The two simple properties of interest to us here are the melting point and the electrical conductivity of the *pure* molten substance.

Recall that the melting point is the temperature at which liquid and solid coexist in equilibrium. The melting point is a characteristic property of a substance. Only rarely will the melting points of two different substances be the same. The melting points for the fluorides of the second-period elements are listed in Table 3.1. Inspection of the data shows that these binary compounds fall into two large classes, based on the relative magnitudes of their melting points;

TABLE 3.1   THE MELTING POINTS OF THE FLUORIDES OF THE SECOND-PERIOD ELEMENTS

	Periodic Group						
	IA	IJA	IIIB	IYB	YB	YIB	YIIB
Compound	LiF	BeF_2	BF_3	CF_4	NF_3	OF_2	F_2
mp, °C	842	800	−126.7	−183.7	−206.6	−223.8	−223
	high melters conductors				low melters nonconductors		

the compounds exhibit either *high melting points* or *low melting points.* The former class of compounds melts well above room temperature, whereas the latter melts considerably below room temperature.

If we had inspected the melting points of many more compounds than are listed in Table 3.1, the results would have been similar, that is, they could be classified as either *high melters* or *low melters.* These terms are *relative* to "room temperature," ~20°C. Compounds with high melting points require greater thermal energy to separate the atomic-sized species of which they are constituted than those with lower melting points. In other words, the chemical species that make up the high melters are held together by stronger forces than those that constitute the lower melters. Thus, the melting points allow us to classify compounds on the basis of the relative magnitude of the forces of attraction between the species that constitute the bulk phase. *Conductivity* is another useful property that aids in understanding the constitution of chemical compounds. The conductivity of a liquid sample (either molten pure substances or a solution of one substance in another) can be determined using the apparatus shown schematically in Figure 3.1.

Although the schematic representation of the apparatus appears simple, the actual equipment can be quite complex—especially if the experiment is to be performed at temperatures much above or below room temperature. Consider, for example, the practical difficulty of measuring the conductivity of pure LiF, which melts at 842°C, or of OF_2, which melts at −223.8°C. We shall not worry too much about the details of doing these experiments; only the results are of interest to us here. Basically, the conductivity experiment involves two electrodes inserted into a liquid sample of the substance under study. The electrodes are connected to a potential source (e.g., a battery) and a device to detect the flow of current in the external circuit; the latter could be a calibrated meter that will quantitatively measure the amount of current that flows or it could simply be a light bulb that glows when a current passes through the external circuit. Irrespective of the nature of the current-indicating device, if a current is detected in the external circuit, an equivalent amount of charge must be flowing between the electrodes immersed in the liquid. Thus, the presence or absence of charged species—also called *ions*—in the liquid sample can be detected by measuring its conductivity. When this experiment is performed on liquid samples of the pure substances listed in Table 3.1, an interesting result, similar to that seen for the melting point, is obtained. The compounds are divided into two classes: conductors and nonconductors. The high-melting substances are conductors of electricity—their melts contain ions—whereas the low-melting substances are nonconductors, or insulators; the conclusion of this experiment is that nonconductors are composed of uncharged species, or molecules. It should be stressed that

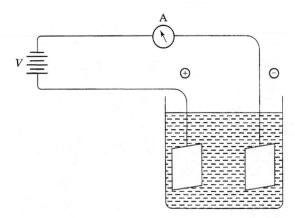

FIGURE 3.1   A schematic diagram of an apparatus that can be used to measure the conductivity of a liquid substance. The current that flows through the circuit upon application of the potential difference *V* is measured by meter A. The magnitude of current flowing is a measure of the rate at which ions in the liquid move to the electrodes.

the conductivity measurements are made on the pure substance in its molten state, *not* on its solutions in water or some other solvent. In the latter instance it is possible that a substance might dissolve in a solvent because a chemical reaction occurs; if the reaction yields ionic species, the solution will, of course, be a conductor. This observation may have no bearing, however, on the presence or absence of ions in the original sample. We discuss the properties of solutions later.

CLASSIFICATION OF CHEMICAL COMPOUNDS

Generally, compounds formed by the alkali and alkaline earth metals (Groups IA and IIA) exhibit high melting points and conduct electricity in the molten state; as a class, such compounds are called *ionic compounds.* The general conclusion is that compounds formed by the alkali and alkaline earth metals are, as a class, ionic. On the other hand, *molecular compounds* exhibit relatively low melting points and are nonconductors in the liquid state. Molecular compounds consist of *molecules* that are attracted to each other by relatively weak forces compared to the forces acting between the ions in ionic compounds. There is no implication in this argument of the relative strength of the forces holding the atoms of a molecule together compared to those holding an ionic substance together. Other data, not discussed here, indicate that the *bonds* holding a molecule together are as strong as the electrostatic forces of attraction between ions; molecular compounds do not contain ions, although the molecules might be polar.

Although most compounds can be classified as molecular (if they have low melting points and are nonconductors) or ionic (high melters and conductors), a few compounds exhibit both the high melting point of ionic compounds and the nonconductivity of covalent compounds; a common example of such a compound is silicon dioxide, SiO_2, the main constituent of glass. Silicon dioxide melts at 1710°C but does not conduct electricity in the liquid state, which indicates that the melt contains no ions. The high melting point is attributed to the fact that the structure of silicon dioxide involves a three-dimensional network of covalently bonded SiO_2 units. In effect, a crystal of silicon dioxide is one giant molecule. Thus, the high melting point of silicon dioxide is a reflection of the energy required to impart relatively free motion to the units of SiO_2, a process that requires some covalent bonds to be broken.

A summary of the classification scheme for chemical compounds based on their constitution appears in Figure 3.2.

THE IONIZATION OF WATER

In the practice of qualitative analysis, we are interested in chemistry—*separations* and *identifications*—as it occurs in aqueous solutions. Accordingly, it is important to understand fully the properties of water in the mixtures (solutions) it forms with different types of substances.

Ordinarily, water is a nonconductor of electricity (or, an *insulator*). However, by using exceedingly sensitive instruments it is possible to detect a current flowing through the purest sample of water that can be produced; essentially, this is an experiment using an apparatus similar to that shown in Figure 3.1. Pure water, when tested with such equipment, shows a minute, but real conductivity indicating the presence of ions. Scientists have never obtained a sample of water that did not exhibit some conductivity. We are forced to conclude that the ions must arise spontaneously from water itself. This and other evidence suggests that some water molecules form ions by the heterolytic cleavage of O–H bonds—that is, bond cleavage to give an uneven

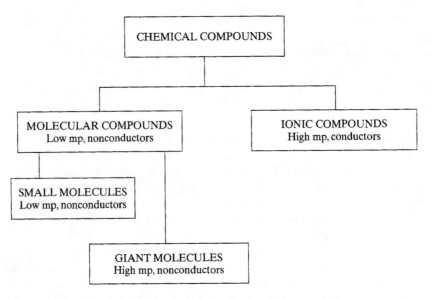

FIGURE 3.2 A general classification scheme for chemical compounds.

distribution of electrons on the two species formed. A water molecule contains two (2) H–O bonds, and if either is to be cleaved to produce ions, one possibility—the correct one—leaves the hydrogen atom with a deficiency of electrons and the oxygen atom with an excess. In this kind of cleavage the hydrogen atom becomes positively charged, H^+. Other evidence suggests that the hydrogen ion—sometimes called the *proton*—is not free in solution, but is associated with another water molecule to form the hydronium ion (H_3O^+); sometimes the fact that the proton is hydrated is ignored and the symbol H^+ is used rather than H_3O^+. The process of ionization is indicated in Eq. (3.1).

$$H_2O + H_2O \leftrightharpoons H_3O^+ + OH^- \tag{3.1}$$

The formation of ions from water [Eq. (3.1)] is an equilibrium process, as described in Chapter 2, that is, water molecules are constantly ionizing, and the ions formed are recombining to reform water molecules. As in all processes at equilibrium, we can write the equilibrium constant for Eq. (3.1) as

$$K = \frac{[H^+][OH^-]}{[H_2O]} = 1.8 \times 10^{-16} \tag{3.2}$$

using the ideas developed previously; notice that the hydrogen ion has been represented for simplicity as H^+ and not H_3O^+. The value for the ionization constant of water has been experimentally determined to be 1.8×10^{-16} at 25°C when the concentration of the species is expressed in moles per liter. A little thought allows us to simplify the equilibrium constant given in Eq. (3.2). Let us consider the molar concentration of water. A liter of water is 1000 g of water, which is the same as $1000/18 = 55.5$ moles of water; in other words, the molar concentration of water in pure water is 55.5 M. Accordingly, we can substitute 55.5 for the quantity $[H_2O]$ in Eq. (3.2):

$$1.8 \times 10^{-16} = \frac{[H^+][OH^-]}{[H_2O]} = \frac{[H^+][OH^-]}{55.5} \tag{3.3}$$

Simplifying Eq. (3.3), we obtain

$$(1.8 \times 10^{-16})(55.5) = 1 \times 10^{-14} = [H^+][OH^-] \tag{3.4}$$

The quantity $[H^+][OH^-]$ is called the *ion product of water* for obvious reasons and is often given the symbol K_w.

$$K_w = [H^+][OH^-] = 1 \times 10^{-14} \tag{3.5}$$

Thus, Eq. (3.5) permits us some quantitative insight into the ionization of water. For example, Eq. (3.1) indicates that as many H^+ as OH^- ions are formed when pure water ionizes; in other words, the concentration of those ions must be equal [Eq. (3.6)] in pure water.

$$[H^+] = [OH^-] \tag{3.6}$$

For both Eqs. (3.5) and (3.6) to be valid, the concentration of both ions must be 10^{-7} M. This conclusion can be obtained simply by substituting Eq. (3.6) into Eq. (3.5) and solving for $[H^+]$.

$$[H^+][H^+] = [H^+]^2 = 1 \times 10^{-14}$$
$$[H^+] = \sqrt{1 \times 10^{-14}} = 1 \times 10^{-7}$$

According to Eq. (3.6), in pure water

$$[OH^-] = [H^+] = 1 \times 10^{-7} \tag{3.7}$$

Thus, we have concluded that in pure water the concentration of both H^+ and OH^- is the same very small number: 1×10^{-7}. It is not surprising that pure water is a very poor conductor of electricity; there are not very many ions present!

Electrolytes

Although pure water is not a good conductor of electricity, it becomes a noticeably better conductor when substances such as H_2SO_4 or $NaCl$ are dissolved in it. Not all substances that dissolve in water, however, give solutions that conduct an electric current to the same extent. Some substances form solutions in water that do not conduct electricity any better than pure water, and these substances are called *nonelectrolytes*. Substances that yield conducting solutions are called *electrolytes,* and the solutions are sometimes referred to as *electrolytic solutions*. The simple experimental arrangement shown in Figure 3.1 can also be used to determine the relative ability of various solutions to conduct electricity. Under equivalent experimental conditions, the conductivity of water is the smallest value obtained for aqueous solutions. Conductivities of all the remaining solutions fall into three broad classes determined by their relative conductivity:

1. Nonelectrolytes, which have a conductivity similar to that of water. Nonelectrolytes contain no ions, other than those arising from the ionization of water.
2. Strong electrolytes, which as a class exhibit the largest conductivity. These solutions contain, essentially, no molecules of the solute.
3. Weak electrolytes, which have a somewhat smaller conductivity than strong electrolytes. These solutions contain a mixture of molecules and ions derived from the solute.

We all know that wire made of certain metals conducts electricity. The passage of an electric current through the wire consists of a flow of electrons that does not bring about a chemical change.

In an analogous manner, the conductivity of solutions arises from the movement of charged particles in the solution, and this process corresponds to the movement of electrons in the wire. The charged particles that give a solution its *conductivity* are not electrons, but *ions;* the passage of an electric current through an electrolytic solution brings about a chemical change, in contrast to the movement of electrons through the wire.

Strong Electrolytes

When substances that consist of a collection of ions in their pure state (ionic compounds) dissolve in water, the ions simply separate in solution. In the solution process, water molecules cluster about ions and screen their charges, which leads to a weaker force of attraction between the dissolved ions. All ionic compounds that dissolve are strong electrolytes. The entire sample of an ionic substance that dissolves exists as ions in solution; there are no molecules of the substance present in these solutions. Examples of strong electrolytes that are ionic in the pure state include KOH, NaOH, and NaCl. All ionic substances would be expected to be stronger electrolytes.

Some covalent compounds also dissolve in water to give solutions of strong electrolytes, which is an interesting fact because covalent compounds contain no ions in the pure state. Thus, HCl gas—a covalent substance—dissolves readily in water to give a solution that is strongly conducting. The only reasonable explanation for this observation is that HCl molecules react with water to form ions [Eq. (3.8)].

$$HCl(g) + H_2O(l) \rightarrow H_3O^+ + Cl^- \qquad (3.8)$$

Various pieces of evidence indicate that no HCl molecules exist in an aqueous solution of HCl gas, showing that the reaction has gone to completion—no equilibrium is established, as indicated by the single arrow in Eq. (3.8). Other important covalent molecules that dissolve in aqueous solution to form strong electrolytes include H_2SO_4 and HNO_3. A list of common strong electrolytes, which also happen to be acids, appears in Table 3.2.

Weak Electrolytes

Although some covalent compounds dissolve in water to give strongly conducting solutions, other covalent substances are weak electrolytes. In both cases water causes these covalent compounds to ionize. The *extent,* or *degree,* of ionization is virtually complete for some covalent compounds and is inherent in the chemical nature of the covalent compound. The terms *weak* and *strong* are used in a *relative* manner to classify all electrolytes. The explanation of the conductance of electricity by solutions of weak electrolytes rests upon a theory suggested by the

TABLE 3.2 COMMON STRONG ACIDS

Formula	Name
HCl	hydrochloric acid
HBr	hydrobromic acid
HI	hydroiodic acid
HNO_3	nitric acid
$HClO_4$	perchloric acid
$HClO_3$	chloric acid
H_2SO_4	sulfuric acid

Swedish chemist Svante Arrhenius (1859–1927) in the year 1887. Arrhenius assumed that the process of dissolving an electrolyte in water brings about the separation of a molecule into ions to at least a small extent. This separation is called *ionization,* or *dissociation,* into ions. Mercuric chloride, $HgCl_2$, is an example of a weak electrolyte, and Eq. (3.9) describes the process occurring in a water solution of this substance.

$$HgCl_2 \rightleftharpoons Hg^{2+} + 2Cl^- \tag{3.9}$$

Again, the use of the two opposing arrows in the ionization equation shows that this is a *reversible* process and that this process is in *dynamic equilibrium.* A solution of mercuric chloride in water contains many un-ionized $HgCl_2$ molecules and only a relatively small number of Hg^{2+} and Cl^- ions, and these are surrounded by solvent molecules. However, the ions constantly recombine to form molecules, and molecules of $HgCl_2$ are in turn constantly undergoing ionization. Thus, the symbolism in Eq. (3.8) is an attempt to describe the fact that $HgCl_2$ molecules are constantly undergoing ionization, while Hg^{2+} and Cl^- ions are recombining. At equilibrium the rate of the forward reaction is equal to the rate of the reverse reaction, and the total number of the various species remains unchanged. Because so few ions are present to act as carriers of the electric current, the solution is a poor conductor, and mercuric chloride is classified as a weak electrolyte. Weak electrolytes are mostly covalent compounds; these are virtually nonconductors, or nonelectrolytes, in the pure state. Other common weak electrolytes include acetic acid (CH_3CO_2H), hydrogen sulfide (H_2S), and ammonia (NH_3). Only upon their dissolution in a solvent such as water do these substances undergo ionization. You should note that the classification of substances on the basis of their ionization in water is dependent upon experimental data. Although this discussion emphasizes water as a solvent, there are many other ionizing solvents—water is merely the most common.

Ionization Constant

As we would expect, solutions of weak electrolytes—such as acetic acid—are chemical systems at equilibrium [Eq. (3.10)].

$$CH_3CO_2H + H_2O \rightleftharpoons CH_3CO_2^- + H_3O^+ \tag{3.10}$$

In the case of acetic acid, the conductivity arises above that of pure water because of the reaction of some acetic acid molecules with water to form hydronium ions (H_3O^+) and acetate ions ($CH_3CO_2^-$). Most of the acetic acid is present as un-ionized molecules. We can write the ionization constant expression for acetic acid in the usual way (see Chapter 2) as

$$K = \frac{[CH_3CO_2^-][H_3O^+]}{[CH_3CO_2H][H_2O]} \tag{3.11}$$

As in the case of pure water we recognize that the concentration of water is a constant. By multiplying both sides of Eq. (3.11) by $[H_2O]$, we obtain

$$K_a = K[H_2O] = \frac{[CH_3CO_2^-][H_3O^+]}{[CH_3CO_2H]} = 1.8 \times 10^{-5} \tag{3.12}$$

where K_a is a new constant because both K and $[H_2O]$ are constants. The subscript "a" in K_a is meant to indicate that K_a is the ionization constant for the substance acting as an *acid.* The value of K_a for acetic acid has been determined experimentally to be 1.8×10^{-5} at 25°C. The low value of the ionization constant indicates that acetic acid does not ionize to an appreciable extent in water; indeed, in a 0.1 M solution of acetic acid only about 2 percent of the molecules are ionized. Acetic acid represents a *class* of substances—weak acids—that are incompletely ionized to yield hydronium ions. The general equilibrium equation for this class of electrolytes is given by

$$HA + H_2O \rightleftharpoons A^- + H_3O^+ \tag{3.13}$$

where A is used to represent the anion of the weak acid, which may be a complex array of atoms. In a manner analogous to that used to obtain the ionization constant for acetic acid [Eq. (3.12)], the ionization constant for Eq. (3.13) is given by

$$K_a = \frac{[A^-][H_3O^+]}{[HA]} \tag{3.14}$$

Each substance in this class exhibits a characteristic low value of K_a. The values of K_a for weak acids are determined from experimental data that provide either the degree of ionization of the electrolyte or the concentration of one of the ions produced in the ionization process. The following examples illustrate the way in which the ionization constant can be calculated from such data.

EXAMPLE 3.1

Conductivity data indicates that the substance MX_2 is a weak electrolyte in aqueous solution. Moreover, the conductivity data reveal that only 1.5 percent of MX_2 undergoes ionization when 0.10 mole is dissolved in sufficient water to make 1 L of solution. Estimate the ionization constant for MX_2.

1. Since MX_2 is a weak electrolyte, the following equilibrium is established:

$$MX_2 \rightleftharpoons M^{2+} + 2X^-$$

2. The equilibrium constant for this process is given by

$$K = \frac{[M^{2+}][X^-]^2}{[MX_2]}$$

3. The number of moles of MX_2 that undergo ionization is given by

$$(0.10)(0.015) = 0.0015 \text{ mole}$$

4. The equation in step 1 indicates the number of moles of M^{2+} or X^- that must be produced for each mole of MX_2 ionized. For 0.0015 moles MX_2 ionizing,

$$\text{Moles } M^{2+} = 0.0015$$
$$\text{Moles } X^- = 2(0.0015) = 0.0030$$

5. Since the volume of solution is 1 L, the molar concentration of the ions is

$$[M^{2+}] = 1.5 \times 10^{-3}$$
$$[X^-] = 3.0 \times 10^{-3}$$

6. The concentration of un-ionized MX_2 is determined by the initial concentration and the amount lost through ionization. Since 0.0015 is much less than 0.10,

$$[MX_2] = 0.10 - 0.0015 \cong 0.1$$

7. Combining the information in steps 5 and 6 with the equilibrium expression in step 2 gives the value of the equilibrium constant:

$$K = \frac{(1.5 \times 10^{-3})(3 \times 10^{-3})}{(0.10)} = 4.5 \times 10^{-7}$$

EXAMPLE 3.2

Assume that an experimental method exists for determining the hydrogen ion concentration in a solution. Furthermore, assume that when the method was applied to a solution prepared by dissolving 0.01 mole of the acid HX in sufficient solution to make 250 mL of solution, the number of moles of H^+ ions was determined to be 1×10^{-4} in this solution. Estimate the value of the ionization constant of the acid HX.

1. The acid HX, when it undergoes ionization, should do so according to the process

$$HX \rightleftharpoons H^+ + X^-$$

2. The ionization constant for HX is given by

$$K_a = \frac{[H^+][X^-]}{[HX]}$$

3. Since 1×10^{-4} mole of H^+ was formed according to the equation in step 1, 1×10^{-4} mole of X^- must also be formed and 1×10^{-4} mole of HX must have been *lost* in the ionization process.

4. The number of moles of HX molecules remaining is

$$0.01 - (1 \times 10^{-4}) \cong 0.01$$

5. The molar concentration of all species can be determined from the information in steps 3 and 4 and the fact that these species are present in 250 mL (0.25 L) of solution:

$$[H^+] = [X^-] = \frac{(1 \times 10^{-4})}{0.25} = 4 \times 10^{-4}$$

$$[HX] = \frac{0.01}{0.25} = 4 \times 10^{-2}$$

6. Substituting the information in step 5 into the equilibrium expression given in step 2,

$$K_a = \frac{(4 \times 10^{-4})(4 \times 10^{-4})}{(4 \times 10^{-2})} = 4 \times 10^{-6}$$

TABLE 3.3 THE IONIZATION CONSTANTS (K_a) OF SOME WEAK ACIDS IN AQUEOUS SOLUTION AT 20°C

Acetic	$HC_2H_3O_2$	1.8×10^{-5}
Arsenic	H_3AsO_4	$K_1 = 2.5 \times 10^{-4}$
		$K_2 = 5.6 \times 10^{-8}$
		$K_3 = 3.0 \times 10^{-13}$
Arsenious	H_3AsO_3	6.0×10^{-10}
Boric	H_3BO_3	6.0×10^{-10}
Carbonic	H_2CO_3	$K_1 = 4.2 \times 10^{-7}$
		$K_2 = 4.8 \times 10^{-11}$
Chromic	H_2CrO_4	$K_1 = 1.8 \times 10^{-1}$
		$K_2 = 3.2 \times 10^{-7}$
Formic	$HCHO_2$	2.1×10^{-4}
Hydrocyanic	HCN	4.0×10^{-10}
Hydrofluoric	HF	6.9×10^{-4}
Hydrosulfuric	H_2S	$K_1 = 1.0 \times 10^{-7}$
		$K_2 = 1.0 \times 10^{-19}$
Hypochlorous	$HClO$	3.2×10^{-8}
Nitrous	HNO_2	4.5×10^{-4}
Oxalic	$H_2C_2O_4$	$K_1 = 3.8 \times 10^{-2}$
		$K_2 = 5.0 \times 10^{-5}$
Phosphoric	H_3PO_4	$K_1 = 7.5 \times 10^{-3}$
		$K_2 = 6.2 \times 10^{-8}$
		$K_3 = 1.0 \times 10^{-12}$
Sulfurous	H_2SO_3	$K_1 = 1.3 \times 10^{-2}$
		$K_2 = 5.6 \times 10^{-8}$

The names of some common weak electrolytes, their formulas, and the values for their ionization constants in aqueous solution are listed in Table 3.3. The larger the value of K_a, the greater the number of ions present compared to molecules, which is equivalent to a greater degree of ionization. Some of the substances listed in Table 3.3 are capable of undergoing multiple ionization processes; in such cases the value for K_a for each step is given.

Note the wide range of values for the ionization constants in this table, indicating that these substances ionize to different extents; however, they are all classified as weak electrolytes.

Even though some reactions do not go to completion because they involve weak electrolytes, it is possible to obtain quantitative information on the species present in such systems if the ionization constant for the weak electrolyte is known. The following example should suffice to illustrate the thought process.

EXAMPLE 3.3

What is the cyanide ion concentration in a solution, which is $0.10\ M$ in HCN?

1. Since HCN is a weak electrolyte (see Table 3.3), the species present in an aqueous solution of this substance are described by:

$$HCN + H_2O \leftrightharpoons H_3O^+ + CN^-$$

2. The equilibrium constant for this system is given by

$$K_a = 4.0 \times 10^{-10} = \frac{[H_3O^+][CN^-]}{[HCN]}$$

where the value of the constant is obtained from Table 3.3.

3. Let us assume that the HCN ionizes to produce x moles per liter of CN^-.

$$[CN^-] = x$$

4. The concentration of all other species can be expressed in terms of x using mass conservation.

$$[H_3O^+] = x$$
$$[HCN] = 0.10 - x$$

5. The unknown quantities in steps 3 and 4 can be substituted into the equilibrium expression to give

$$K_a = 4.0 \times 10^{-10} = \frac{[H_3O^+][CN^-]}{[HCN]} = \frac{(x)(x)}{0.10 - x}$$

6. The expression in step 5 can be solved exactly using the quadratic formula (Appendix I) or it can be simplified by assuming that the amount of HCN lost through ionization (x) is small in comparison with the total amount of HCN present $(0.10\ M)$

$$[HCN] = 0.10 - x \cong 0.10$$

which gives

$$4.0 \times 10^{-10} = \frac{x^2}{0.10}$$

7. Solving for x gives

$$x^2 = 4.0 \times 10^{-11}$$
$$x = 6.3 \times 10^{-6} = [CN^-]$$

Note that it is certainly true that 6.3×10^{-6} is small compared to 0.10, which makes our assumption in step 6 valid.

ACIDS AND BASES

Study of the multitude of known chemical compounds is considerably simplified because many of them can be grouped into a much smaller number of *classes of compounds* on the basis of similarities in their properties. Many different classes of compounds are recognized, each of which embraces hundreds of individual chemicals. To be included in a given classification, a compound must exhibit certain characteristics common to all other substances within the class. The compounds belonging to a particular group may show considerable differences in their specific properties; these, however, are differences in degree rather than in kind.

Acids

All substances designated as *acids* have certain properties in common; these properties are discussed here as they refer to acids in water solution since it is the solvent most commonly employed in the laboratory. Aqueous solutions of acids taste sour, cause certain substances (called *indicators)* to change color, react with active metals to form hydrogen, and neutralize bases. *Indicators* are substances that are sensitive to the presence of either acids or bases and exhibit different colors in different environments. For example, litmus is a natural dyestuff that can exhibit either a blue or a red color; if the blue form of litmus is put into an acid solution, the color immediately changes to red, whereas the red form turns blue in basic solution. The most distinctive chemical property of acids is that their water solutions react with active metals, such as zinc, to liberate hydrogen.

The properties that we have ascribed to acids are based on the reaction of aqueous solutions of these substances, and for our purpose an *acid* is defined as *a substance that increases the hydronium ion concentration in water* when it dissolves in that solvent. Most of the familiar acids in the pure anhydrous (water-free) form are liquids or gases at ordinary temperatures, indicating the covalent bonding that exists in these substances. Hydrogen chloride, hydrogen bromide, and hydrogen sulfide are gases, while pure nitric acid (hydrogen nitrate) and pure sulfuric acid (hydrogen sulfate) are liquids; many organic acids, such as benzoic acid, are readily volatile solids at room temperature. All these substances are poor conductors of electricity in the pure liquid state and do not exhibit any of the properties characteristic of acids. If these covalent, hydrogen-containing substances are dissolved in water, however, they give solutions that conduct an electric current and exhibit acidic properties. Thus, the reactions of hydrogen chloride and hydrogen nitrate with water are described by Eq. (3.8), reproduced below, and Eq. (3.15), respectively.

$$HCl(g) + H_2O \rightarrow H_3O^+ + Cl^- \tag{3.8}$$

$$HNO_3(l) + H_2O \rightarrow H_3O^+ + NO_3^- \tag{3.15}$$

An aqueous solution of hydrogen chloride is properly described as a solution of the ionic substance hydronium chloride, since there are no HCl molecules present in solution. The term *hydrochloric acid* is often used to denote aqueous solutions of hydrogen chloride gas.

It can be shown by various methods that the properties of acids are due to the presence of hydronium ions in solution and that the addition of an acid to water involves a reaction with the solvent. The preceding discussion does not imply that all hydrogen-containing compounds behave as acids or that compounds that do not contain hydrogen cannot act as acids. For example, methane (CH_4) is not acidic in water, but sulfur trioxide (SO_3) gives solutions that have acidic properties. To the extent that methane dissolves in water, these solutions contain only methane molecules. In the latter case, sulfur trioxide reacts with water to form sulfuric acid [Eq. (3.16)], which then ionizes to give an acidic solution [Eq. (3.17)].

$$SO_3 + H_2O \rightarrow H_2SO_4 \tag{3.16}$$

$$H_2SO_4 + 2H_2O \rightarrow 2H_3O^+ + SO_4^{2-} \tag{3.17}$$

Many nonmetal oxides dissolve in water to form acids; these oxides are called *acid anhydrides.* Other examples of acid anhydrides are carbon dioxide (CO_2) and phosphorous pentoxide (P_4O_{10}).

$$CO_2 + H_2O \rightarrow H_2CO_3 \tag{3.18}$$

$$P_4O_{10} + 6H_2O \rightarrow 4H_3PO_4 \tag{3.19}$$

Acids may be conveniently classified in terms of the number of different elements that they contain (Table 3.4). Common *binary acids* include HCl, HBr, and H_2S. The molecules of acids most commonly used in the laboratory and industry contain the elements hydrogen and oxygen, together with one other element that is largely nonmetallic in character. Compounds of this type are called *ternary acids;* some of the most common are H_2SO_4 (sulfuric acid), H_2CO_3 (carbonic acid), HNO_3 (nitric acid), H_3PO_4 (phosphoric acid), and $HC_2H_3O_2$ (acetic acid). In the last-named compound, only one of the four hydrogen atoms present in the molecule can be replaced by a metal, and this is emphasized by the manner in which the formula is written— $HC_2H_3O_2$, *not* $C_2H_4O_2$ or $H_4C_2O_2$.

Not all acidic substances produce the same number of hydronium ions per mole when they react with water. As mentioned earlier in our discussion of strong electrolytes, some acids, such as HCl, undergo complete reaction, so that in solution there are no *molecules* of HCl present; all the HCl molecules react to form the equivalent number of hydronium ions and chloride ions [Eq. (3.8)]. On the other hand, substances such as hydrocyanic acid (HCN) and acetic acid undergo only partial ionization; this is indicated by the double arrow in Eqs. (3.20) and (3.10), respectively.

$$HCN + H_2O \rightleftharpoons H_3O^+ + CN^- \tag{3.20}$$

Acids that are completely ionized (or nearly so) are called *strong acids,* whereas those that are only partially ionized are called *weak acids*; the relative strength of weak acids is indicated by the value of the ionization constant (Table 3.3), which is a measure of the extent of ionization.

Bases

Compounds belonging to the class known as *bases* possess properties decidedly different from those of acids. Bases that dissolve in water form solutions that have a bitter taste and feel slippery or soapy, and most of them cause pronounced irritation when they come into contact with the skin. Solutions of bases are also capable of causing the colors of indicators to change, the change being the reverse of that produced by acids. For example, the red form of litmus turns

TABLE 3.4   SOME BINARY AND TERNARY ACIDS

Binary Acids		Ternary Acids	
HCl	Hydrochloric acid	HCN	Hydrocyanic acid
HBr	Hydrobromic acid	HNO_2	Nitrous acid
HI	Hydroiodic acid	HNO_3	Nitric acid
HF	Hydrofluoric acid	H_2SO_3	Sulfurous acid
H_2S	Hydrosulfuric acid	H_2SO_4	Sulfuric acid
		H_3PO_3	Phosphorous acid
		H_3PO_4	Phosphoric acid

blue when placed in a basic solution. *Bases* are substances that dissolve in water to yield solutions containing hydroxide ions in excess of the number present in pure water.

In much the same way that a replaceable hydrogen ion is the component common to many acids, the hydroxide ion (OH⁻) is the group common to most, but not all, bases. Potassium hydroxide (KOH) and sodium hydroxide (NaOH) are examples of bases that contain hydroxide ions and are appreciably soluble in water. These substances, which contain hydroxide ions in the solid state, increase the concentration of OH⁻ in solution when they are dissolved in water [Eqs. (3.21) and (3.22)].

$$Na^+OH^- \rightarrow Na^+ + OH^- \tag{3.21}$$

$$K^+OH^- \rightarrow K^+ + OH^- \tag{3.22}$$

KOH and NaOH are already ionized in the solid state, and in the presence of water the ions separate. In addition, some substances that do not contain hydroxide ions can be considered as bases because they react with water to form hydroxide ions in solution. Examples of this type of base are calcium oxide and potassium oxide.

$$CaO + H_2O \rightarrow Ca^{2+} + 2OH^- \tag{3.23}$$

$$K_2O + H_2O \rightarrow 2K^+ + 2OH^- \tag{3.24}$$

The names of bases consist of the word *hydroxide,* together with the name of the element or radical in combination with the hydroxide group. In addition to the terms *base* and *hydroxide,* these compounds (and, more particularly, KOH and NaOH) are often referred to by the general term *alkali.*

Most of the metal hydroxides mentioned in this section are strong bases because they are completely ionized in solution. There is, however, one commonly encountered weak base. An aqueous solution of ammonia (commonly called *ammonium hydroxide* or *aqueous ammonia*) contains relatively few ammonium ions (NH₄⁺) and hydroxide ions; consequently aqueous ammonia is classified as a *weak base.*

$$NH_3 + H_2O \leftrightarrows NH_4^+ + OH^- \tag{3.25}$$

The ionization constant for ammonia (K_b) is given by

$$K_b = \frac{[NH_4^+][OH^-]}{[NH_3]} = 1.76 \times 10^{-5} \tag{3.26}$$

Neutralization

Acids react with bases in a process called *neutralization.* The extent of the reaction between acids and bases is governed entirely by the autoionization of water [Eq. (3.1), which is reproduced below].

$$H_2O + H_2O \leftrightarrows H_3O^+ + OH^- \tag{3.1}$$

An acid solution contains more hydronium ions than are present in pure water and a basic solution contains more hydroxide ions. By applying Le Chatelier's principle (Chapter 2) to an aqueous solution that has been made acidic, we determine that the solution still contains some hydroxide ions, but the number is less than that in pure water. This is perhaps an alternative—but indirect—method of describing an acid solution. The equilibrium expressed by Eq. (3.1) indicates that if hydronium ions are added to the system, the position of the equilibrium shifts, in accordance with Le Chatelier's principle, to remove the added hydronium ions. To do this, hydronium ions must react with hydroxide ions to form un-ionized water molecules. The new position of the equilibrium would then correspond to a solution containing a smaller concentration of hydroxide ions than is present in pure water. An analogous process occurs if hydroxide ions are added instead. This means that if the concentration of one of the ionic species (hydronium ions or hydroxide ions) is increased, the concentration of the other decreases. In fact, the ion product of water [Eq. (3.4), which is reproduced below] permits us to determine precisely the hydroxide ion concentration in a solution containing any amount of hydronium ion.

$$[H^+][OH^-] = 1 \times 10^{-14} \tag{3.4}$$

Rearranging Eq. (3.4) we obtain

$$[OH^-] = \frac{1 \times 10^{-14}}{[H_3O^+]} \tag{3.27}$$

Conversely, the hydronium ion concentration in the presence of any concentration of hydroxide ion can be calculated from Eq. (3.28).

$$[H_3O^+] = \frac{1 \times 10^{-14}}{[OH^-]} \tag{3.28}$$

It is clear from the expression of the ion product of water [Eq. (3.4)] that the concentrations of H_3O^+ and OH^- are inextricably linked: The concentration of OH^- increases as the concentration of H_3O^+ decreases, and *vice versa*, if their product is to remain constant at 1×10^{-14}. Thus, mixing an acidic solution with a basic solution involves bringing together two aqueous solutions that, if there were no reaction, would give a solution with an ion product greater than that of water. This situation, however, does not correspond to the real world, and the concentrations change by the reaction of H_3O^+ and OH^- to form water until the correct value of the ion product is achieved. In essence, we apply the Le Chatelier argument to Eq. (3.1). The general discussion of neutralization can be made quantitative for a specific case; let us consider, briefly, the neutralization of HCl and NaOH.

An understanding of acid–base reactions requires that the nature of acidic and basic solutions and of the ionic species responsible for the corresponding sets of characteristic properties be kept clearly in mind. Suppose, for example, that 1 L of 1 *M* hydrochloric acid is to be mixed with 1 L of 1 *M* sodium hydroxide.[1] By definition, the solutions contain 1 mol weight of HCl (1.009 + 35.453 = 36.461 g) and 1 mol weight of NaOH (22.990 + 15.999 + 1.008 = 39.998 g), respectively. Since HCl is a strong acid, it is completely ionized [Eq. (3.8), reproduced below]

$$HCl(g) + H_2O \rightarrow H_3O^+ + Cl^- \tag{3.8}$$

---

1. The symbol 1 *M* means *one molar*; it represents the concentration of a solution in units of moles of dissolved species per liter of solution (see Chapter 4).

and the solution contains 1 mol of H_3O^+ and 1 mol of Cl^-. Similarly, the solution of the strong base NaOH contains 1 gram-molecular weight of Na^+ and OH^- [Eq. (3.21), reproduced below].

$$Na^+OH^- \rightarrow Na^+ + OH^- \tag{3.21}$$

The hydronium ion concentration in the HCl solution is greater than that of pure water and the solution is acidic. The sodium hydroxide solution contains hydroxide ion in excess of the concentration found in pure water, so this solution is basic. Incidentally, the solution of NaOH is not completely devoid of H_3O^+. The equilibrium in Eq. (3.1) indicates that some H_3O^+ must exist in this solution, and the concentration of this species can be calculated from Eq. (3.28). Thus, a 1 M NaOH solution contains 1×10^{-14} M H_3O^+, that is,

$$[H_3O^+] = \frac{1 \times 10^{-14}}{[OH^-]} = \frac{1 \times 10^{-14}}{1} = 1 \times 10^{-14} \tag{3.29}$$

When these two solutions are mixed, *the resulting mixture does not have the properties of either an acid or a base* and is, therefore, said to be *neutral*. Evaporation of the water leaves 1 gram-molecular weight of solid NaCl. These are experimental observations and can only mean that the acidic solution and the basic solution have reacted in a manner such that the ions responsible for the acidic and basic properties have been eliminated. The acid–base reaction must involve the reaction of hydronium ions with hydroxide ions:

$$H_3O^+ + OH^- \rightarrow 2H_2O \tag{3.30}$$

which is the reverse of the process occurring in the ionization of water [Eq. (3.1)]. This reaction is equivalent to stating that the two ionic species have combined to form a covalent compound that, because of its slight degree of ionization, furnishes only very small and equal concentrations of hydronium and hydroxide ions. Since neither H_3O^+ nor OH^- predominates, the resulting solution can exhibit neither the property of acidity nor that of basicity. The solid NaCl remaining after evaporation of the water is easily explained in light of the preceding discussion. The sodium ions and the chloride ions were present in the basic and acidic solutions, respectively, before mixing. They did not take part in the reaction and are still present in the solution after the reaction has occurred. The overall reaction can be represented as

$$(H_3O^+ + Cl^-) + (Na^+ + OH^-) \rightarrow 2H_2O + Na^+ + Cl^- \tag{3.31}$$
$$\text{acidic} \qquad\qquad \text{basic} \qquad\qquad \text{neutral}$$
$$\text{solution} \qquad\qquad \text{solution} \qquad\qquad \text{solution}$$

The pH Scale. It should be noted from the previous discussion that the hydronium ion concentration in aqueous solutions can vary enormously. In 1 M hydrochloric acid the hydronium ion concentration is 1 M; in pure water the concentration of H_3O^+ is 1×10^{-7} M (0.0000001 M); in 1 M NaOH the concentration of H_3O^+ is 1×10^{-14} M (0.00000000000001 M). Rather than attempt to deal with such widely varying numbers, chemists have devised a more convenient scale for comparing hydronium ion concentrations. The pH of a solution is defined as

$$pH = \log\frac{1}{[H^+]} = -\log[H^+] \tag{3.32}$$

Originally, pH was defined in terms of the *hydrogen ion* concentration. Although it is well known that hydrogen ions do not exist as such in aqueous solutions, the original terminology is still retained. Mathematically, taking the logarithm of a set of numbers has the effect of compressing them to a smaller scale. Thus, according to Eq. (3.32), the pH values of a 1 M HCl solution, a neutral solution, and a 1 M NaOH solution are 0, 7, and 14, respectively. The pH of any aqueous solution can be calculated from Eq. (3.32) if the hydronium ion concentration is known. The results of such calculations for solutions with a wide range of hydrogen ion concentrations are summarized in Table 3.5. Solutions with a pH lower than 7 are acidic and solutions with a pH greater than 7 are basic. A neutral solution has a pH of 7. Solutions can exist with pH values outside the range 0–14; however, much of the chemistry of aqueous solutions occurs within the 1–14 pH range.

If, for some reason, we might be interested in acid–base phenomena from the standpoint of the base (the substance which increases the $[OH^-]$), the pH concept can be carried over to this species. Thus, the pOH of a solution is defined as

$$POH = \frac{1}{[OH^-]} = -\log[OH^-] \tag{3.33}$$

Since $[H^+]$ and $[OH^-]$ in an aqueous solution are interrelated through Eq. (3.5), pH and pOH have an equivalent interrelationship, namely,

$$pH + pOH = 14 \tag{3.34}$$

In fact, the idea of the *p-function* has found extensive use where a series of numbers exhibit a very wide range of values, as in the case of the values of the ionization constants of weak acids (see Table 3.3).

TABLE 3.5 THE RELATIONSHIPS AMONG THE HYDRONIUM ION CONCENTRATION, THE HYDROXIDE ION CONCENTRATION, AND pH IN AQUEOUS SOLUTIONS

$[H_3O^+]^a$	$[OH^-]^a$	pH	
1 or 1×10^0	0.00000000000001 or 1×10^{-14}	0	
0.1 or 1×10^{-1}	0.0000000000001 or 1×10^{-13}	1	
0.01 or 1×10^{-2}	0.000000000001 or 1×10^{-12}	2	Acid range
0.001 or 1×10^{-3}	0.00000000001 or 1×10^{-11}	3	
0.0001 or 1×10^{-4}	0.0000000001 or 1×10^{-10}	4	
0.00001 or 1×10^{-5}	0.000000001 or 1×10^{-9}	5	
0.000001 or 1×10^{-6}	0.00000001 or 1×10^{-8}	6	
0.0000001 or 1×10^{-7}	0.0000001 or 1×10^{-7}	7	Neutral
0.00000001 or 1×10^{-8}	0.000001 or 1×10^{-6}	8	
0.000000001 or 1×10^{-9}	0.00001 or 1×10^{-5}	9	
0.0000000001 or 1×10^{-10}	0.0001 or 1×10^{-4}	10	Basic range
0.00000000001 or 1×10^{-11}	0.001 or 1×10^{-3}	11	
0.000000000001 or 1×10^{-12}	0.01 or 1×10^{-2}	12	
0.0000000000001 or 1×10^{-13}	0.1 or 1×10^{-1}	13	
0.00000000000001 or 1×10^{-14}	1 or 1×10^0	14	

^{a} The brackets denote concentration of the enclosed species in moles per liter (M).

Thus, the pK_a of a weak acid is defined as

$$pK_a = -\log K_a \qquad (3.35)$$

Using this definition of pK_a, the pK_a of acetic acid ($K_a = 1.8 \times 10^{-5}$) is 4.7.

Salts

This rather lengthy description of a simple acid–base reaction is intended to show that the reaction of *any acid* and *any base* involves the formation of water from the hydronium ions and the hydroxide ions that are present in the respective solutions.

$$H_2SO_4 + Ba(OH)_2 \rightarrow 2H_2O + BaSO_4(s) \qquad (3.36)$$

It is apparent that the formation of a salt from the negative ion of the acid and the positive ion of the base is a fortuitous circumstance. In some instances the salt that is formed in an acid–base reaction is insoluble in water and separates as a precipitate. Thus, the insoluble salt barium sulfate, the product of the reaction between sulfuric acid and barium hydroxide, precipitates from the reaction mixture.

In view of the facts outlined above, neutralization can now be appropriately defined as *the formation of water by the reaction between the hydronium ions of an acid and the hydroxide ions of the base.* Classifications of chemical compounds are rarely absolute. Some compounds can be classified both as bases and as salts; they are called *basic salts.* Similarly, compounds called *acid salts* have the characteristics of both acids and salts. Calcium hydrogen carbonate is a typical example of an acid salt, and an even more familiar example is $NaHCO_3$, or sodium hydrogen carbonate (sodium bicarbonate). The existence of acid salts depends on the fact that acids containing more than one replaceable hydrogen ion per molecule can be neutralized in a stepwise fashion. The ionization of the weak acid H_2CO_3 is represented by the following equilibria.

$$\textit{Step 1:} \quad H_2CO_3 + H_2O \rightleftharpoons H_3O^+ + HCO_3^- \qquad (3.37)$$
$$\textit{Step 2:} \quad HCO_3^- + H_2O \rightleftharpoons H_3O^+ + CO_3^{2-} \qquad (3.38)$$

To prepare the acid salt $NaHCO_3$, sufficient NaOH is needed to react with the H_3O^+ available from step 1.

$$(H_3O^+ + HCO_3^-) + (Na^+ + OH^-) \rightarrow 2H_2O + Na^+ + HCO_3^- \qquad (3.39)$$

Sodium hydrogen carbonate contains an ion capable of further ionization to provide another H_3O^+ (step 2). The latter can combine with the OH^- from another gram-molecular weight of NaOH to lead to the formation of the *normal salt* (nonacid) sodium carbonate.

$$(Na^+ + H_3O + CO_3^{2-}) + (Na^+ + OH^-) \rightarrow 2H_2O + 2Na^+ + CO_3^{2-} \qquad (3.40)$$

A combination of the two preceding equations represents the complete neutralization of the original carbonic acid in a single equation.

$$H_2CO_3 + 2NaOH \rightarrow 2H_2O + Na_2CO_3 \qquad (3.41)$$

By analogy with acid salts, basic salts contain hydroxide groups that can be neutralized with acids. The following are examples of the neutralization of some common basic salts.

$$Pb(OH)Cl + HCl \rightarrow H_2O + PbCl_2 \tag{3.42}$$

$$Mg(OH)Cl + HCl \rightarrow H_2O + MgCl_2 \tag{3.43}$$

$$Sb(OH)_2Cl + 2HCl \rightarrow 2H_2O + SbCl_3 \tag{3.44}$$

Stepwise Ionization.   The isolation of acid salts and other experimental evidence suggests that polyprotic acids undergo a stepwise ionization process. For example, aqueous solutions of H_3PO_4 contain all the possible phosphorus-containing species indicated in Eqs. (3.45)–(3.47).

$$H_3PO_4 \rightleftharpoons H^+ + H_2PO_4^- \tag{3.45}$$

$$H_2PO_4^- \rightleftharpoons H^+ + HPO_4^{2-} \tag{3.46}$$

$$HPO_4^{2-} \rightleftharpoons H^+ + PO_4^{3-} \tag{3.47}$$

We might be tempted to write the overall reaction for the ionization of H_3PO_4 as

$$H_3PO_4 \rightleftharpoons 3H^+ + PO_4^{3-} \tag{3.48}$$

but Eq. (3.48) does not indicate all the phosphorus-containing species we know to be present in solutions of H_3PO_4. Thus, Eqs. (3.45)–(3.47) provide a more accurate description of a solution of phosphoric acid. Since each of these equations is an equilibrium process, we can write the equilibrium constant expression in the usual way [Eqs. (3.49)–(3.51)].

$$K_1 = \frac{[H_2PO_4^-][H^+]}{[H_3PO_4]} \tag{3.49}$$

$$K_2 = \frac{[HPO_4^{2-}][H^+]}{[H_2PO_4^-]} \tag{3.50}$$

$$K_3 = \frac{[PO_4^{3-}][H^+]}{[HPO_4^{2-}]} \tag{3.51}$$

The equilibrium constant for the overall ionization reaction [Eq. (3.48)] is given by

$$K = \frac{[PO_4^{3-}][H^+]^3}{[H_3PO_4]} \tag{3.52}$$

and it can be shown that the relationship among the stepwise equilibrium constants [Eqs. (3.49)–(3.51)] and the overall equilibrium constant [Eq. (3.52)] is given by Eq. (3.53).

$$K_1 \cdot K_2 \cdot K_3 = K \tag{3.53}$$

since

$$\frac{[H_2PO_4^-][H^+]}{[H_3PO_4]} \cdot \frac{[HPO_4^{2-}][H^+]}{[H_2PO_4^-]} \cdot \frac{[PO_4^{3-}][H^+]}{[HPO_4^{2-}]} = \frac{[PO_4^{3-}][H^+]^3}{[H_3PO_4]} \tag{3.54}$$

The values of the individual stepwise ionization constants have been established for most of the common polyprotic acids (Table 3.3). As we discussed qualitatively in Chapter 2, hydrogen sulfide is an important polyprotic acid in our analysis scheme since it is a common source of S^{2-}, which precipitates certain metal ions from solution. The stepwise ionization constants for H_2S are given by Eqs. (3.55) and (3.56).

$$H_2S \rightleftharpoons H^+ + HS^- \qquad (K_1 = 1.0 \times 10^{-7}) \tag{3.55}$$

$$HS^- \rightleftharpoons H^+ + S^{2-} \qquad (K_2 = 1.0 \times 10^{-19}) \tag{3.56}$$

The value for the overall ionization for H_2S [Eq. (3.57)] is given by Eq. (3.58).

$$H_2S \rightleftharpoons 2H^+ + S^{2-} \tag{3.57}$$

$$K = \frac{[H^+]^2[S^{2-}]}{[H_2S]} = K_1K_2 = 1 \times 10^{-26} \tag{3.58}$$

The concentration of S^{2-} in a solution of H_2S is controlled by the hydrogen ion concentration. Solving Eq. (3.58) for $[S^{2-}]$, we obtain

$$[S^{2-}] = \frac{[1 \times 10^{-26}][H_2S]}{[H^+]^2} \tag{3.59}$$

It should be apparent from Eq. (3.59) that the $[S^{2-}]$ in solution is governed by $[H^+]$; an increase in H^+ (decrease in pH) will cause $[S^{2-}]$ to decrease, while making the solution more basic (decreasing $[H^+]$ or increasing the pH of the solution) will cause $[S^{2-}]$ to increase. We make important use of this principle in the separation of the copper-arsenic group (Procedure 5, Chapter 10) from the aluminum-nickel group (Procedure 15, Chapter 11). The sulfides of the copper-arsenic group are precipitated by the sulfide ion concentration developed from the ionization of H_2S in pure water [Eq. (3.1)], but this concentration of sulfide ion is not sufficiently large to precipitate the sulfides of the aluminum-nickel group. The application of Le Chatelier's principle to the H_2S equilibrium indicates that the S^{2-} concentration will increase if a base is added to a solution containing H_2S. Thus, the sulfides in the aluminum-nickel group are precipitated by H_2S from a solution containing ammonia. The addition of the base to an H_2S solution shifts the equilibrium to the right because H^+ reacts with OH^-. This Le Chatelier shift also increases the S^{2-} concentration.

Hydrolysis of Salts

Because acetic acid is a weak electrolyte (acid), it affects the nature of the species present in salts of acetic acid. By implication, all soluble salts are completely ionized and, if the ions present do not contain easily ionizable hydrogen ions (such as H_2PO_4), we would not expect salts to affect the pH of solutions. Salts should just dissolve to form solutions of their constituent ions. So it is with solutions of, for example, NaCl and other salts of strong acids and strong bases. The salts of weak acids, such as $NaCH_3CO_2$, are another matter. Experimentally, the dissolution of $NaCH_3CO_2$ in pure water yields a basic solution, for example, as indicated by an indicator experiment. As we might expect, the dissolution of $NaCH_3CO_2$ yields hydrated sodium ions and acetate ions.

$$NaCH_3CO_2 \rightarrow Na^+ + CH_3CO_2^- \tag{3.60}$$

Recall that water undergoes self-ionization [Eq. (3.1)]. Thus, an aqueous solution of $NaCH_3CO_2$ contains acetate ions and hydrogen ions. Since acetic acid is a weak acid [see Eq. (3.10)]—which is another way of saying that acetate ions and hydrogen ions would prefer to exist as the undissociated CH_3CO_2H molecules rather than as individual ions—the two ions combine to form molecules of acetic acid. As hydrogen ions are consumed in this process, the water equilibrium [Eq. (3.1)] shifts in accordance with Le Chatelier's principle to maintain a constant ion product [Eq. (3.5)]. In doing so, the hydroxide ion concentration increases, which makes the solution basic in accord with the experimental observation. In effect, the overall reaction is given by Eq. (3.61),

$$CH_3CO_2^- + H_2O \rightleftharpoons CH_3CO_2H + OH^- \tag{3.61}$$

which shows the formation of OH^-. The equilibrium constant for the equilibrium in Eq. (3.61) is given by

$$K = \frac{[OH^-][CH_3CO_2H]}{[CH_3CO_2^-][H_2O]} \tag{3.62}$$

As is the case with most equilibrium expressions that involve water, since the concentration of water $[H_2O]$ is constant, we can combine this value with the ionization constant [Eq. (3.63)] to give a new constant K_H,

$$K_H = K[H_2O] = \frac{[OH^-][CH_3CO_2H]}{[CH_3CO_2^-]} \tag{3.63}$$

and K_H is called the *hydrolysis constant*. Hydrolysis is the general term for the reaction of any chemical species with water. The numerical value of K_H can be obtained from standard data by some simple arithmetic; multiply the right-hand side of Eq. (3.63) by the ratio $[H^+]/[H^+]$ (which, since it is another way of writing unity, does not change the equation).

$$K_H = \frac{[H^+][OH^-][CH_3CO_2H]}{[CH_3CO_2^-][H^+]} \tag{3.64}$$

Note in Eq. (3.64) that the product $[H^+][OH^-]$ is the ion product of water K_w [Eq. (3.5)] and the remaining factor is the reciprocal of the ionization constant for acetic acid [Eq. (3.12)]. Thus, K_H is given by

$$K_H = \frac{K_w}{K_a} \tag{3.65}$$

Since we know the value of K_a for acetic acid (Table 3.3) and the ionic product of water, we can calculate the hydrolysis constant of acetate ion.

$$K_H = \frac{1 \times 10^{-14}}{1.8 \times 10^{-5}} = 5.6 \times 10^{-10} \tag{3.66}$$

A knowledge of the hydrolysis constant for an ion permits us to calculate the pH of a solution of salts of that ion.

EXAMPLE 3.4

Calculate the pH that results from a solution by dissolving 0.05 mole of $NaCH_3CO_2$ in sufficient water to prepare 0.25 L of solution.

1. The molar concentration of salt is

$$\frac{0.05 \text{ mole}}{0.25 \text{ liter}} = 0.2 \text{ mole/liter} = 0.2 \ M$$

2. Since the salt is completely ionized,

$$NaCH_3CO_2 \rightarrow Na^+ + CH_3CO_2^-$$

The initial concentration (before hydrolysis) of the ions in this solution is

$$[Na^+] = [CH_3CO_2^-] = 0.2 \ M$$

3. Acetate ion hydrolyzes according to

$$CH_3CO_2^- + H_2O \rightleftharpoons CH_3CO_2H + OH^-$$

4. For this, the hydrolysis constant expression is

$$K_H = \frac{[CH_3CO_2H][OH^-]}{[CH_3CO_2^-]}$$

5. The value of K_H is given by

$$K_H = \frac{K_w}{K_a} = \frac{(1 \times 10^{-14})}{(1.8 \times 10^{-5})} = 5.6 \times 10^{-10}$$

6. Let x be the concentration of $CH_3CO_2^-$, which hydrolyzes according to step 3. Then the concentrations of all the species in step 3 become

$$[CH_3CO_2H] = x$$
$$[OH^-] = x$$
$$[CH_3CO_2^-] = 0.2 - x$$

7. Substituting the equilibrium quantities deduced in step 6 into step 4 and using the value for K_H gives

$$5.6 \times 10^{-10} = \frac{(x)(x)}{0.2 - x}$$

8. Since step 7 is one equation in one unknown (x), it is theoretically solvable. We can use the solution of the quadratic equation shown in Appendix I or we can try a previous strategy to simplify the equation, namely, assume that the amount of acetate hydrolyzed is small compared to the initial amount present $0.2 \gg x$:

$$0.2 - x \cong 0.2$$

Thus, the equation in step 7 becomes

$$5.6 \times 10^{-10} = \frac{x^2}{0.2}$$
$$x^2 = (5.6 \times 10^{-10})(0.2) = 1.12 \times 10^{-10}$$
$$x = 1.06 \times 10^{-5}$$

9. The pH now can be calculated. Since $x = [OH^-]$, the hydrogen ion concentration can be calculated from the ion product of water:

$$1 \times 10^{14} = [H^+][OH^-] = [H^+][1.06 \times 10^{-5}]$$
$$[H^+] = \frac{1 \times 10^{-14}}{1.06 \times 10^{-5}} = 0.94 \times 10^{-9}$$

By definition

$$pH = -\log[H^+] = -\log(9.4 \times 10^{-10})$$
$$pH = -[0.97 - 10.00] = 9.03$$

The analysis just given for the hydrolysis of the acetate ion applies for the hydrolysis of any salt containing the anion of a weak acid HX. Thus, the alkali metal salt of HX dissolves to form ions,

$$MX \rightarrow M^+ + X^- \tag{3.67}$$

and the anion hydrolyzes [Eq. (3.68)] to form a basic solution.

$$X^- + H_2O = HX + OH^- \tag{3.68}$$

The hydrolysis constant expression is given by

$$K_H = \frac{K_w}{K_a} = \frac{[HX][OH^-]}{[X^-]} \tag{3.69}$$

where K_a is the ionization constant of the weak acid (Table 3.3). A similar analysis can be made for the hydrolysis of the salt of a weak base (ROH) and a strong acid HY. When the salt RY dissolves, it does so completely in ionized form.

$$RY \rightarrow R^+ + Y^- \tag{3.70}$$

The R^+ cations then undergo hydrolysis to form the weak base and a solution that contains $[H^+]$ in excess of pure water.

$$R^+ + H_2O = ROH + H^+ \tag{3.71}$$

The equilibrium expression is

$$K_H = \frac{K_w}{K_b} = \frac{[ROH][OH^+]}{[R^+]}$$

Again the value of K_b can be obtained from the ionization constant of the weak base.

Hydrolysis of Metal Ions

A large number of metal ions become hydrated in aqueous solution, and these solutions are acidic. For example, solutions of $Al(NO_3)_3$, $Cr(NO_3)_2$, $FeCl_3$, $ZnCl_2$, or $MgCl_2$ exhibit all the properties of acids. The source of the hydronium ion in these systems is best understood from the point of view of the hydration of the metal cation. Thus, when $Al(NO_3)_3$ dissolves in water, hydrated aluminum ions are formed [Eq. (3.72)].

$$Al(NO_3)_3 + 6H_2O \rightarrow Al(H_2O)_6^{3+} + 3NO_3^- \tag{3.72}$$

The hydrated metal ion reacts with excess water to produce an acid solution [Eq. (3.73)].

$$Al(H_2O)_6^{3+} + H_2O \rightleftharpoons Al(H_2O)_5(OH)^{2+} + H_3O^+ \tag{3.73}$$

The formation of the hydrated metal ion [Eq. (3.72)] occurs because the metal ion is highly charged, and it attracts the unshared electron pairs on the water molecule (see Chapter 5 for a detailed discussion of metal ion complexation). Water molecules so associated with highly charged metal ions are more acidic than normal water molecules because the charge on the cation draws electron density to it, which increases the polarization of the O–H bond; this sequence of events leads to a situation where the hydrogen on the complexed water molecule is more readily removed as a proton [Eq. (3.73)] than the proton of an uncomplexed water molecule. Intuitively, we might expect water to interact most strongly with metal cations that have a high charge density—that is, a small radius and/or a large positive charge. The charge density (the ratio of ionic charge to ionic radius) for several representative ions is given in Table 3.6. As a matter of reference, the data for Na^+, which does not form hydrates, are included. All the multiply charged cations apparently have sufficiently high charge densities to form hydrates that undergo proton transfer in water.

As with most weak acids it is possible to write an ionization constant for the dissociation process; thus, the equilibrium shown as Eq. (3.74) has an ionization constant of the form

$$K_a = \frac{[Al(H_2O)_5(OH)^{2+}][H_3O^+]}{[Al(H_2O)_6^{3+}]} \tag{3.74}$$

and it has the value shown in Table 3.6. Note that metal cation hydrates have the potential to induce successive ionization of protons. For example, after the first proton is removed from the hydrated aluminum ion [Eq. (3.73)], it is possible to remove two more protons [Eqs. (3.75) and (3.76)].

$$Al(H_2O)_5(OH)^{2+} + H_2O \rightleftharpoons Al(H_2O)_4(OH)_2^+ + H_3O^+ \tag{3.75}$$

$$Al(H_2O)_4(OH)_2^+ + H_2O \rightleftharpoons Al(H_2O)_3(OH)_3 + H_3O^+ \tag{3.76}$$

TABLE 3.6 THE CHARGE DENSITY AND HYDROLYSIS CONSTANTS
FOR SOME TYPICAL METAL IONS

Metal Ion	Ionic Charge/Ionic Radius	Cation Hydrate	K_1
Na^+	1.0	—	—
Cu^{2+}	2.8	$Cu(H_2O)_4^{2+}$	3.0×10^{-8}
Mg^{2+}	3.1	$Mg(H_2O)_6^{2+}$	3.8×10^{-12}
Al^{3+}	6.7	$Al(H_2O)_6^{3+}$	7.2×10^{-6}
Fe^{3+}	4.8	$Fe(H_2O)_6^{3+}$	8×10^{-11}

The final aluminum-containing species in Eq. (3.76) is hydrated aluminum hydroxide, which is insoluble in water. The end result of the hydrolysis of highly charged metal cations is often the formation of an insoluble hydrated hydroxide. Often it is necessary to acidify solutions of highly charged metal cations to keep these species in solution. In the presence of excess acid, Le Chatelier's principle dictates that the equilibrium in Eqs. (3.73), (3.75), and (3.76) be shifted to the left.

Quantitative statements concerning the species in solutions of hydrated metal cations can be made if we know the value of the ionization constant, as in other weakly ionized systems.

EXAMPLE 3.5

What is the pH of an aqueous 0.1 M $AlCl_3$ solution?

1. $AlCl_3$ dissolves in water to produce hydrated aluminum ions, which then undergo hydrolysis.

$$Al(H_2O)_6^{3+} + H_2O \rightleftharpoons Al(H_2O)_5(OH)^{2+} + H_3O^+$$

2. The equilibrium constant for the process described in step 1 is

$$K = 7.2 \times 10^{-6} = \frac{[Al(H_2O)_5(OH)^{2+}][H_3O^+]}{[Al(H_2O)_6^{3+}]}$$

The value of this equilibrium constant appears in Table 3.6.

3. Since we do not know the extent of this hydrolysis, let us assume the molar concentration of H_3O^+ at equilibrium can be represented by x.

$$[H_3O^+] = x$$

4. The concentration of all the remaining species can be established using mass conservation.

$$[Al(H_2O)_5(OH)^{2+}] = x$$
$$[Al(H_2O)_6^{3+}] = 0.1 - x$$

5. Placing the concentration of all species present into the equilibrium expression gives

$$\frac{[Al(H_2O)_5(OH)^{2+}][H_3O^+]}{[Al(H_2O)_6^{3+}]} = \frac{(x)(x)}{0.10 - x} = 7.2 \times 10^{-6}$$

6. Since the hydrolysis constant is small, we can assume that the amount of $Al(H_2O)_6^{3+}$ lost through ionization is negligible and

$$[Al(H_2O)_6^{3+}] = 0.10 - x \cong 0.10$$

7. The expression in step 5 becomes

$$\frac{x^2}{0.1} = 7.2 \times 10^{-6}$$

or

$$x = 8.4 \times 10^{-4} = [H_3O^+]$$

8. The pH is given by

$$pH = -\log[H_3O^+] = -\log(8.4 \times 10^{-3}) = 2.18$$

The detection of bismuth, antimony, and tin in the copper-arsenic group (Chapter 10) is based on the formation of insoluble hydrolysis products formed by these ions in water.

Buffers

As indicated earlier in this chapter, not all acids and bases are completely ionized (Table 3.3). For example, aqueous solutions of HCN contain an appreciable concentration of HCN *molecules* and a relatively small concentration of hydronium ions and cyanide ions [Eq. (3.21)]. In other words, the chemical nature of HCN is such that a mixture of CN^- and H_3O^+ exists preferentially as un-ionized HCN molecules rather than as the separate ions. Thus, a solution of HCN can serve as a source of H_3O^+ even though the concentration of this species is very low. The addition of a substance that reacts with hydronium ions would bring about ionization of some of the HCN molecules, forming more H_3O^+ (and CN^-) to replace the hydronium ions that were removed according to Le Chatelier's principle. This process could obviously continue until all the HCN molecules have reacted. Thus, a solution of HCN *resists* attempts to remove H_3O^+.

Similarly, a solution containing cyanide ions can react with H_3O^+ added to the solution from an external source. For example, if hydronium ions (in the form of a solution of HCl) are added to a solution of sodium cyanide (NaCN), hydronium ions and cyanide ions would react to form HCN. This process occurs because HCN is *weakly* ionized, which is another way of saying that the position of the equilibrium [Eq. (3.21)] favors the molecular form (HCN) rather than the corresponding ions. Thus, a solution of NaCN will *resist* attempts to increase the concentration of H_3O^+ when this species is added from an external source.

In the light of this discussion, it should not be surprising to learn that a mixture of HCN and NaCN will resist attempts to bring about a change in the H_3O^+ concentration of that solution by either the *addition* of an acid or the *removal* of H_3O^+ by a base. Such mixtures are called *buffers*. A mixture of any weak acid or weak base and its salt in water will form a buffer solution.

EXAMPLE 3.6

In the separation of the aluminum-nickel group (Chapter 11) the solution is buffered with a mixture of NH_4Cl and NH_4OH to keep the hydroxides of some of the elements of the next group (mainly Mg^{2+}) from precipitating. As an example of the buffering action of this type of mixture, calculate the concentration of OH^- and the pH of a solution that is 0.20 M in aqueous NH_3 and 0.10 M in NH_4Cl.

1. Since NH_4Cl is a strong electrolyte (a salt) and NH_3 is a weak base, the equilibrium we must consider is

$$NH_3 + H_2O \rightleftharpoons NH_4^+ + OH^-$$

2. The equilibrium constant for this process is given by

$$K_b = 1.8 \times 10^{-5} = \frac{[NH_4^+][OH^-]}{[NH^3]}$$

where the value of K_b is 1.8×10^{-5}.

3. Let us assume that NH_3 ionizes to produce x moles per liter of OH^-.

$$[OH^-] = x$$

4. The concentration of all the other species can be expressed in terms of x using mass conservation.

$$[NH_3] = 0.20 - x$$
$$[NH_4^+] = 0.10 + x$$

In the latter case recall that there are two sources of NH_4^+ one obtained from the complete ionization of NH_4Cl (0.10), and the other from the ionization of NH_3 (x).

5. The quantities in steps 3 and 4 can be substituted into the equilibrium expression to give

$$K_b = 1.8 \times 10^{-5} = \frac{[NH_4^+][OH^-]}{[NH_3]} = \frac{(0.10 + x)(x)}{0.20 - x}$$

6. The expression in step 5 can be solved exactly using the quadratic formula (Appendix I) or it can be simplified by assuming the amount of NH_4^+ gained through the ionization of NH_3 is small compared with the total amount of NH_4^+ already present (0.10). If this assumption is true, it follows that x is small compared to 0.20.

$$[NH_4^+] = (0.10 - x) \cong 0.10$$
$$[NH_3] = (0.20 - x) \cong 0.20$$

which gives

$$1.8 \times 10^{-5} \cong \frac{0.10}{0.20}x$$

7. Solving for x gives

$$x \cong 1.8 \times 10^{-5} \times \frac{0.20}{0.10} \cong 3.6 \times 10^{-5}$$

Note that 3.6×10^{-5} is indeed small compared with 0.10 and 0.20, and it is obvious that our assumption in step 6 is valid.

8. Since $[OH^-] = 3.6 \times 10^{-5}$, we find that pOH = 4.44 and pH = 9.56.

TABLE 3.7   VARIOUS BUFFER MIXTURES AND THE pH RANGE OF THEIR BUFFERING ACTION

Buffer Mixture	pH Range
Citric acid-sodium citrate	3.0–6.2
Acetic acid-sodium acetate	3.6–5.6
KH_2PO_4-K_2HPO_4	6.9–8.0
Boric acid-borax (sodium borate)	8.7–9.2
Sodium carbonate-sodium bicarbonate	9.1–10.7

The solution in Example 3.6 is buffered because it will resist a change in $[OH^-]$ (or $[H^+]$). If acid is added, H^+ will be consumed by reaction with the NH_3 present (essentially $0.20\ M$) according to

$$NH_3 + H_3O^+ \rightleftharpoons NH_4^+ + H_2O \tag{3.77}$$

Added bases react with the NH_4^+: present (essentially $0.10\ M$).

$$NH_4^+ + OH^- \rightarrow NH_3 + H_2O \tag{3.78}$$

It is possible to prepare buffer solutions that will maintain a particular hydronium ion concentration (or pH) by judicious choice of the weak acid or weak base (Table 3.7).

If the amounts of an acid and its salt present in a mixture and the strength of the acid are known, it is possible to calculate the pH at which the solution will be buffered. In fact, in many aspects of chemistry, buffer solutions are sufficiently important that tables giving the proportions of the specific components necessary to prepare solutions within the pH ranges of various buffer systems are available, eliminating the need to make the necessary calculations each time a buffer solution is prepared.

Consideration of the effects of *removing* hydroxide ions from a solution of a weak base and *adding* hydroxide ions to a solution containing the salt of the same weak base leads to the conclusion that mixtures of weak bases and their salts (such as NH_4OH and NH_4Cl) can also serve as buffers.

PROBLEMS

3.1 Indicate whether $0.10\ M$ aqueous solutions of the following substances would be expected to be acidic, basic, or neutral. Give equations describing the nature of the processes occurring when these substances dissolve in water.

    a. NH_4Cl

    b. CH_3CO_2H

    c. SO_2

    d. NH_3

    e. KBr

    f. CH_3OH

    g. $NaCN$

    h. $C_{12}H_{22}O_{11}$

    i. Rb_2O

    j. $NaHSO_3$

3.2 Write the equations for the stepwise neutralization of H_3AsO_4.

3.3 A 0.1 M solution of a weak acid, HX, has a pH of 4.3. Calculate the ionization constant of the acid.

3.4 What is the $[OH^-]$ of the solution described in Problem 3?

3.5 A weak acid HX is 4 percent ionized at 25°C. For a 0.1 M solution of HX, calculate each of the following.

    a. The concentration of all species present at 25°C.

    b. The ionization constant of HX at 25°C.

3.6 Calculate the pH of the solution described in Problem 5.

3.7 Estimate the value of the hydrolysis constant for CN^-.

3.8 Estimate the pH of the solutions described in Problem 3.l(a), (d), (g), and (j).

3.9 Calculate the $[OH^-]$ of a 0.1 M aqueous solution of NH_3.

3.10 Calculate the pH of the solution described in Problem 3.9.

3.11 Calculate the $[H^+]$ in a liter of 0.1 M CH_3CO_2H containing 0.2 mole $NaCH_3CO_2$. Determine the pH of this solution.

3.12 How many moles of KCN must be dissolved in a liter of 0.3 M HCN to yield a solution with $[H^+]$ of $1 \times 10^{-5}\ M$?

3.13 Calculate the pH of a 0.1 M solution of potassium formate, $KCHO_2$.

3.14 Calculate the molar concentration of all species present in the following mixtures.

    a. 0.5 L of 1.0 M HNO_3 and 100 mL of 15 M NH_4OH.

    b. 25 mL of 0.1 M HCl and 10 mL of 0.15 M NH_3.

    c. 100 mL of 0.01 M HCN and 50 mL of 0.02 M KOH.

    d. 10 mL of 0.1 M H_2S and 20 mL of 0.05 M NaOH.

3.15 Calculate the pH of the following solutions.

    a. 0.5 L of 0.05 M formic acid $(HCHO_2)$.

    b. 0.5 L of 0.10 M HCN containing 0.75 M KCN.

    c. A 0.05 M solution of aqueous NH_3.

    d. A 0.10 M solution of $NaNO_2$.

3.16 Establish the concentration of all species present in 0.5 L of a 0.1 M solution of nitrous acid containing 0.05 mole of $NaNO_2$.

3.17 What is the pH of a solution, which is 0.05 M in Cu^{2+}?

3.18 Estimate the concentration of hydrated Mg^{2+} ions $[Mg(H_2O)_6^{2+}]$ in a solution buffered at pH = 6.0.

3.19 What is the sulfide ion concentration in a solution containing 0.1 M H_2S at a pH of 5.0?

3.20 What would the sulfide ion concentration be for the solution described in Problem 3.19 if the pH were changed to 8.0?

Solution Phenomena

Since the scheme for qualitative analysis presented in this text (see Figure 1.2) involves the separation and identification of ions by precipitation, it will be useful to summarize the nature of aqueous solutions and the principles that govern the solubility of species dissolved in such solutions.

A *solution* is a homogeneous mixture with variable composition. Ordinarily, most mixtures that we may encounter are *heterogeneous*; that is, the components of the mixture can be distinguished visually, with the aid of the microscope if necessary. The components of a solution, however, cannot be seen with the most powerful microscope known. Thus, solutions are intimate and random mixtures of particles on the molecular level and appear to be *homogeneous*. The components of a solution never separate spontaneously, although it is possible for them to develop strong density gradients; solutions pass through the finest filters unchanged. The components of a solution distribute themselves in a completely random manner, given sufficient time, much as the contents of a burst gas cylinder distribute themselves throughout a room. For example, a lump of sugar in a glass of water dissolves, and eventually the sugar molecules can be found evenly distributed throughout the water even when mechanical stirring is not employed. Initially the solution near the surface of the solid sugar has a higher proportion of sugar molecules than the solution in the far reaches of the container. Eventually, however, the entire volume of water has a constant distribution of sugar molecules throughout. This phenomenon is similar to the process of diffusion that occurs with gases at a somewhat greater rate, and we must reach the same conclusion concerning the nature of this process in solution: The molecules of sugar—as well as water molecules—must be in constant motion. In other words, the general ideas of the kinetic-molecular theory also apply to the liquid state. In the case of solutions (or any liquids) the molecules present do not move far before they encounter other molecules, which is consistent with the fact that diffusion in liquids is considerably slower than diffusion in gases.

A special nomenclature has evolved for the components of a solution. The *solvent* is the major constituent of the solution and the *solute* is the minor component. If 10 g of alcohol are dissolved in 100 g of water, alcohol is the solute and water is the solvent. If 10 g of water are dissolved in 100 g of alcohol, however, the roles, by definition, are reversed. We should stress that not all substances are completely *miscible*—capable of forming homogeneous liquid phases in all proportions. We will have more to say concerning the solubility of different substances in each other shortly.

The most commonly encountered and most useful solutions are those of substances dissolved in liquids, but as many types of aqueous solutions exist as there are different combinations of the three states of matter. Of all the liquid solvents used in chemistry laboratories, in industry, and at home, water is the most commonly employed and is the best of the inorganic solvents.

The dissolution of a solute in a solvent sometimes corresponds to a single physical process in which solvent molecules separate the constituent particles of the solute. Thus, when solid sodium chloride dissolves in water the solvent molecules separate the Na^+ and Cl^- ions, which are already present in the solid substance. Gaseous hydrogen chloride (HCl) dissolves in both benzene and water. Solutions in benzene contain HCl *molecules* such as those existing in the gaseous state; a benzene solution of HCl is essentially a physical mixture of two kinds of molecules. On the other hand, solutions of HCl in water conduct an electric current, which—as we saw in Chapter 3—is due to the presence of H_3O^+ and Cl^- ions; no HCl molecules exist in aqueous solutions of hydrogen chloride. A more accurate description of the dissolution of HCl in water recognizes that a *reaction* occurs between these substances [Eq. (4.1)] and that the reaction products, H_3O^+ and Cl^-, are soluble in water.

$$HCl\,(g) + H_2O \rightarrow H_3O^+ + Cl^- \tag{4.1}$$

Thus, we need to make a clear distinction between the ordinary physical process of dissolution of a solute and the formation of a solution because a chemical reaction occurs between the solute and the solvent leading to soluble products. Unfortunately, this difference is not always recognized and confusion often results. For example, we may read that "iron dissolves in hydrochloric acid" [Eq. (4.2)]:

$$Fe + 2HCl \rightarrow H_2 + FeCl_2 \tag{4.2}$$

A more proper description of the process is that iron *reacts* with hydrochloric acid, which is an aqueous solution of HCl, to form products that are soluble. The solution contains iron (II) chloride, $FeCl_2$, and not metallic iron, as could be demonstrated experimentally by evaporating the liquid remaining after the iron dissolved.

CONCENTRATION UNITS

Many chemical reactions are conducted in solutions of one kind or another, and it is often necessary to know the amount of solute in the reaction mixture. Since the required amount of solute can usually be dissolved in varying amounts of solvent, a standard method for expressing these quantities is necessary. The *concentration* of a solution is defined as the amount of solute present in a given quantity of solvent. Care must be used at this point because chemists often speak of *concentrated* solutions (conc. HCl), *dilute* solutions, *very dilute* solutions, and so on. These designations give only a rough, relative idea of the concentration of a solution. Such designations are little more than qualitatively useful, but they are, nevertheless, widely employed.

Perhaps the most common way to express the concentration of a solution is on the basis of the weight of solute in a given weight of solvent (also called *weight percent*). For example, a solution can be prepared by dissolving 4.50 g of KCl in 25 g of H_2O. The concentration of this solution could be expressed as 4.50 g KCl per 25 g H_2O, 0.18 g KCl per 1 g H_2O, 18 g KCl per 100 g H_2O, or 180 g KCl per 1000 g H_2O. A solution of the same concentration could be prepared by dissolving 9.00 (2 × 4.50) g of KCl in 50 (2 × 25) g of water or by dissolving

2.25 (0.5×4.50) g of KCl in 12.5 (0.5×25) g of water. Thus, a statement of the concentration of a solution does not imply anything concerning the total amount of solute or solvent in the mixture, but merely gives a commonly accepted standard ratio of solute to solvent in terms of some convenient (and sometimes arbitrary) units. Sometimes the percent concentration of a solution is stated. As might be expected, the percent concentration is the weight of solute present expressed on the basis of 100 g of solution [Eq. (4.3)].

$$\text{Percent concentration} = \frac{(\text{weight of solute}) \times 100}{\text{weight of solute} + \text{weight of solvent}} \tag{4.3}$$

$$= \frac{(\text{weight of solute}) \times 100}{\text{weight of solution}}$$

In our original example the percent concentration of KCl would be determined as

$$\frac{4.50 \text{ g KCl} \times 100}{(25 \text{ g H}_2\text{O} + 4.50 \text{ g KCl})} = 15.3\% \text{ KCl}$$

Because the weight of a liquid (solvent) is not as conveniently determined experimentally as its volume, a more practical unit of concentration involves the volume of solvent or of the solution. It should be recognized that there is a difference between the volume of solution and the volume of the solvent used to prepare a solution. For example, a solution of KCl may contain 1.25 g of KCl per 1000 mL (or 1 L) of solution or 1.25 g KCl per 1000 mL of solvent. These are not exactly the same solutions because the volume of the solution specified in the former case includes the volume of the solute particles.

Not surprisingly, chemists find it convenient to express the amount of solute in terms of the number of *moles* present rather than the number of grams. Recall that when the amount of solute is expressed in moles and the volume of solution in liters, a special name—the *molarity*—is given to the concentration unit; thus, a 1 molar (abbreviated 1 *M*) solution contains 1 mole of solute per 1 L of solution. Expressing the concentration of a solution on a molar basis provides an indirect way of conveniently counting the number of molecules of solute, since 1 mole of any substance contains the Avogadro number (6.02×10^{23}) of molecules. Thus, a 0.25 *M* solution of, for example, sugar contains 1.5×10^{23} ($0.25 \times 6.02 \times 10^{23}$) molecules per liter of solution. This does not imply that we have a liter of solution or 0.25 mole in the sample, but implies only that the ratio of solute to solution, expressed in terms of moles of solute and liters of solution, is 0.25.

EXAMPLE 4.1

Calculate the molarity of a solution prepared by dissolving 2.45 g of H_2SO_4 in sufficient water to make 80 mL of solution.

1. Molarity is defined as

$$\frac{\text{moles of solute}}{1 \text{ L of solution}}$$

2. Calculate the number of grams of solute per liter:

$$\frac{2.45 \text{ g } H_2SO_4}{80 \text{ mL of solution}} = \frac{0.0306 \text{ g } H_2SO_4}{1 \text{ mL of solution}} = \frac{31 \text{ g } H_2SO_4}{1 \text{ L of solution}}$$

3. Convert weight of solute in 1 L to moles of solute in 1 L:

$$\text{Molar mass of } H_2SO_4 = 2(1) + 32 + 4(16) = 98$$

$$\frac{31 \text{ g } H_2SO_4}{1 \text{ L of solution}} \times \frac{1 \text{ mole } H_2SO_4}{98 \text{ g } H_2SO_4} = \frac{0.32 \text{ mole } H_2SO_4}{1 \text{ L of solution}}$$

4. Therefore, the molarity of the solution is 0.32 *M*.

If we know the volume and concentration of a solution, we indirectly know the amount of solute present [Eq. (4.4)].

$$\text{Concentration} \times \text{volume} = \text{amount of solute}$$

$$\frac{\text{Amount of solute}}{\text{Volume of solution}} \times \text{volume of solution} = \text{amount of solute} \qquad (4.4)$$

EXAMPLE 4.2

What weight of KCl is required to prepare 250 mL of a solution containing 0.025 *M* Cl^-?

1. Since KCl is completely ionized in water, the number of moles of KCl required for this problem is also the number of moles of Cl^-.
2. Calculate the number of moles of KCl required:

$$\text{Moles KCl} = 0.025 \text{ mole/L} \times 0.250 \text{ L} = 0.0063 \text{ moles}$$

3. Calculate the weight of KCl:

$$\text{Molecular weight KCl} = 39.1 + 35.5 = 74.6 \text{ g/mole}$$

$$\text{Weight of KCl} = 0.0063 \text{ moles} \times (74.6 \text{ g/mole}) = 0.47 \text{ g}$$

EXAMPLE 4.3

How much solute is present in 30 mL of a 0.6 *M* solution?

1. 30 mL = 0.030 L
2. 0.6 *M* = 0.6 mole solute per 1 L of solution
3. 0.6 mole solute per 1 L of solution × 0.030 L = 0.02 mole solute

SOLUBILITY

In the formation of the majority of aqueous solutions, there is a maximum amount of solute that can dissolve in a given quantity of solvent to form a stable solution under a particular set of conditions. This maximum amount is called the *solubility* of the solute and can be expressed in any convenient concentration unit; such a solution is said to be *saturated*. For example, 100 g of pure water at 25°C can dissolve 151 g of LiI to form a stable saturated solution, but the same quantity of water under the same conditions dissolves only 0.673 g of $PbCl_2$. In some cases there is no upper limit to the amount of solute that a given quantity of solvent can dissolve; such substances are said to be *miscible* in all proportions. For example, a mixture of any two gaseous substances is homogeneous and ethyl alcohol is completely miscible with water.

The solubility of the solute is frequently confused with the rate of solution of that solute. Recall that 100 g of water at 25°C will ultimately dissolve 151 g of LiI; however, this amount of solute dissolves more rapidly if it is finely divided than if it is a single crystal. In both instances when a stable solution is attained and the solution is saturated, 151 g of LiI will have dissolved at 25°C; however, the *rate* of solution will be different. The more rapid dissolution of a finely powdered sample as compared to a single crystal of the same weight results from the fact that a larger surface area is exposed to the action of the solvent. The rate of dissolution can also be increased by stirring (which brings solvent that is less than saturated into more uniform contact with the solute), by increasing temperature, or by the application of both techniques. If the solute is a gas, its solubility can be increased by increasing its pressure; however, an increase in temperature will decrease the solubility of a gaseous solute.

Dissolution: An Equilibrium Process

At first glance, the previous discussion of the process of solution of a solute might suggest that a static condition prevails when a saturated solution has been achieved. It is relatively simple, however, to show experimentally that a *dynamic equilibrium* (Chapter 2) exists between the solute dissolved in solution and excess undissolved solid in contact with the saturated solution. In other words, particles (molecules, atoms, or ions, depending on the solute) on the solid surface are continually going into solution, but since the solution is already saturated, an equal number of particles in solution redeposit on the surface of the excess solid solute.

A state of equilibrium exists when two opposing processes occur at the same rate; such a situation is often depicted by two opposing arrows, as in Eq. (4.5).

$$\text{Solid solute + solvent} \rightleftharpoons \text{dissolved solute} \tag{4.5}$$

With this concept of an equilibrium process, the definition of a saturated solution is concisely and precisely made as a solution that is in equilibrium with an excess of the solute.

Several experimental proofs are available to show that a condition of dynamic equilibrium exists between a saturated solution and excess, undissolved solute. One involves a simple observation that broken crystals become well formed after they have been in contact with a saturated solution. Suppose we prepare a saturated solution of a substance that exhibits an easily recognizable crystalline form—for example, NaCl, which crystallizes in the form of cubes. Take an imperfect crystal (e.g., one with a corner missing), submerge it in the saturated solution, and allow it to remain in contact with the solution; after a period of time the imperfection of the crystal disappears and the crystal is once again its usual cubic form. Moreover, the weight of the crystal has not changed in the process, although careful measurements might show that the cube has slightly shorter edges than the original. We are forced to conclude that a rearrangement of the crystal has occurred during the time it was in contact with the saturated solution. Two possible explanations come to mind. The particles of the NaCl crystal have rearranged themselves by sliding around on the surface until the cubic form was attained. If this were so, then we might expect the broken crystal to become perfect by simply standing in the air; however, such a process has never been observed to occur. If you look at the grains of salt in a shaker, you will note the presence of many broken cubes; if broken cubes of NaCl spontaneously became perfect cubes, we should not expect to see broken cubes in any sample of NaCl because they would be constantly repaired. The other possible—and correct—explanation is that NaCl is dissolving in the saturated solution at the same rate as NaCl is redepositing from solution [recall Eq. (4.5)]. In other words, a state of equilibrium has been established. At equilibrium there is no net transfer of solute, but the dynamic process results in the redistribution of solid NaCl.

An experiment that is more difficult to perform, but one that is perhaps more convincing, involves the use of radioactive tracers. Suppose we prepared a saturated solution of NaCl, as in the preceding experiment, and introduced into it a crystal of NaCl containing radioactive chlorine. After a time, we would find that the undissolved solute crystal had lost some of its radioactivity and the saturated solution gained an equivalent amount of radioactivity. That is, there has been exchange of solute between the saturated solution and the excess solid solute.

The Dependence of Solubility on Experimental Conditions

In general, three major factors—pressure, temperature, and the nature of the solute and the solvent—individually or in combination influence the solubility of the solute in a solvent. All of these factors may not be equally important for a specific case.

Pressure. Commonly observed phenomena support the fact that an increase in pressure increases the solubility of a gaseous solute in a liquid solvent. For example, removal of the cap from a bottle of soda water invariably causes the liquid to effervesce. Soda water consists of carbon dioxide dissolved in water under considerable pressure; when the cap is removed, the pressure on the liquid is reduced to atmospheric pressure and carbon dioxide leaves the solution. Since carbon dioxide leaves the solution at a lower pressure, it follows that the solubility of carbon dioxide in water is dependent on the pressure of the gas above the liquid. As a rule, the solubility of gases increases with increasing pressure.

Pressure generally has little or no effect upon the solubility of solid or liquid solutes in liquid solvents.

Temperature. In general, a change in temperature affects the solubility of gaseous solutes differently than it does the solubility of solid solutes. The solubility of a gas in a liquid solvent decreases with increasing temperature. Most of us have observed the persistent evolution of gas bubbles (after the initial effervescence) when a cold, opened bottle of soda water is allowed to warm to room temperature, or the bubbles (of air) that form in a heated pan of water before the boiling point is reached.

With relatively few exceptions the solubility of solids in liquids increases with temperature; the solubilities of several typical substances are shown graphically in Figure 4.1. In some instances the increase in solubility is very large; for example, the solubility of potassium nitrate is approximately 31 g per 100 g of water at 25°C and about 83 g per 100 g of water at 50°C. On

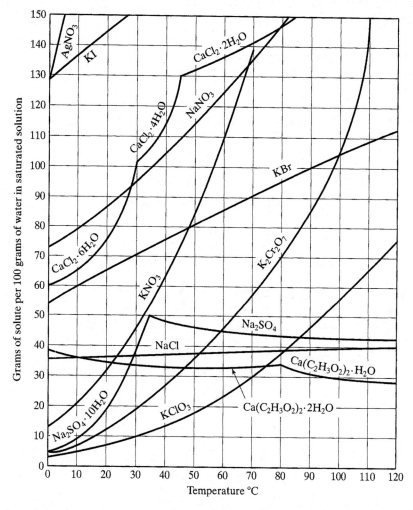

FIGURE 4.1   The temperature dependence of the solubility of several typical substances.

the other hand, the solubility of some solutes, such as ordinary salt (NaCl), shows very little dependence on temperature.

Often such differences in solubility at various temperatures are used to advantage in the preparation, isolation, or purification of substances by the process of crystallization.

Nature of Solute and Solvent.   Numerical values representing the solubility of solutes in solvents differ greatly, depending on the nature of the substances involved. For example, NaCl and other ionic substances are appreciably soluble in water but are, for all practical purposes, insoluble in organic solvents such as benzene, ether, and saturated hydrocarbons, while many covalent organic compounds, such as naphthalene, are insoluble in water. This is not to imply, however, that all organic substances are insoluble in water and all ionic compounds are soluble. Such observations have led to the old generalization concerning solubilities: like dissolves like.

Solubility of Substances in Water

To this point our discussion has not been focused on particular solutes and/or solvents but on general principles. Since the qualitative analysis scheme described here is centered on aqueous chemistry, it would be useful to consider some of the special characteristics of water as a solvent.

Solubility of Polar Substances in Water.   Since the O–H bond is polar and the water molecule is bent (**1**), it possesses a dipole moment, the negative end of which is on the oxygen atom.

$$
\begin{array}{c}
\delta- \\
O \\
H \qquad H \\
\delta+ \\
\mathbf{1}
\end{array}
$$

In addition, water is a highly associated liquid, the individual molecules being held together by *hydrogen bonds*. This type of bond is, perhaps, best described as an electrostatic attraction between the positive end of the polar hydrogen–oxygen covalent bond and the unbonded electron pairs on the oxygen atom in another water molecule (Figure 4.2). Because of the geometry of the water molecule and the fact that each oxygen atom can form two hydrogen bonds, the association of water molecules leads to an open structure rather than one in which the molecules are packed as efficiently as possible; hydrogen bonding accounts for the expansion of water when it freezes.

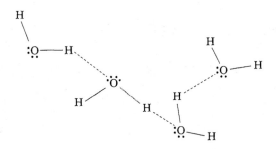

FIGURE 4.2   A schematic representation of hydrogen bonding in liquid water.

It is reasonable to imagine that if a substance is to dissolve in water, there must exist a fairly strong interaction between the water molecules and solute molecules, when compared to the hydrogen-bonded interactions and the interaction between the solute molecules (or ions). In other words, when the solute dissolves, the energy required to break the hydrogen-bonded structure of water must be compensated by a gain in energy because of the solute–solvent interactions present in the solution (Figure 4.3).

Water–solute interactions can occur primarily because the water molecule is polar (**1**) and there is the possibility that either the positive or negative end of the water dipole can become electrostatically attracted to a corresponding charge in the solute species that becomes solvated in the process. This suggests that polar solute molecules should interact with polar water molecules to

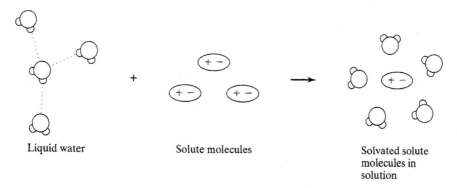

Liquid water   Solute molecules   Solvated solute
                                                                    molecules in
                                                                    solution

FIGURE 4.3   The solvation of polar molecules by water involves the breakup of the water
structure as well as the structure of the bulk polar solute.

form solutions and that nonpolar substances should interact less readily, leading to poor solubility; that is, "like dissolves like." Note that the negative ends of the water molecules associate
with the positive ends of the polar solute molecules and the positive ends of water molecules are
associated with the negative end of the solute molecule.

For the most part, these predictions are verified with real systems; thus, methanol (**2**) and
acetone (**3**), which are both polar molecules, are completely miscible with water—they mix in
all proportions to form a homogeneous mixture.

2

3

4

On the other hand, benzene (**4**) and other nonpolar molecules are markedly less soluble in
water. Thus, we can see the importance of molecular geometry and bond polarity in determining the magnitude of the electrostatic interaction between molecules and, indirectly, the solubility of substances in water. The flat, symmetrical benzene molecule is constituted of atoms
that possess a small electronegativity difference; benzene molecules will not interact strongly

with a polar water molecule, leading to low solubility. On the other hand, the less symmetrical methanol (**2**) and acetone (**3**) molecules, incorporating atoms with larger electronegativity differences, are more soluble in water.

Solubility of Ionic Substances in Water.   We would expect ionic substances to dissolve readily in water to form solutions because ions are charged particles that should electrostatically attract the corresponding end of the water dipole (**1**). Since ions carry a full charge, the attractive forces for ion-water interactions are larger than those for polar molecule-water interactions. To form a solution of an ionic substance (Figure 4.4), energy must be expended to break up the hydrogen-bonded structure of water, as well as the crystal lattice of the ionic solid.

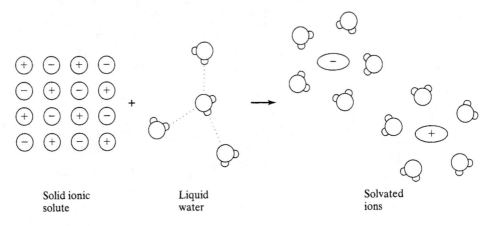

| Solid ionic solute | Liquid water | Solvated ions |

FIGURE 4.4   A schematic representation of the dissolution of an ionic substance in water to form a solution of solvated ions.

This energy must come from the solvation of the ions by water molecules. For the most part, ionic substances are reasonably soluble in water, indicating that sufficient energy is generally obtained from the solvation of the ions to overcome the energy of the crystalline lattice. On the other hand, there are some ionic substances that are not very soluble in water, indicating that the crystal lattice energy is too great to be overcome by the ionic solvation energy. Again, note the arrangement of the polar solvent molecules around the ions; the negative ends point toward the positive ions and the positive ends associate with the negative ions.

Qualitative Solubility Rules

Although an understanding of the principles that govern the interactions between solvent molecules and dissolved species is in hand, the process of quantitative prediction of solubilities is not well developed; the following qualitative observations describing the solubility of ionic compounds have been developed empirically.

1. The nitrates, chlorates, and acetates of all metals are soluble in water. Silver acetate is sparingly soluble.
2. All sodium, potassium, and ammonium salts are soluble in water.
3. The chlorides, bromides, and iodides of all metals except lead, silver, and mercury(I) are soluble in water. Mercuric iodide, HgI_2, is insoluble in water, while $PbCl_2$, $PbBr_2$, and

PbI_2 are soluble in hot water. The water-insoluble chlorides, bromides, and iodides are also insoluble in dilute acids.

4. The sulfates of all metals except lead, mercury(I), barium, and calcium are soluble in water. Silver sulfate is slightly soluble. The water-insoluble sulfates are also insoluble in dilute acids.

5. The carbonates, phosphates, borates, sulfites, chromates, and arsenates of all metals except sodium, potassium, and ammonium are insoluble in water, but soluble in dilute acids. Also, $MgCrO_4$ is soluble in water; $MgSO_4$ is slightly soluble in water.

6. The sulfides of all metals except barium, calcium, magnesium, sodium, potassium, and ammonium are insoluble in water; BaS, CaS, and MgS are sparingly soluble.

7. The hydroxides of sodium, potassium, and ammonium are very soluble in water. The hydroxides of calcium and barium are moderately soluble. The oxides and hydroxides of all other metals are insoluble.

These solubility rules have practical application in the qualitative analysis scheme described in Chapters 9 through the end of this book. For example, the practical effect of Rule 2 is that the cations Na^+, K^+, and NH_4^+ are left to the end of the analysis scheme (Figure 1.1) after all other cations have been removed by the formation of insoluble compounds formed with certain anions. Chloride ions are used to separate Pb^{2+} and Hg^{2+} (Rule 3), while sulfide ions (at two different concentrations) and hydroxide ions are used to separate all other metals (Rule 6). Neither sulfide ions nor hydroxide ions will precipitate Na^+, K^+ and NH_4^+ (Rule 2 and Rule 7). However, sulfate ions are used to separate Ba^{2+} from the other cations in the barium-magnesium group (Rule 4).

These qualitative solubility rules are of great practical importance in the analysis of salts or mixtures of salts (Chapter 14). Anions in a salt mixture often can be eliminated on the basis of cation analysis. Thus, a soluble salt mixture containing silver cannot also contain chloride, bromide, or iodide ions (Rule 3), carbonate, phosphate, borate, sulfite, chromate, or arsenate (Rule 4), or hydroxide (Rule 7), but it may contain sulfate (Rule 4). The salt mixture probably contains nitrate, chlorate, or acetate (Rule 1). There are numerous examples in Part III of the application of these solubility rules.

EQUILIBRIUM PROCESSES INVOLVING SPARINGLY SOLUBLE IONIC SUBSTANCES

Earlier in this chapter we indicated that a state of equilibrium exists between a saturated solution of a solute and excess solid solute; in other words, the solute is dissolving as rapidly as it is precipitating. Such a system is susceptible to a very useful quantitative treatment. Let us consider the sparingly soluble substance calcium carbonate, $CaCO_3$, which is one of the main constituents of limestone. Solid $CaCO_3$ that has come to equilibrium with water is involved with two processes—dissolution and reprecipitation—as described by Eq. (4.6).

$$CaCO_3 (s) \leftrightharpoons Ca^{2+} + CO_3^{2-} \tag{4.6}$$

Since $CaCO_3$ is ionic, we write the products as Ca^{2+} and CO_3^{2-} ions rather than as $CaCO_3$, which would imply molecules. Recalling our discussion of the factors that affect the rate of dissolution of a solute, we would expect that the rate of dissolution of solid $CaCO_3$ would depend on the surface area of the solid. Equation (4.7) expresses this relation in mathematical form:

$$\text{Rate of dissolution} = k_1 \times (\text{area}) \tag{4.7}$$

where k_1 is a proportionality constant, often known as a *rate constant*. The rate at which $CaCO_3$ deposits from solution should be proportional to the surface area available for deposition and the number of times a Ca^{2+} ion collides with a CO_3^{2-} ion at the surface, but the number of Ca^{2+} collisions at the surface depends on the number of these ions in solution; in a similar manner, the number of collisions of CO_3^{2-} with the surface depends on the concentration of CO_3^{2-} ions in solution. Recalling that molar quantities are related to the number of particles present, and using brackets, [ ], to represent molar concentration, the rate of precipitation of $CaCO_3$ is given by Eq. (4.8), where k_2 is another proportionality constant.

$$\text{Rate of precipitation} = k_2 \times (\text{area})[Ca^{2+}][CO_3^{2-}] \tag{4.8}$$

At equilibrium the rate of dissolution is equal to the rate of precipitation [Eq. (4.9)].

$$\text{Rate of dissolution} = \text{rate of precipitation} \tag{4.9}$$

Substituting Eqs. (4.8) and (4.7) into Eq. (4.9) gives

$$k_1 \times (\text{area}) = k_2 \times (\text{area})[Ca^{2+}][CO_3^{2-}] \tag{4.10}$$

Equation (4.10) can be rewritten as

$$\frac{k_1 \times (\text{area})}{k_2 \times (\text{area})} = [Ca^{2+}][CO_3^{2-}] \tag{4.11}$$

The left side of Eq. (4.11) consists of a ratio of two constants (k_1 and k_2), which should be another constant, and the area factors cancel. This new constant is commonly called the *solubility product constant* (K_{sp}); we write Eq. (4.11) as

$$K_{sp} = \frac{k_1 \times (\text{area})}{k_2 \times (\text{area})} = [Ca^{2+}][CO_3^{2-}] \tag{4.12}$$

In general, the solubility product expression for any sparingly soluble ionic substance can be written directly from its formula rather than carrying out the detailed arguments used to develop the equivalent of Eq. (4.12). Thus, given any sparingly soluble electrolyte M_aX_b, we can write the equation that corresponds to its dissolution [Eq. (4.13)].

$$M_aX_b \text{ (s)} \rightarrow aM^{b+} + bM^{a-} \tag{4.13}$$

Since it is ionic, we know from the formula that M^{b+} and X^{a-} ions are formed in the ratio of a to b. The solubility product for this process is written as

$$K_{sp} = [M^{b+}]^a[X^{a-}]^b \tag{4.14}$$

Note that the molar concentrations of the ions are raised to a power represented by the coefficients in the formula. The examples shown in Table 4.1 will suffice to illustrate these ideas.

Numerical values for the solubility product of a substance are easily calculated from its solubility. Let us use $CaCO_3$ as an example for this process.

TABLE 4.1   SOLUBILITY PRODUCT EXPRESSIONS FOR SELECTED SOLUTES

Substance	Dissolution in Water	Solubility Product Expression
AgCl	$AgCl\ (s) \rightleftharpoons Ag^+ + Cl^-$	$K_{sp} = [Ag^+][Cl^-]$
CaF_2	$CaF_2\ (s) \rightleftharpoons Ca^{2+} + 2F^-$	$K_{sp} = [Ca^{2+}][F^-]^2$
$Al(OH)_3$	$Al(OH)_3\ (s) \rightleftharpoons Al^{3+} + 3\ OH^-$	$K_{sp} = [Al^{3+}][OH^-]^3$
$BaSO_4$	$BaSO_4\ (s) \rightleftharpoons Ba^{2+} + SO_4^{2-}$	$K_{sp} = [Ba^{2+}][SO_4^{2-}]$
Bi_2S_3	$Bi_2S_3\ (s) \rightleftharpoons 2Bi^{3+} + 3S^{2-}$	$K_{sp} = [Bi^{3+}]^2[S^{2-}]^3$

EXAMPLE 4.4

The solubility of $CaCO_3$ is 0.012 g/L at 20°C. Determine the solubility product of $CaCO_3$ at 20°C.

1. Calculate the *molar* solubility of $CaCO_3$.
   a. The molar mass of $CaCO_3 = 40 + 12 + 3(16) = 100$
   b. 0.012 g/L $\times$ 1 mole/100 g $= 0.00012\ M$
2. Calculate the concentration of ions in solution. Since

$$CaCO_3\ (s) = Ca^{2+} + CO_3^{2-}$$

we know

$$[Ca^{2+}] = [CO_3^{2-}] = 0.00012\ M$$

3. Calculate the solubility product:

$$K_{sp} = [Ca^{2+}][CO_3^{2-}]$$

$$K_{sp} = (0.00012)(0.00012) = 1.4 \times 10^{-8}$$

The solubility products of many sparingly soluble substances have been determined; Table 4.2 contains a selection of a few of these sparingly soluble ionic substances.

TABLE 4.2   THE SOLUBILITY PRODUCTS[a] FOR SOME COMMON SALTS AT 25°C

Aluminum hydroxide	$[Al^{3+}] \times [OH^-]^3$	5×10^{-33}
Barium carbonate	$[Ba^{2+}] \times [CO_3^{2-}]$	1.6×10^{-9}
Barium chromate	$[Ba^{2+}] \times [CrO_4^{2-}]$	1.2×10^{-10}
Barium oxalate	$[Ba^{2+}] \times [C_2O_4^{2-}]$	1.5×10^{-8}
Barium sulfate	$[Ba^{2+}] \times [SO_4^{2-}]$	1.5×10^{-9}
Bismuth sulfide	$[Bi^{3+}]^2 \times [S^{2-}]^3$	1×10^{-70}
Cadmium hydroxide	$[Cd^{2+}] \times [OH^-]^2$	2×10^{-14}
Cadmium sulfide	$[Cd^{2+}] \times [S^{2-}]$	8×10^{-28}
Calcium carbonate	$[Ca^{2+}] \times [CO_3^{2-}]$	6.9×10^{-9}

Continued

TABLE 4.2 *Continued*

Calcium chromate	$[Ca^{2+}] \times [CrO_4^{2-}]$	1×10^{-4}
Calcium oxalate	$[Ca^{2+}] \times [C_2O_4^{2-}]$	1.3×10^{-9}
Calcium sulfate	$[Ca^{2+}] \times [SO_4^{2-}]$	2.4×10^{-5}
Chromium hydroxide	$[Cr^{3+}] \times [OH^-]^3$	7×10^{-31}
Cobalt sulfide	$[Co^{2+}] \times [S^{2-}]$	6×10^{-22}
Copper (II) hydroxide	$[Cu^{2+}] \times [OH^-]^2$	1.6×10^{-19}
Copper (II) sulfide	$[Cu^{2+}] \times [S^{2-}]$	6×10^{-37}
Iron (II) hydroxide	$[Fe^{2+}] \times [OH^-]^2$	2×10^{-15}
Iron (III) hydroxide	$[Fe^{3+}] \times [OH^-]^3$	6×10^{-38}
Iron (II) sulfide	$[Fe^{2+}] \times [S^{2-}]$	6×10^{-19}
Lead carbonate	$[Pb^{2+}] \times [CO_3^{2-}]$	1.5×10^{-13}
Lead chloride	$[Pb^{2+}] \times [Cl^-]^2$	1.6×10^{-5}
Lead chromate	$[Pb^{2+}] \times [CrO_4^{2-}]$	2×10^{-16}
Lead iodide	$[Pb^{2+}] \times [I^-]^2$	8.3×10^{-9}
Lead sulfate	$[Pb^{2+}] \times [SO_4^{2-}]$	1.3×10^{-8}
Lead sulfide	$[Pb^{2+}] \times [S^{2-}]$	3×10^{-28}
Magnesium carbonate	$[Mg^{2+}] \times [CO_3^{2-}]$	4×10^{-5}
Magnesium hydroxide	$[Mg^{2+}] \times [OH^-]^2$	8.9×10^{-12}
Magnesium oxalate	$[Mg^{2+}] \times [C_2O_4^{2-}]$	8.6×10^{-5}
Manganese hydroxide	$[Mn^{2+}] \times [OH^-]^2$	2×10^{-13}
Manganese sulfide	$[Mn^{2+}] \times [S^{2-}]$	3×10^{-11}
Mercury (I) chloride	$[Hg_2^{2+}] \times [Cl^-]$	1.1×10^{-18}
Mercury (II) sulfide	$[Hg^{2+}] \times [S^{2-}]$	4×10^{-33}
Nickel hydroxide	$[Ni^{2+}] \times [OH^-]^2$	1.6×10^{-16}
Nickel sulfide	$[Ni^{2+}] \times [S^{2-}]$	1×10^{-22}
Silver arsenate	$[Ag^+]^3 \times [AsO_4^{3-}]$	1×10^{-23}
Silver bromide	$[Ag^+] \times [Br^-]$	5×10^{-13}
Silver carbonate	$[Ag^+]^2 \times [CO_3^{2-}]$	8.2×10^{-12}
Silver chloride	$[Ag^+] \times [Cl^-]$	2.8×10^{-10}
Silver chromate	$[Ag^+]^2 \times [CrO_4^{2-}]$	1.9×10^{-12}
Silver iodate	$[Ag^+] \times [IO^{3-}]$	3×10^{-8}
Silver iodide	$[Ag^+] \times [I^-]$	8.5×10^{-17}
Silver phosphate	$[Ag^+]^3 \times [PO_4^{3-}]$	1.8×10^{-18}
Silver sulfide	$[Ag^+]^2 \times [S^{2-}]$	6×10^{-37}
Silver thiocyanate	$[Ag^+] \times [CNS^-]$	1×10^{-12}
Tin (II) sulfide	$[Sn^{2+}] \times [S^{2-}]$	1×10^{-26}
Zinc hydroxide	$[Zn^{2+}] \times [OH^-]^2$	5×10^{-17}
Zinc sulfide	$[Zn^{2+}] \times [S^{2-}]$	3×10^{-23}

[a] Some data taken from R. J. Meyers, *J. Chem. Ed.*, 1986, 687: 63.

The solubility product also gives us an indication of the minimum amount of reagent required to precipitate a given species.

EXAMPLE 4.5

At what pH does $Zn(OH)_2$ just begin to precipitate in a 0.20 M solution of $ZnCl_2$?

1. When precipitation starts, the solution is saturated with respect to $Zn(OH)_2$ and the following equilibrium is established:

$$Zn(OH)_2 \text{ (s)} \leftrightharpoons Zn^{2+} + 2OH^-$$

for which

$$K_{sp} = [Zn^{2+}][OH^-]^2 = 5 \times 10^{-17} \text{ (from Table 4.2)}$$

2. In a 0.20 M $ZnCl_2$ solution,

$$[Zn^{2+}] = 0.20$$

3. Substituting the results in step 2 into the expression for K_{sp} in step 1 permits us to calculate $[OH^-]$ at incipient $Zn(OH)_2$ precipitation.

$$5 \times 10^{-17} = (0.20)[OH^-]^2$$

$$[OH^-]^2 = \frac{5}{0.20} \times 10^{-17} = 2.5 \times 10^{-13}$$

$$[OH^-] = \sqrt{2.5 \times 10^{-13}} = 5 \times 10^{-7}$$

4. Calculate the pH from step 3.

$$pOH = -\log[OH^-] = -\log 5.0 \times 10^{-7}$$

$$pOH = 6.3$$

$$pH = 14 - 6.3 = 7.7$$

In other words, at a pH of 7.7, $Zn(OH)_2$ starts to precipitate from a 0.20 M Zn^{2+} solution.

A knowledge of solubility products can be useful in several ways. Qualitatively we can say that the smaller the value of K_{sp} for a substance, the less soluble is the substance. Also, quantitative estimates of the solubility of a substance can be made from a knowledge of its solubility product, as shown in the following example.

EXAMPLE 4.6

Calculate the molar solubility of CuS.

1. When CuS dissolves, the following equilibrium is established:

$$CuS \text{ (s)} \leftrightharpoons Cu^{2+} + S^{2-}$$

2. Let s be the molar solubility of CuS.

3. When equilibrium is established, s moles per liter of CuS dissolve, but, since we assume CuS is completely ionized, all the dissolved CuS appears as the corresponding ions.

4. The concentration of ions is given by

$$[Cu^{2+}] = [S^{2-}] = s\ M$$

5. The solubility product expression for CuS is given by

$$K_{sp} = [Cu^{2+}][S^{2-}]$$

6. Combining steps 4 and 5,

$$K_{sp} = (s)(s) = s^2$$

7. According to Table 4.2,

$$K_{sp} = 4 \times 10^{-35}$$

8. Combining steps 6 and 7,

$$s^2 = 4 \times 10^{-35}$$

9. Solving for s (the solubility of CuS),

$$s = \sqrt{4 \times 10^{-35}}$$
$$= 6.3 \times 10^{-18}\ M$$

In other words, the molar solubility of CuS is 6.3×10^{-18} moles/L.

Knowledge of the solubility product of a substance also permits us to establish its solubility in a solution containing a common ion. For example, the solubility of TlCl at 25°C is $1.3 \times 10^{-2}\ M$. Using the arguments developed previously (see Example 4.4 above) for the equilibrium established between solid TlCl and its ions in solution [Eq. (4.15)], its solubility product can be calculated at 1.7×10^{-4}.

$$TlCl\ (s) \leftrightarrows Tl^+ + Cl^- \tag{4.15}$$

Now consider the solubility of TlCl in a solution containing $0.10\ M\ Cl^-$; this experimental situation is described by establishing Eq. (4.15) in a solution containing $0.1\ M$ KCl. How much TlCl would dissolve in this solution? When equilibrium is established, some TlCl dissolves according to Eq. (4.15); let us call the number of moles of TlCl that dissolve in a liter of solution s. If equilibrium has been established, the solubility product expression yields Eq. (4.16).

$$K_{sp} = [Tl^+][Cl^-] = 1.7 \times 10^{-14} \tag{4.16}$$

The total chloride ion concentration at equilibrium must be the amount present initially (from KCl that is completely ionized) plus the amount that is formed from the TlCl dissolving [Eq. (4.17)].

$$[Cl^-] = (0.10) + s \tag{4.17}$$

Since Tl^+ ions arise only because of the $TlCl$ that dissolves, the concentration of Tl^+ is given by

$$[Tl^+] = s \tag{4.18}$$

Combining Eqs. (4.16), (4.17), and (4.18) gives

$$(0.10 + s)\,(s) = 1.7 \times 10^{-4} \tag{4.19}$$

Equation (4.19) contains only one unknown quantity and hence can be solved for s; rearranging this equation gives Eq. (4.20), which is a quadratic equation in the unknowns.

$$0.10\,s + s^2 = 1.7 \times 10^{-4} \tag{4.20}$$

Although it is possible to solve the quadratic equation by a relatively simple method, as shown in Appendix I, a more direct approach to the solution is to attempt to simplify Eq. (4.20). Notice that Eq. (4.20) can be simplified if we assume that the chloride ion concentration that arises from the dissolution of $TlCl$ (s) is small compared to the chloride ion concentration of the original solution (0.1 M); this appears to be a reasonable assumption since $TlCl$ is not very soluble in pure water to begin with. We shall have occasion to check the validity of this assumption shortly. Mathematically our assumption is equivalent to writing

$$s \ll 0.10 \tag{4.21}$$

It follows that

$$[Cl] \cong 0.10$$

Combining this and Eq. (4.18) with Eq. (4.16), we obtain

$$(0.10)s = 1.7 \times 10^{-4} \tag{4.22}$$
$$s = 1.7 \times 10^{-3}$$

Thus, the solubility of $TlCl$ in 0.10 M KCl is 1.7×10^{-3} M. The assumption expressed in Eq. (4.21) is certainly valid, since 0.0017 is considerably smaller than 0.10. Note that $TlCl$ is considerably less soluble (1.7×10^{-3} M) in solutions containing Cl^- ions than in pure water (1.3×10^{-2} M). A similar effect would be noted if we attempted to dissolve $TlCl$ in a solution containing excess Tl^+ ions. This decrease in solubility (compared to that in pure water) of a substance in a solution containing an ion found in the solute is called the *common ion effect*, which itself is a "restatement" of Le Chatelier's principle. The common ion effect reflects a good rule of thumb in laboratory practice when one wishes to precipitate a compound from solution. An excess of the precipitating reagent will cause more of the substance to precipitate from solution than if we added the exact—stoichiometric—amount. Another example of a problem involving the common ion effect follows.

EXAMPLE 4.7

Calculate the concentration of Ag^+ and CrO_4^{2-} and the solubility of Ag_2CrO_4 in a solution prepared by mixing an excess of solid Ag_2CrO_4 and 0.1 mole of K_2CrO_4 and sufficient water to make 1 L of solution. The value of K_{sp} for Ag_2CrO_4 is 1.9×10^{-12}.

1. A part of the Ag_2CrO_4 dissolves and at equilibrium the following equation obtains:

$$Ag_2CrO_4 \rightleftharpoons 2Ag^+ + CrO_4^{2-}$$

2. The solubility product expression for Ag_2CrO_4 is

$$K_{sp} = [Ag^+]^2[CrO_4^{2-}] = 1.9 \times 10^{-12}$$

3. When the K_2CrO_4 dissolves, it forms a solution that is $0.1\ M$ in CrO_4^{2-}. If s is the molar solubility of Ag_2CrO_4 in this solution, then the concentrations of the Ag^+ and CrO_4^{2-} ions at equilibrium are

$$[Ag^+] = 2s$$

$$[CrO_4^{2-}] = s + 0.1$$

4. Substituting the concentration of the ions established in step 3 into the expression for K_{sp} in step 2 gives

$$1.9 \times 10^{-12} = (2s)^2(s = 0.10)$$

5. The expression given in step 4 represents an exact solution to the problem. If we could solve for s, we would have the solubility of Ag_2CrO_4 in the mixture; through step 3, the concentrations of the ions of interest could be calculated. Unfortunately, since the equation in step 4 is a cubic equation, it is very difficult to solve it exactly. We can, however, make a simplifying assumption: The contribution of Ag_2CrO_4 to the final CrO_4^{2-} concentration is negligible. Mathematically we express this as

$$s \ll 0.1$$

Thus, the Ag^+ and CrO_4^{2-} ion concentrations become

$$[Ag^+] = 2s \text{ (unchanged)}$$

$$[CrO_4^{2-}] \cong 0.1$$

6. Under this assumption the solubility product expression becomes

$$1.9 \times 10^{-2} = (2s)^2(0.1)$$

Solving for s,

$$(2s)^2 = 1.9 \times 10^{-11}$$

$$s^2 = \frac{(1.9 \times 10^{-11})}{4}$$

$$s = \sqrt{4.75 \times 10^{-12}} = 2.8 \times 10^{-6}\ M$$

7. Note that the assumption is valid since

$$2.8 \times 10^{-6} \ll 0.1$$

Finally, the relative magnitudes of the solubility products of two substances can form the basis for the separation of two or more cations. For example, Zn^{2+} can be separated from Cu^{2+} because of the differences in the solubility product of the corresponding sulfides.

EXAMPLE 4.8

A solution contains Zn^{2+} and Cu^{2+}, each at 0.020 M. If this solution is made 0.10 M in H_3O^+ and saturated with H_2S, can Zn^{2+} and Cu^{2+} be separated?

1. The overall ionization constant for H_2S is given by

$$H_2S = 2H^+ + S^{2-}$$

   for which (see Chapter 3)

$$K = \frac{[H^+]^2[S^{2-}]}{[H_2S]} = 1 \times 10^{-26}$$

2. A saturated H_2S solution is about 0.10 M in H_2S at room temperature. Thus, $[S^{2-}]$ in a 0.10 M H_3O^+ solution at room temperature is given by

$$[S^{2-}] = \frac{1 \times 10^{-26} \times [H_2S]}{[H^+]^2} = \frac{1 \times 10^{-26} \times 0.10}{(0.10)^2} = 1 \times 10^{-25}$$

3. The solubility product of ZnS is (Table 4.2)

$$K_{sp} = 3 \times 10^{-23} = [Zn^{2+}][S^{2-}]$$

4. In a solution containing 0.020 M Zn^{2+}, 0.10 M H_3O^+, and saturated in H_2S, the quantity $[Zn^{2+}][S^{2-}]$ is

$$(0.020)(1 \times 10^{-25}) = 2 \times 10^{-27}$$

   which is less than the solubility product of ZnS, as given in step 3. Thus, Zn^{2+} would not precipitate under the conditions described.

5. Similarly, the question is posed whether CuS would precipitate under these conditions given the fact (Table 4.2) that the solubility product of this substance is

$$K_{sp} = 6 \times 10^{-37} = [Cu^{2+}][S^{2-}]$$

6. In a solution containing 0.020 M Cu^{2+}, 0.10 M H_3O^+, and saturated in H_2S, the quantity $[Cu^{2+}][S^{2-}]$ is

$$(0.020)(1 \times 10^{-25}) = 2 \times 10^{-27}$$

   which is greater than the solubility product of CuS as given in step 5. Thus, CuS would precipitate under the conditions described, and a separation of Cu^{2+} and Zn^{2+} would be effected.

Frequently, the ions that constitute a precipitate can also participate in other equilibria. Thus, precipitates incorporating the anions of weak acids, such as CN^-, CrO_4^{2-}, OH^-, and S^{2-}, are more soluble in acid solution because of the competing reactions of the anions with H^+.

EXAMPLE 4.9

What $[H^+]$ is needed to dissolve completely 0.0010 mole of AgCN in 1 L of solution?

1. The pertinent process involved in the dissolution is given by

$$AgCN + H^+ = Ag^+ + HCN$$

2. The solubility product for AgCN is

$$K_{sp} = 1.6 \times 10^{-14} = [Ag^+][CN^-]$$

when all the AgCN dissolves

$$[Ag^+] = 0.0010 \; M$$

and $[CN^-]$ can be calculated as

$$[CN^-] = \frac{K_{sp}}{[Ag^+]} = \frac{1.6 \times 10^{-14}}{0.0010} = 1.6 \times 10^{-11}$$

3. The ionization of HCN is given by

$$HCN = H^+ + CN^-$$

for which

$$K_a = \frac{[H^+][CN^-]}{[HCN]} = 4.0 \times 10^{-10}$$

When all the AgCN dissolves, the concentration of all the species containing CN^- is given by

$$[CN^-] + [HCN] = 0.0010 \; M$$

Since HCN is weakly ionized, $[CN^-]$ is negligible compared to $[HCN]$, and

$$[HCN] \cong 1.0 \times 10^{-3}$$

4. We can establish $[H^+]$ from the ionization constant of HCN, together with steps 2 and 3.

$$K_a = \frac{[H^+][CN^-]}{[HCN]} = 4.0 \times 10^{-10}$$

$$[H+] = \frac{4.0 \times 10^{-10} \, [HCN]}{[CN^-]}$$

$$\cong \frac{4.0 \times 10^{-10}(1.0 \times 10^{-3)}}{(1.6 \times 10^{-11})}$$

$$\cong 2.5 \times 10^{-2} \; M$$

Note that this is the *final* concentration of H^+. The concentration of H^+ that must be added is

$$[H^+] + [HCN] = 0.025 + 0.001 = 0.026 \, M$$

COLLOIDAL DISPERSIONS

An understanding of the factors affecting the formation and stability of colloids is useful in the study of qualitative analysis. The schemes of separation and identification invariably involve the formation of precipitates. Thus, an understanding of how to produce well-formed precipitates, or what to do if well-formed precipitates do not form, is essential.

If a sample of sand is vigorously stirred with water, a *suspension* results and the sand quickly settles to the bottom of the container when stirring is discontinued. The most readily apparent difference between a suspension and a solution is that the components of the former tend to separate under the influence of gravity, while those of the latter do not. The particles of a suspension consist of large, easily visible clusters, but solute particles in a true solution consist of individual molecules (or ions) and cannot be observed with the most powerful microscope. A *colloidal dispersion* is an intermediate type of mixture that has some of the properties of both suspensions and solutions. Matter in the colloidal state is in the form of very small particles that, individually, are invisible to the unaided eye, but they are larger than molecules or ions. However, a colloidal dispersion often has a cloudy or hazy appearance.

If the sand we used to describe the formation of a suspension previously was very finely ground and the resulting powder stirred into water, we would observe that some of the particles would settle, but much more slowly than before. A continuation process of grinding the solid particles more and more finely ultimately results in the production of particles so small that they would not settle out when suspended in water. They could not be considered to be dissolved, however, because discrete surface boundaries separate the particles—which may consist of thousands of molecules or ions—from the liquid medium in which they exist.

By implication, the size of colloidal particles lies between that of the solute molecules in true solutions and the relatively large particles in suspensions that slowly settle upon standing. It is impossible to define a colloidal particle in terms of some discrete size, because the properties of the resulting dispersions vary with the size of the particles. An ill-defined range of sizes can be considered for the sake of discussion, however. Colloidal particles are usually considered to have diameters between 1 and 100 nm. Even with the best light microscope fitted with an oil immersion objective, colloidal particles are too small to be seen, since the lower limit of resolution of such optical systems is about 100 nm. The limits on the colloidal-size range are somewhat indefinite, but unfortunately, chemists have acquired the bad habit of designating the diameter of colloidal particles. Colloidal particles are, perhaps, best described in terms of behavior and properties rather than size or weight.

Properties of Colloids

Colloidal dispersions exhibit a unique set of properties, none of which is characteristic of a true solution. In contrast with true solutions, colloidal materials have essentially no effect on the vapor pressure, boiling point, or freezing point of the dispersion medium.

The Tyndall Effect.   Colloidal particles scatter light; an incident beam of light is reflected or scattered in a random way by colloidal particles. The light-scattering ability of colloidal particles is called the *Tyndall effect* after the English scientist John Tyndall, who first described the phenomenon. A true solution containing solute molecules homogeneously mixed with solvent molecules does not scatter light, and this simple observation can be used to distinguish between colloidal dispersions and true solutions.

Brownian Motion.   If instead of attempting to look directly at a colloidal particle through a light microscope, a colloidal particle is viewed with a high-powered microscope placed at right angles to a light beam, a random motion of the particle is observed. This motion, first observed by the Scottish botanist Robert Brown in a suspension of plant pollen grains in water, is called *Brownian motion*. Briefly, Brownian motion arises because the molecules of the dispersion medium are in constant and random motion, bombarding the colloidal particles in a random manner. The bombardment is unbalanced since it is highly improbable that a given colloidal particle could be hit from opposite directions at the same time by precisely the same number of solvent molecules traveling at the same velocity; there occurs a net imbalance of forces on the colloidal particle to which it responds. Brownian motion is an important factor in keeping colloidal particles from settling out of the dispersion medium. Brownian motion is not observed in ordinary suspensions because the suspended particles are too large in comparison with the forces exerted by the impacts of solvent molecules.

Adsorption.   One of the important properties of colloidal dispersions is the large total surface area that the dispersed phase exhibits. Some colloidally dispersed phases have been estimated to exhibit a total surface area in excess of 1 acre/g of colloid.

Ordinarily, the fact that molecules can exert considerable forces of attraction for other molecules is not readily discernible, but the large surface area of colloidal dispersions makes this phenomenon easier to observe. By using simple experiments, it is possible to show that substances dissolved in the dispersion medium may be concentrated on the surface of the colloidal particles. This adherence to a surface is called *adsorption* (not absorption). Colloids are highly selective in the adsorption of dissolved species, which can be either molecules or ions. Some types of colloidal particles can adsorb only negative ions on their surfaces, and others can adsorb only positive ions. Since all particles of a given colloid adsorb ions of the same charge, these particles repel each other; this is another factor that tends to keep colloids from settling out of the dispersion medium or clustering together to form larger particles.

Colloidal dispersions containing particles on which ions are adsorbed conduct an electric current, the charged colloidal particles migrating toward the electrode bearing a charge opposite to that of the adsorbed ions. Since all the colloidal particles of a dispersion have the same charge, all the particles migrate toward the same electrode. There are many practical consequences of this phenomenon; for example, the adsorption of ions on finely dispersed precipitates, which gives each particle the *same* charge, leads to difficulty in coagulating a precipitate to make it more easily separable from the solution.

An important consideration in the separation schemes in qualitative analysis involves techniques to neutralize the charge of the ions adsorbed on precipitates. Although adsorption may complicate the separation of cations, this phenomenon can be employed to advantage in some cases. For example, aluminum ion (Al^{3+}) is identified by the fact that $Al(OH)_3$ is insoluble and it adsorbs the dye *aluminon* from solution to form a characteristic red gelatinous mass referred to as a *lake*; $Al(OH)_3$, itself, is a white gelatinous substance that often takes on a translucent appearance, which may be difficult to observe. Along similar lines, Mg^{2+} is identified as $Mg(OH)_2$, which forms a blue lake with another dye.

Methods of Colloid Formation

From a broad point of view, all methods for the preparation of colloids are classified as either *dispersion* or *condensation* processes. In dispersion methods, bulk matter is broken down into smaller and smaller units until it finally becomes colloidal in size. Such methods are not particularly of interest in qualitative analysis, where colloids are most often formed through condensation methods. Condensation methods start with dissolved substances in the ionic or molecular form, and these species combine into aggregates of colloidal size. Processes of this type usually involve relatively rapid chemical reactions because a slow rate of reaction favors the formation of larger crystals, which do not remain dispersed. Chemical reactions that occur in solution to yield an insoluble product often form colloidal dispersions. Thus, the rapid addition of hydrogen sulfide to arsenious acid produces a yellow colloidal suspension of arsenious sulfide [Eq. (4.23)].

$$2H_3AsO_3 + 3H_2S \rightarrow As_2S_3 + 6H_2O \tag{4.23}$$

The resulting arsenious sulfide particles provide a typical example of a colloid that adsorbs negative ions. In a similar manner, reddish brown, colloidal ferric hydroxide, which adsorbs positive ions, can be prepared by the reaction between ferric chloride and water [Eq. (4.24)].

$$FeCl_3 + 3H_2O \rightarrow Fe(OH)_3 \, (s) + 3H^+ + 3Cl^- \tag{4.24}$$

This particular type of reaction in which a substance reacts with water is called *hydrolysis*, or *a hydrolysis reaction*, and the substance that reacts with water is said to be *hydrolyzed*. Many of the precipitates formed during the operations of qualitative analysis give finely divided precipitates that show all the properties of the colloidal state. The precipitates are often so finely divided that they do not settle when centrifuged; they cannot be removed by filtration because the particles are so small that they pass through the pores of the filter paper. The substances that cause the most trouble in this respect are the hydroxides and sulfides of the metals and occasionally elemental sulfur when it is formed (as a by-product).

Cold water tends to cause certain insoluble substances to be dispersed in a colloidal condition, the sulfides being especially susceptible. In this finely dispersed state, the particles remain in suspension. To avoid this dispersion, hot wash water may be used; also, the wash water may contain a small amount of ammonium acetate or some other electrolyte (for example, see the last paragraph of Procedure 5, Chapter 10).

Colloids frequently form when they are not wanted, and it is important to know how to destroy colloids. Before chemists arrived at a reasonable understanding of the factors responsible for the stability of colloids, efforts to destroy them were restricted to the unsatisfactory trial-and-error approach. With the gradual accumulation of new information concerning the nature of colloids, it became apparent that the properties of colloids are primarily dependent on their large exposed surfaces. Once chemists came to understand the role of adsorption upon surfaces and the effect of such factors as adsorbed ions and protective films, it became relatively easy to devise methods to counteract the factors that cause colloids to be stable.

Counteracting Colloid Dispersions. Some colloids may be coagulated merely by boiling them. The increase in temperature increases the kinetic energy of the colloidal particles, causing them to move more rapidly and to collide more frequently and with a greater force until they coagulate into larger aggregates that gradually settle out (precipitate).

Centrifugation provides another method for separating neutral colloidal particles from the dispersing medium. Particles in solutions or suspensions are, in effect, placed in an increased gravitational field by the centrifuge; the higher the speed (revolutions per second) at which the centrifuge spins the samples, the greater the centripetal force on the particles. Particles that would not settle out or would do so only very slowly under the force of the earth's gravity often separate more rapidly on centrifugation. There is a direct relationship between the mass of a particle and the speed required to cause it to separate.

Electrically charged colloidal particles repel each other; this type of colloid can be coagulated by neutralizing the charges, either electrically at an electrode or by adding a large amount of an electrolyte. When an electrolyte is added, the charges of the adsorbed ions are neutralized, and the colloidal particles no longer repel each other. Upon collision of the uncharged particles, larger aggregates form and precipitate. Precipitating reagents used in qualitative analysis are electrolytes; consequently, the excess of the precipitant helps coagulate the precipitate. In certain cases, as in the precipitation of the copper-arsenic group (see Procedure 5, Chapter 10) and the aluminum-nickel group (see Procedure 15, Chapter 11), the presence of a strong electrolyte ($NH_4CH_3CO_2$ or NH_4Cl) helps ensure coagulation of the precipitate.

PROBLEMS

4.1 A solution is prepared by dissolving 15.4 g of NaCl in 94.6 g of water. What is the percent concentration of this solution?

4.2 If the density of the solution described in Problem 4.1 is 1.009 g/mL, what is its molarity?

4.3 Determine the molarity of a solution prepared by dissolving 4.62 g HNO_3 in sufficient water to make 90 mL of solution.

4.4 If the solution prepared in Problem 4.3 is 5.0 percent nitric acid, determine its density.

4.5 What weight of solute is present in the following?

  a.   50 mL of 0.056 M H_2SO_4

  b.   23 mL of 0.18 M NaCl

  c.   700 mL of 0.2 M KOH

  d.   1 mL of 15 M NH_3

4.6 What weight of solute is necessary to prepare the following solutions?

  a.   50 g of a 0.025 percent solution of H_2SO_4

  b.   25 mL of a 0.15 M solution of NaCl

  c.   150 mL of a 1.55 M solution of HCl

4.7 Write the solubility product expressions for the following substances.

  a.   AgI

  b. $CaSO_4$

  c. $Ca(OH)_2$

  d. LaF_3

  e. $La_2(SO_4)_3$

4.8 Lanthanum molybdate, $La_2(MoO_4)_3$, has a solubility of 0.0017 g per 100 mL of solution at 25°C. Calculate the solubility product for this substance.

4.9 Given the solubilities of the following solutes in grams per 100 g of solvent, calculate their solubility products.

    a. $Yb_2(C_2O_4)_3$ has a solubility of 0.00033.

    b. SnS_2 has a solubility of 0.00002.

4.10 Calculate the molar solubility for the following substances.

    a. PbI_2

    b. Ag_2S

    c. $Fe(OH)_3$

4.11 Calculate the molar solubility of PbF_2 in a solution containing 0.1 M NaF.

4.12 What is the Ag^+ concentration in a saturated solution of AgCl containing 0.1 M HCl?

4.13 Calculate the pH of a solution saturated in $Al(OH)_3$.

4.14 A solution formed by equilibrating water with solid Ag_3PO_4 was found to contain 4.8×10^{-5} mole per liter of Ag^+. Calculate the solubility product of Ag_3PO_4.

4.15 What molar concentration of Ag^+ must be presented to start the precipitation of AgCl from a solution containing 1×10^{-3} M Cl^-?

4.16 Pure $Ca(OH)_2$ was equilibrated with water, and the resulting saturated solution was observed to have a pH of 12.3. Calculate the solubility product of $Ca(OH)_2$.

4.17 Calculate the molar solubility of $Fe(OH)_3$ in a solution with pH = 9.3.

4.18 What is the concentration of Zn^{2+} in a solution saturated with $Zn(OH)_2$ at a pH of 11.1?

4.19 A solution contains 1×10^{-2} M Ag^+. How many drops of 0.10 M NaOH are required to just start the precipitation of AgOH? Assume there are 20 drops per milliliter and the solubility product of AgOH is 1.52×10^{-8}.

4.20 Consider the information in Problem 4.19. How many drops of NaOH are required to decrease the molar concentration of Ag^+ to 1×10^{-6}?

4.21 What is the pH of the final solution described in Problem 4.20?

Complex Compounds

LEWIS THEORY OF CHEMICAL COMBINATION

In Chapter 3, we developed the experimental difference between two classes of compounds: ionic compounds, which have high melting points and are good conductors of electricity in the pure molten state; and covalent compounds, which possess low melting points and are relatively poor electrical conductors. Gilbert N. Lewis formulated a simple theory of chemical combination involving the outer electrons of an atom by which elements could achieve a stable electronic arrangement. Since compounds of the rare gases were unknown to Lewis, he assumed that the electronic arrangements of rare gas atoms were the most stable that could be attained. Accordingly, it was logical to assume that when compounds formed, the constituent atoms achieved a stable—or rare gas—electronic configuration. According to Lewis, this result could be achieved either by a complete loss or gain of electrons or by a sharing of electrons.

Ionic Compounds

Recall from Chapter 3 that the high melting and conducting compounds are constituted of ions, hence, their classification as ionic compounds. The Lewis Theory describes how ions can be formed from atoms.

Elements that have one or two electrons more or less than a rare gas atom can achieve a stable electronic configuration by either losing or gaining electrons. Thus, for example, atoms such as fluorine and oxygen, which are near the end of their respective periods, can gain one or two electrons, respectively.

$$F(1s^2 2s^2 2p^5) + e^- \rightarrow [F(1s^2 2s^2 2p^6)]^- \tag{5.1}$$

$$O(1s^2 2s^2 2p^4) + 2e^- \rightarrow [O(1s^2 2s^2 2p^6)]^{2-} \tag{5.2}$$

Oxygen and fluorine thus can be expected to exist as anions (combined with cations) in many of their compounds. Conversely, atoms that are near the beginning of their respective periods, such as sodium and magnesium, can achieve a rare gas electronic structure (that of neon) by a loss of one and two electrons, respectively.

$$Na(1s^22s^22p^63s^1) \rightarrow [Na(1s^22s^22p^6)]^+ + e^- \qquad (5.3)$$

$$Mg(1s^22s^22p^63s^2) \rightarrow [Mg(1s^22s^22p^6)]^{2+} + 2e^- \qquad (5.4)$$

In attaining the neon configuration, these atoms acquire a $1+$ and $2+$ charge. The combining powers (or valences) of these cations are 1 and 2, respectively, because it takes one and two counter ions, respectively, of charge 1^- to form neutral compounds. In essence, the Lewis theory states that ionic compounds are formed because the constituent atoms have gained or lost electrons to acquire the nearest rare gas electronic structure; the valences of ionic species are related to their charges. Lewis was thus able to account for the stoichiometry and the ionicity of the compounds formed by certain groups of elements in the periodic arrangement.

Covalent Bonds

Lewis proposed the concept of the covalent bond to account for the properties of molecular compounds. The formation of a covalent bond occurs when two atoms share a pair of electrons. In the formalism of electron counting, the bond is most often formed between two atoms that each contribute one electron to the bond. These ideas are applied to the nonionic fluorides of the period 2 elements in Figure 5.1. Although it is impossible to distinguish the original source of the electrons in these systems, the electrons from the central atom are designated by x, whereas those on other atoms are indicated by $\cdot$; you should remember that this convention is merely a "bookkeeping" convenience. When covalent bonds are formed in this series of compounds, *each* atom in *all* the molecules (except BF_3) achieves a rare gas structure.

Since neutral species are formed, there is little association between the molecules except in very special situations. These covalently bonded substances exhibit relatively low melting points and boiling points; since ions are not present, these substances do not conduct an electric current in the pure molten state (Chapter 3).

Coordinate-Covalent Bonds

Although the Lewis theory leads to a simple correlation of many observations, there are still obvious difficulties. For example, the covalent compounds of boron cannot achieve a rare gas structure

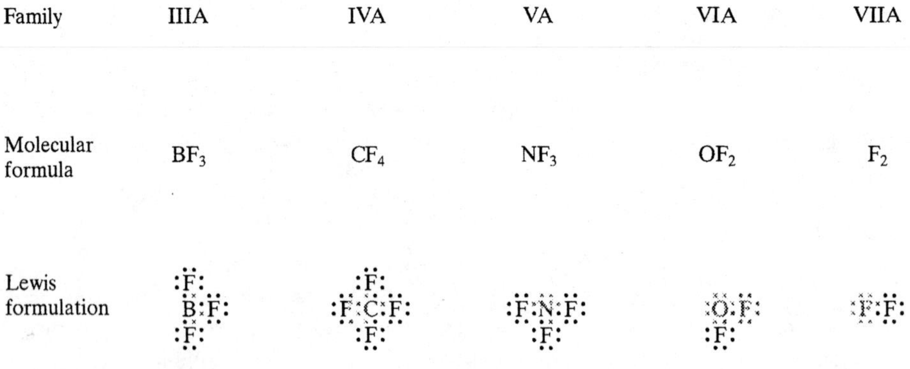

Family	IIIA	IVA	VA	VIA	VIIA
Molecular formula	BF_3	CF_4	NF_3	OF_2	F_2
Lewis formulation					

FIGURE 5.1   The Lewis formulation for the covalent fluorides of the period 2 elements.

electronic configuration because boron has only three valence electrons available for covalent bond formation (see Figure 5.1). This electron deficiency of boron compounds is manifested in their ability to react with substances that possess non-bonded electron pairs to form products in which the boron atom has achieved a rare gas configuration (Figure 5.2). The formation of the BF_4^- ion and H_3NBF_3 involves the formation of a covalent (a shared-electron pair) bond in which the bonding electron pair is supplied by one atom—namely, F^- in the case of BF_4^- and the nitrogen atom in H_3N in the case of H_3NBF_3. Lewis called this kind of bond a *coordinate-covalent* bond. Thus, although some atoms cannot achieve a rare gas structure in the formation of simple covalent compounds, the chemistry of these substances reflects this electron deficiency. Conversely, some species—like NH_3 or F^-—possess electron pairs that are not originally involved in bond formation, but can be involved under certain conditions.

The species that accepts the electron pair in such a bond is called a *Lewis acid*, and that which donates the electron pair is a *Lewis base*. Virtually all the representative elements in period 3 and higher can form compounds that behave as Lewis acids, such as $MgCl_2$, $AlCl_3$, ICl_3, $SiCl_4$, and PF_5 (see Table 5.1). In addition, compounds of the elements that occur before carbon in period 2—such as $BeCl_2$ and BF_3—also behave as Lewis acids. Lewis bases, on the other hand, are species that carry pairs of electrons not initially involved in bonding; such species are usually anions (such as the halide ions) or simple neutral molecules such as amines (NR_3), water (H_2O or HOH), and its derivatives (ROH, ROR).

For the purposes of brevity, R generally means a carbon-containing group attached to another atom; for example, R could be a methyl group (CH_3), in which case ROH and R_2O mean methyl alcohol (CH_3OH) and dimethyl ether [$(CH_3)_2O$], respectively. Table 5.1 contains some examples of Lewis acid–base reactions involving the representative elements.

Transition metal ions also act as Lewis acids, forming compounds that can be imagined to contain coordinate-covalent bonds. For example, anhydrous copper sulfate, $CuSO_4$, is a colorless substance, but dissolution in water gives a sky-blue solution; in liquid ammonia, a purple solution forms. If potassium chloride is added to an aqueous solution of $CuSO_4$, a bright green

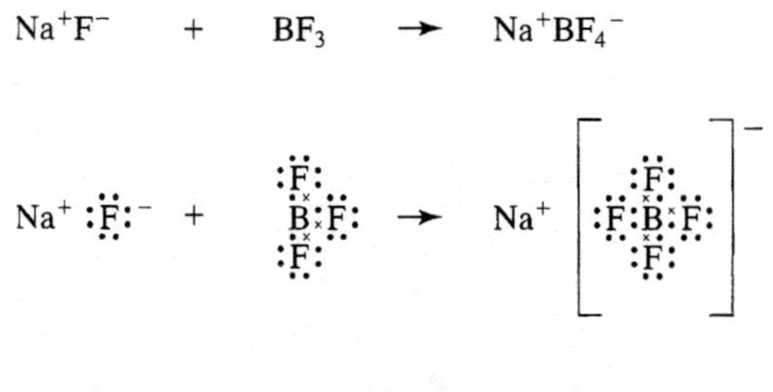

FIGURE 5.2 Atoms that have a deficiency of electrons, in a Lewis sense, can form additional electron-pair bonds. The product may be ionic or it may be a neutral species.

TABLE 5.1 EXAMPLES OF LEWIS ACID–BASE INTERACTIONS AMONG SPECIES OF THE REPRESENTATIVE ELEMENTS

Group	Period	Lewis Acid	Lewis Base	Coordinately Bonded Species
VII	5	ICl_3	Cl^-	ICl_4^- $(KICl_4)$
VI	4	$SeCl_4$	Cl^-	$SeCl_6^{2-}$ (K_2SeCl_6)
V	3	PF_5	F^-	PF_6^- (KPF_6)
IV	3	SiF_4	F^-	SiF_6^{2-} (K_2SiF_6)
III	3	$AlCl_3$	NH_3	$AlCl_3 \cdot NH_3$
			Cl^-	$AlCl_4^-$ $(KAlCl_4)$
III	2	BF_3	F^-	BF_4^- (KBF_4)
			R_2O	F_3BOR_2
II	2	$BeCl_2$	Cl^-	$BeCl_4^{2-}$ (K_2BeCl_4)
			R_3N	$Cl_2Be\,(NR_3)_2$
II	3	$MgCl_2$	Cl^-	$MgCl_4^{2-}$ (K_2MgCl_4)
			NH_3	$Mg(NH_3)_6^{2+}$ $[Mg(NH_3)_6](ClO_4)_2$

solution forms. It is possible to isolate compounds with formulations such as $CuSO_4 \cdot 4H_2O$, $CuSO_4 \cdot 4NH_3$, and $CuCl_2 \cdot 2KCl$ from these solutions.[1] The first two compounds appear to be "solvates" and the third, a "double salt." However, a better description of these compounds, based on a variety of data, involves species in which the copper ion acts as a Lewis acid toward H_2O, NH_3, and Cl^-, which function as Lewis bases, viz. $[Cu(H_2O)_4]SO_4$, $[Cu(NH_3)_4]SO_4$, and $K_2[CuCl_4]$. The species enclosed in brackets behave as units that persist in chemical reactions unchanged and are called *complex ions,* or *coordination complexes.* The aqueous chemistry of transition metal ions is dominated by their ability to act as electron-pair acceptors toward Lewis bases, which are often called *ligands* in this context. As might be expected, the number of ligands associated with various transition metal ions and their arrangement in space is characteristic of the metal ion. Since most qualitative analysis schemes are conducted in water (which is, itself, a Lewis base) with reagents that possess unshared electron pairs, it should not be surprising to learn that complex ions play an important role in such schemes of analysis.

The formulation of complex ions containing metal ions can be traced to Alfred Werner, who suggested that all metals possess two types of valence: primary and secondary. The primary valence is the normal ionic valence associated with the metal and can be satisfied only by anions. The secondary valence can be satisfied by either anions or neutral species. Just as each metal has a characteristic primary valence, it also possesses a characteristic secondary valence; thus, in the example of cobalt complexes that possess an empirical formula such as $CoCl_3 \cdot 5NH_3$, cobalt has a primary valence of $3+$ and a secondary valence of 6. The suggestion that anions can satisfy either primary or secondary valences arose from the reactions of complex compounds such as $CoCl_3 \cdot 5NH_3$ [Eqs. (5.5) and (5.6)].

$$CoCl_3 \cdot 5NH_3 + 2AgNO_3 \rightarrow Co(NO_3)_2Cl \cdot 5NH_3 + 2AgCl \qquad (5.5)$$

$$CoCl_3 \cdot 5NH_3 + H_2SO_4 \rightarrow Co(SO_4)Cl \cdot 5NH_3 + 2HCl \qquad (5.6)$$

---

1. The common form of "hydrated" copper sulfate is $CuSO_4 \cdot 5H_2O$, in which one of the water molecules is hydrogen-bonded in lattice. The other four water molecules form coordinate covalent bonds to Cu^{2+} ions through the lone electron pairs on the oxygen atom.

It is significant that the cobalt-containing products in Eqs. (5.5) and (5.6) still retain one chlorine atom. The products are characterized by a Co : Cl : NH$_3$ ratio of 1:1:5, which suggests that the complex ion $[Co(NH_3)_5Cl]^{2+}$ is present in these compounds. Thus, cobalt exhibits a primary valence of 3+ (although the charge on the complex ion is 2+) and a secondary valence of 6. The older term *secondary valence* has given way to the modern term *coordination number*, which is used in much the same sense as it is in crystallography for describing the number of nearest neighbors.

Those ligands that satisfy the characteristic coordination number of a transition metal ion are said to occupy the *coordination sphere* of the ion, which is usually depicted by brackets in the line formula of the compound, such as $[Co(NH_3)_5Cl]^{2+}$. A list of some typical transition metal coordination compounds is given in Table 5.2. Coordination numbers of 4 and 6 are the most prevalent; however, species with lower and higher coordination numbers have been identified. A survey of the coordination numbers and the corresponding geometrical arrangements

TABLE 5.2   EXAMPLES OF KNOWN COORDINATION NUMBERS FOR SOME TRANSITION METALS AND SOME REPRESENTATIVE ELEMENTS

Coordination Number	Geometry	Examples	
		Transition Metals	Representative Elements
2	Linear	$Ag(NH_3)_2^+$; $Hg(NH_3)_2^{2+}$; $AuCl_2^-$; $M(CN)_2^-$; $M = Cu^+, Ag^+, Au^+$	$BeCl_2(g)$
3	Trigonal planar	$Ag(R_3P)_3^+$, HgI_3^-, $Ag(R_2S)_3^+$, $Cu(R_3P)_3^+$, $Au(AsR_3)_3^+$	BCl_3, ICl_3
4	Tetrahedral	$ZnCl_4^{2-}$, $CdCl_4^{2-}$, HgI_4^{2-}, $FeBr_4^-$	BF_4^-
	Planar	$PtCl_2(NH_3)_2$, $Ni(CN)_4^{2-}$	ICl_4^-
5	Trigonal bipyramid	$Fe(CO)_5$, $Cu(terpy)Cl_2$,^{a} $Ni(R_3P)_2Br_3$	PF_5
	Square pyramid	$NiBr_2 \cdot triArs$,^{b} $Ni(PR_3)_2Br_3$	BrF_5
6	Octahedral	$Cr(CN)_6^{3-}$, $Co(NH_3)_6^{3+}$, $Cr(CO)_6$	PF_6^-, SiF_6^{2-}
7	Pentagonal bipyramid	ZrF_7^{3-}, HfF_7^{3-}, UF_7^{3-}	IF_7
	Distorted trigonal prism	NbF_7^{2-}, TaF_7^{2-}	—
	Distorted octahedron	$NbOF_6^{3-}$	—
8	Dodecahedron (bis-diphenoid)	$Mo(CN)_8^{4-}$	—
	Square antiprism	TaF_8^{3-}, $Zr(acac)_4$^{c}	—
9	Distorted trigonal prism	$Nd(H_2O)_9^{3+}$, UCl_9^{6-}, $La(OH)_9^{6-}$, ReH_9^{2-}	—

^{a}terpy = 2, 2′, 2″-terpyridine.
^{b}triArs = $(CH_3)_2AsC_2H_4As(CH_3)C_2H_4As(CH_3)_2$.
^{c}acac = acetylacetonate.

of the ligands that have been observed for transition metal species indicates that 6 and 4 are the most common coordination numbers exhibited by the transition metal cations. The 6-coordinate species are octahedral; some — such as CuX_6^{n+} — are distorted; the 4-coordinate species exist as either tetrahedral or square planar complexes. In some instances tetrahedral complexes can also be distorted. Generally, the transition elements in their common oxidation states exhibit high coordination numbers at the beginning of a series, with the lower coordination numbers being attained at the end of the series. The heavier metals in any family can attain higher coordination numbers than the lighter members. It is not unusual to find a metal in the same oxidation state exhibiting different coordination numbers. Complexes with odd coordination numbers are not usually found except in unusual circumstances.

Ligands

One of the best methods for classifying complex compounds is on the basis of the nature of the ligands involved. Generally speaking, the ligands supply all the electron density for the covalent bond formed to the transition metal atom. Thus, it is not surprising that the ligand atoms directly attached to the transition metal atom are those found at the extreme right of the periodic table, where the number of electrons available exceeds the number of normal covalent bonds the ligand atom can form. The ligand atoms can be associated with neutral molecules, such as the trivalent Group V derivatives $MR_3(M = N, P, As)$, where the substituents R can be organic groups, hydrogen, or (less often) halogens, as well as similar divalent derivatives of the Group VI elements, R_2M. In each of these cases, the ligand atom has at least one pair of unshared electrons [(**1**) and (**2**)].

Other neutral ligands include oxidized derivatives of the Group V elements (R_3MO), binary oxides (CO and NO), and organic derivatives containing the CN moiety (RCN and RNC). Even organic compounds containing unsaturated systems (i.e., molecules containing multiple bonds), such as C_2H_4 and C_6H_6, can act as ligands toward some transition metal atoms.

The most important ligand in qualitative analysis is water, which has two pairs of nonbonding electrons, making it a good Lewis base. The solution of transition metal ions in water invariably involves the formation of aquo complexes. Thus, the predominant species present in aqueous solutions of aluminum ion (Al^{3+}) and ferric iron (Fe^{3+}) are the corresponding octahedral aquo complexes [Eqs. (5.7) and (5.8)].

$$Al^{3+} + 6\,H_2O \rightarrow Al(OH_2)_6^{3+} \tag{5.7}$$

$$Fe^{3+} + 6\,H_2O \rightarrow Fe(OH_2)_6^{3+} \tag{5.8}$$

As described in Chapter 4, such hydrated, highly charged ions invariably undergo hydrolysis to give acidic solutions [Eqs. (5.9) and (5.10)].

$$Al(OH_2)_6^{3+} + H_2O \rightleftharpoons Al(OH_2)_5(OH)^{2+} + H_3O^+ \tag{5.9}$$

$$Fe(OH_2)_6^{3+} + H_2O \rightleftharpoons Fe(OH_2)_5(OH)^{2+} + H_3O^+ \tag{5.10}$$

The complexed ions can be kept in solution in the presence of acid, which shifts equilibria in Eqs. (5.9) and (5.10) to the left. If such solutions get slightly basic, complete hydrolysis occurs, forming the generally insoluble hydrated hydroxides $Al(OH)_3(OH_2)_3$ and $Fe(OH)_3(OH_2)_3$, respectively.

It should not be surprising that virtually all the anions can act as ligands, since these species carry a negative charge, implying that an unshared pair of electrons is available; however, a few anions—such as tetraalkyl borates (BR_4^-)—must be excluded from this generalization. Theoretically, some anions have two or more possible sites for coordination. For example, the Lewis formulations for CN^- (**3**), SCN^- (**4**), and NO_2^- (**5**) show two possible coordination sites.

$$:C\equiv N:$$
3

$$:\ddot{S}{-}C\equiv N:$$
4

$$\underset{\textstyle 5}{\overset{\textstyle \ddot{N}}{\cdot\ddot{O}\diagup\quad\diagdown\cdot\ddot{O}:}}$$

In each instance, complex compounds in which the metal is coordinated to one site or the other have been isolated. This point is considered in more detail in the next section.

It should be recalled that some of the transition metals can exhibit oxidation states as high as 8+. In some ways the oxy anions, in which the transition metals exhibit their highest oxidation states, can be considered to be composed of a highly charged cation surrounded by oxide (O^{2-}) ligands.

In this view the oxidation states M^{7+}, M^{6+}, and M^{8+} give rise to the tetrahedral complex oxy anions $MO_4^-(MnO_4^-, ReO_4^-)$, $MO_4^{2-}(CrO_4^{2-}, ReO_4^{2-}, MoO_4^{2-}, WO_4^{2-})$, and $MO_4^{3-}(VO_4^{3-})$. Of course, it is not correct to describe these species as collections of the corresponding ions because the high polarizing power of atoms in high oxidation states leads to significant covalent bonding; however, this point of view is often useful in certain discussions.

Amphoterism.   Among the complex ions encountered in the study of qualitative analysis, $Zn(OH)_4^{2-}$, $Sn(OH)_4^{2-}$, $Cr(OH)_4^-$, and $Al(OH)_4^-$ are of particular interest because their formation illustrates the *amphoteric* character of the hydroxides of the metals involved. Using zinc as an example, if a solution of sodium hydroxide is added drop by drop to a solution of a soluble zinc salt, the white precipitate $Zn(OH)_2$ will first form because zinc hydroxide is the least soluble substance in the solution (Chapter 4). If the addition of the sodium hydroxide solution is continued, the solid $Zn(OH)_2$ will gradually dissolve until, finally, a water-clear solution is obtained. This clear solution is found to contain the complex ion $Zn(OH)_4^{2-}$, the formation of which is given by Eq. (5.11).

$$Zn(OH)_2 + 2OH^- \rightleftharpoons Zn(OH)_4^{2-} \quad \text{(tetrahydroxozincate ion)} \qquad (5.11)$$

In this process OH^- acts as a ligand (Lewis base) toward the zinc ion.

The $Zn(OH)_2$ that is precipitated when a solution of a zinc salt is treated with NaOH will also dissolve in any strong acid in accordance with Eq. (5.12).

$$Zn(OH)_2 + 2H^+ \rightleftharpoons Zn(aq)^{2+} + 2H_2O \qquad (5.12)$$

There is nothing unusual about the reaction in Eq. (5.12); $Zn(OH)_2$ is the basic hydroxide of a metal, and basic hydroxides react with acids to form salts and water. The solubility of $Zn(OH)_2$ in excess sodium hydroxide is, however, unusual. It means that $Zn(OH)_2$ will react with either a strong acid or a strong base. A substance, usually a hydroxide, that reacts with either a strong acid or a strong base is said to be *amphoteric.*

The hydroxides of aluminum, chromium, tin, lead, and antimony, like $Zn(OH)_2$, are amphoteric: They also react with strong acids in the manner typical of the basic hydroxides of metals. They react with strong bases to form stable complex ions. Although there is some question about the details of the composition of some of these hydroxo complexes, we shall represent all of them by the same general formula, $M(OH)_4^{n-}$ [$Zn(OH)_4^{2-}$, $Al(OH)_4^-$, $Cr(OH)_4^-$, $Sn(OH)_4^{2-}$, $Pb(OH)_4^{2-}$, and $Sb(OH)_4^-$].

If a solution of a strong acid is added drop by drop to a solution containing any one of the above complex hydroxo ions, the hydroxide of the metal is completely precipitated when the neutral point is reached. On further addition of acid, the hydroxide dissolves to form a solution containing the cation of the metal. This behavior indicates the existence of an equilibrium of the type represented by Eq. (5.13).

$$Cr(OH)_4^- + H^+ \leftrightharpoons Cr(OH)_3 + H_2O \tag{5.13}$$

The hydrogen ions (acid) that are added remove OH^- ions and, thereby, reduce the rate of reaction to the right. This causes the equilibrium to be shifted to the left, which causes $Cr(OH)_3$ to precipitate. When the excess OH^- ions have been neutralized, the acid dissolves the basic hydroxide in the conventional manner, as indicated by Eq. (5.14).

$$Cr(OH)_3 + 3H^+ \leftrightharpoons Cr^{3+} + 3H_2O \tag{5.14}$$

If solutions of the complex hydroxo ions formed by dissolving the amphoteric hydroxides in sodium hydroxide are concentrated by evaporation, solid salts of compositions represented by the formulas Na_2ZnO_2, Na_2SnO_2, and $NaAlO_2$ will crystallize out. This indicates that the solutions may have contained the ions ZnO_2^{2-}, SnO_2^{2-}, and AlO_2^-. The presence of these ions can be accounted for by equilibria of the type shown in Eq. (5.15).

$$Sn(OH)_4^{2-} \leftrightharpoons SnO_2^{2-} + 2H_2O \tag{5.15}$$

The question will arise as to why, among the many metal hydroxides, the six discussed above should be amphoteric, while the others are not. We shall find upon examination of the periodic table that, with the exception of chromium, the metals in question lie along the border between the characteristically nonmetallic elements in the upper right section of the table and the more characteristically metallic elements that occupy the rest of the table. Accordingly, these hydroxides—although predominantly basic—are also somewhat acidic in character.

The separation of aluminum, chromium, and zinc ions from iron, cobalt, nickel, and manganese ions depends on the fact that the hydroxides of the first three elements listed are amphoteric, but the hydroxides of the last four are not. If a solution containing a mixture of the cations of these seven metals is treated with an excess of sodium hydroxide, the hydroxides of iron, cobalt, nickel, and manganese, which are not amphoteric, will form precipitates, while the amphoteric hydroxides, $Al(OH)_3$, $Cr(OH)_3$, and $Zn(OH)_2$, will dissolve in excess NaOH to give the soluble $Al(OH)_4^-$, $Cr(OH)_4^-$, and $Zn(OH)_4^{2-}$ ions (see Procedure 16, Chapter 11).

NOMENCLATURE

A complex ion is always named as a derivative of the central ion (the cation). The constituent parts of the complex ion are named in the following order: anion, neutral molecule, and central ion (cation); within these kinds of ligands, the order is alphabetical. For ligands that are anions with names ending in *ide*, the *ide* changes to *o*: chloride becomes chloro, cyanide becomes cyano, hydroxide becomes hydroxo, and so on. For ligands that end in *ate* or *ite*, the *e* changes to *o*: sulfate becomes sulfato, sulfite becomes sulfito, cyanate becomes cyanato, and so forth. Some neutral ligand molecules have special names; NH_3 is *ammine*, H_2O is *aquo*, and CO is *carbonyl*. Most neutral molecules go by their usual names.

In a negative complex ion, the name used to designate the central ion always ends in *ate*, and it is generally derived from the name from which the symbol of the element is derived. The number of each kind of ligand is designated in the name with the appropriate Greek prefix. Thus, $Ag(CN)_2^-$ is the dicyano*argentate* ion and $Pb(OH)_4^{2-}$ is the tetrahydroxo*plumbate* ion. In a positive complex ion the common name of the central ion is used. Hence, $Ag(NH_3)_2^+$ is the diammine*silver* ion and $PbCl^+$ is the monochloro*lead* ion.

A Roman numeral enclosed in parentheses shows the oxidation state of the central element. In accordance with the rules above:

$Fe(H_2O)_2Cl_4^-$	is the tetrachlorodiaquoferrate(III) ion.
$FeCl^{2+}$	is monochloroiron(III).
$Sn(OH)_4^{2-}$	is tetrahydroxostannate(II).
$Sn(OH)_6^{2-}$	is hexahydroxostannate(IV).
$Co(NH_3)_5Cl^{2+}$	is pentamminemonochlorocobalt(III).
$CuCl_4^{2-}$	is tetrachlorocuprate(II).
$Cu(NH_3)_4^{2+}$	is tetramminecopper(II).
$Ag(S_2O_3)_2^{3-}$	is dithiosulfatoargentate(I).
$Fe(CN)_6^{4-}$	is hexacyanoferrate(II).

THEORETICAL CONSIDERATIONS

Ever since it was recognized that some transition metal compounds exhibit a capacity to react with ligands above and beyond that expected on the basis of the normal valence rules, there has been an active interest in the theoretical explanation of these effects. Werner's recognition of the existence of two types of valences—primary and secondary—was an important organizational aspect of coordination chemistry. The consequence of this suggestion was far-reaching in terms of structural considerations, even though no adequate theory existed for predicting such valences or the nature of the bonding between ligands and transition metal atoms. Several theoretical arguments are now available that attempt to describe the structure of transition metal coordination compounds and their bonding. A single theory that encompasses all aspects of this problem does not exist; rather several theories, successful in different aspects of coordination chemistry, are available. We describe these theories here in a semihistorical manner because such an exposition contributes to an understanding of the experimental basis upon which they were modified and leads to an appreciation of why certain terms or concepts are still found in the modern literature. Again, we must remember that a theory need not be entirely correct from a fundamental point of view to be useful in correlating the available data.

Sidgwick Theory

The Lewis interpretation of the nature of the normal covalent bond and its adaptation to coordinate-covalent bond formation in species such as $BF_3 \cdot NH_3$ (6) was extended to transition metal complexes by Sidgwick and Lowry.

$$
\begin{array}{ccc}
F\diagdown & & \diagup H \\
F\!\longrightarrow\!B & :\;N\!\longleftarrow\!\!H \\
F\diagup & & \diagdown H
\end{array}
$$

6

   In their collective view, the primary valences of Werner were the ordinary electrovalences associated with ionic compounds and hence could only be satisfied by an appropriate number of counter ions. Secondary valences, on the other hand, were satisfied by the formation of coordinate-covalent bonds between the transition metal atom and the ligand, which may, or may not, be charged. Thus, anions can satisfy both primary and secondary valences, whereas neutral ligands can satisfy only secondary valences. Coordinate-covalent bonds are sometimes called *dative,* or *semipolar, bonds.*

   According to Sidgwick, transition metal atoms acquire sufficient electrons in forming coordinate-covalent bonds to achieve the next rare gas configuration. In this respect the Sidgwick model followed one of the postulates of the Lewis theory of covalent bond formation for the representative elements described earlier in this chapter. The *effective atomic number* (EAN) of a metal atom in a complex ion represents the total number of electrons associated with that atom in the complex ion. In the Sidgwick theory, transition metal ions form complexes so that they achieve an effective atomic number approaching that of the next rare gas if at all possible. The following example illustrates the way in which the effective atomic number of a complex ion, $[Co(NH_3)_6]Cl_3$, is determined.

Species	Number of Electrons
Co^{3+}	$(Z - 3) = 27 - 3 = 24$
6 NH_3 groups	$6 \times 2 = \underline{12}$
EAN of Co in complex	36

   The effective atomic number of cobalt in this complex is 36, which is the same as the atomic number of the next rare gas, krypton. Thus, cobalt, which cannot attain a rare gas configuration with its primary valence, can achieve this structure by forming six coordinate-covalent bonds. Sidgwick suggested that the six ligands in complexes of this kind are disposed in an octahedral arrangement about the central metal ion. The effective atomic number of a transition metal atom in a complex is generally near that of the next rare gas, although there are numerous examples where complex species have lower $[Ti(H_2O)_6^{4+}]$ or higher $[Ni(NCS)_6^{4+}]$ effective atomic numbers. The EAN correlates reasonably well with the fact that the elements at the end of the transition series possess low coordination numbers for their normal valence states; those at the beginning of the transition series exhibit the higher coordination numbers required to achieve a rare gas configuration. In addition, the observation that the higher valence state of an element generally exhibits the higher coordination number is reflected in the EAN of the metal in these complexes.

Valence Bond Theory

Although valence bond theory was not developed specifically for coordination compounds, it has had some success when applied to such systems. The valence bond description of a normal covalent chemical bond recognizes that the electrons in the uncombined atoms exist in atomic orbitals and that the bond is formed when half-filled orbitals overlap to form an area of high electron density between the bound nuclei; this process is depicted in Figure 5.3. The coordinate-covalent bond formed between a Lewis acid and a ligand involves the overlap of an empty orbital on the acid with a filled orbital on the ligand. Valence bond theory was originally applied to the bonding in complex ions in an attempt to describe the nature of the bonds formed, the stereochemistry of the complexes, and their magnetic properties.

As is the case with the valence bond description of the compounds of the representative elements, atomic orbitals are generally not the most efficient for bonding; hybrid orbitals, which are mathematical mixtures of atomic orbitals, must be used to provide maximum overlap. We saw earlier in this chapter that two coordination numbers—4 and 6—are commonly observed for complex ions.

Moreover, three structural types (square planar, tetrahedral, and octahedral) are commonly found among these species. Although three hybridization schemes are necessary to account for the structures of most of the complex species known, hybrid orbitals can be constructed for virtually all the geometric arrangements of ligands about the transition metal atom. It should be recognized that structures with coordination numbers greater than 4, as well as the square planar structure, require hybrid orbitals incorporating a d component. Recall from your earlier studies of atomic electronic structure that transition metal atoms have d orbitals as a part of their valence shell. Thus, the valence bond description of complex ions involves using empty hybrid orbitals with an appropriate geometry on the metal atom overlapping with filled orbitals on the ligand.

Normal Covalent Bond Formation

Lewis
formulation   A·   +   ·B   →   A:B

Valence-bond
description

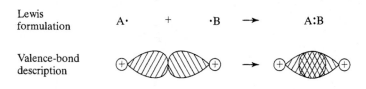

Coordinate-Covalent Bond Formation

Lewis
formulation   M   +   :L   →   M:L

Valence-bond
description   M

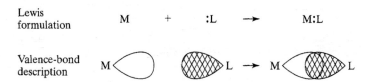

FIGURE 5.3   A comparison and pictorial representation of coordinate-covalent bond formation. The orbitals depicted are generalized; half-filled orbitals are shaded, empty orbitals are clear, and doubly filled orbitals are cross-hatched.

An outline of the application of valence bond theory to complex ions is best illustrated using an example, $Co(H_2O)_6^{3+}$. The ion Co^{3+} has an argon structure plus six $3d$ electrons, and it might be expected to exhibit the configuration shown in (**7**) in the gaseous state if Hund's rule is obeyed; note that the higher-energy $4s$ and $4p$ orbitals are empty.

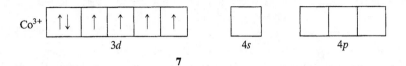

7

The ion $Co(H_2O)_6^{3+}$ contains an octahedral arrangement of ligands that requires vacant hybrid orbitals of the type d^2sp^3, but (**7**) shows electrons present in the lowest-lying d orbitals available, $3d$. Thus, to create a vacant set of hybrid orbitals with the appropriate symmetry, two electrons must be promoted in energy. In this case the electrons can be either paired in the $3d$ shell or placed in an orbital of higher energy than $4p$. Of these two possibilities, the former requires less energy, giving the excited configuration shown in (**8**). The empty $3d$, $4s$, and $4p$ orbitals are now available to form

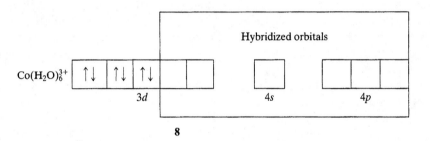

8

d^2sp^3 hybrids that overlap with the filled ligand orbitals to form six metal-ligand bonds (**9**).

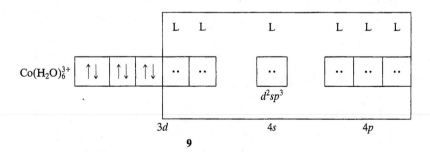

9

Thus, the conclusion of the valence bond method is that $Co(H_2O)_6^{3+}$ should possess no unpaired electron spins and, therefore, be diamagnetic, as it is experimentally observed to be. Other examples of the application of valence bond arguments to bonding in different complexes appear in Table 5.3.

Crystal Field Theory

An electrostatic approach to an understanding of the properties of complex species—crystal field theory—was suggested by Hans Bethe in 1929, but it did not find use by chemists until

TABLE 5.3 EXAMPLES OF VALENCE BOND DESCRIPTIONS OF SOME COMMON COORDINATION SPECIES

Species	Geometry	Valence Bond Electron Description
		$3d$ $4s$ $4p$
$Cr(H_2O)_6^{3+}$	Octahedral	[↑][↑][↑][••][••] [••] [••][••][••] — d^2sp^3
$Fe(CN)_6^{4-}$	Octahedral	[↑↓][↑↓][↑↓][••][••] [••] [••][••][••] — d^2sp^3
$Ni(CN)_6^{2-}$	Square planar	[↑↓][↑↓][↑↓][↑↓][••] [••] [••][••] — dsp^2
$Cu(NH_3)_4^{2+}$	Square planar	[↑↓][↑↓][↑↓][↑↓][••] [••] [••][••][↑] — dsp^2
$Ni(NH_3)_4^{2+}$	Tetrahedral	[↑↓][↑↓][↑↓][↑][↑] [••] [••][••][••] — sp^3
$MnCl_4^{2-}$	Tetrahedral	[↑][↑][↑][↑][↑] [••] [••][••][••] — sp^3
$ZnCl_4^{2-}$	Tetrahedral	[↑↓][↑↓][↑↓][↑↓][↑↓] [••] [••][••][••] — sp^3

about 25 years later. Crystal field theory considers the effect of the electrostatic fields of the ligands on the energy of the d orbitals of the metal.

To illustrate the principles of the crystal field theory, consider an octahedral complex species, ML_6^{n+}. The central metal ion in the gas phase contains a degenerate set of five d orbitals that can be described conveniently with reference to the normal Cartesian coordinate system.

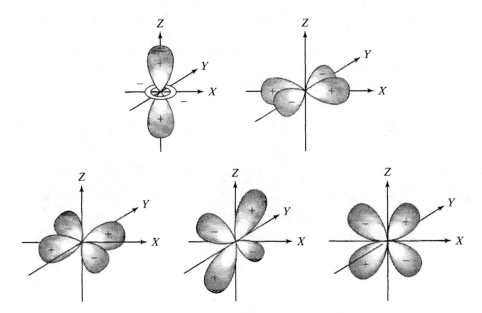

FIGURE 5.4   A representation of d orbitals for a gaseous transition metal ion.

The boundary surfaces of the d orbitals and their arrangement in space are shown in Figure 5.4. We can imagine the influence of the six ligands on the d orbitals by placing the ligands on the Cartesian axes at infinity and moving them toward the metal atom until the experimentally observed metal-ligand internuclear distance is attained; this is shown schematically in Figure 5.5. Recall that ligands are either anions or polar species, the negative ends of which are coordinated to the metal ion. Under this geometry the metal comes under the influence of a negative electric field. If the electric field arising from the ligands were spherically symmetrical (i.e., a field of

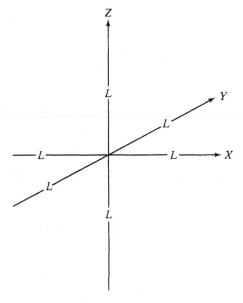

FIGURE 5.5   In the crystal field approach for an octahedral complex the ligands are placed on the conventional Cartesian coordinate system centered on the metal atom.

equal intensity at all points that are the same distance from the metal atom), we would expect that the energy of all the d orbitals would be raised to the same extent; that is, it would require a higher energy to maintain an electron in an orbital in the presence of a negative field than without a field, as shown in Figure 5.6.

The six ligands generate an octahedral field, however, that interacts differently with the d orbitals than does a spherical field. The orbitals, which lie along the Cartesian axes, $d_{x^2-y^2}$ and d_{z^2}, are perturbed more than the orbitals that lie between these axes, d_{xy}, d_{xz}, and d_{yz}. Thus, in an octahedral field, the five degenerate d orbitals split into two groups; the orbitals that lie along the axes form one group, whereas the orbitals that lie between the axes form another. This result, achieved in an intuitive manner, can be more rigorously obtained by applying group theory to the symmetry of the d orbitals. The two sets of d orbitals present in an octahedral field are identified several different ways in the literature. In the literature, the higher-energy set has been called e_g, d_y, or γ_3; the corresponding labels for the set at lower energy, t_{2g}, d_ε, or γ_5. The energy difference between the two sets of orbitals that arise from an octahedral field is measured in terms of the parameter Δ_0 or 10 Dq. Under the influence of a purely electrostatic field the average energy of the perturbed orbitals is zero. Thus, the higher e_g levels must be 6 Dq, or 3/5 greater in energy than the average value. In other words, if the original set of d orbitals in the gaseous ion contained five electrons, an octahedral perturbation of the levels would increase the energy of the electrons in the two e_g orbitals by 2×6 Dq (or $2 \times 3/5$ Δ_0). The gain in energy when electrons occupy the t_{2g} orbitals compared to the energy they would have if they occupied orbitals corresponding to the average energy is called the *crystal field stabilization energy* (CFSE). The crystal field stabilization energy is equal in magnitude to 10 Dq (or 2/5 Δ_0) per electron.

Crystal field effects on the d orbitals of transition metals surrounded by any number of ligands also can be deduced using similar arguments. In the case of tetrahedral complexes, the pattern for the splitting of the d orbitals is the reverse of that found in octahedral complexes, as

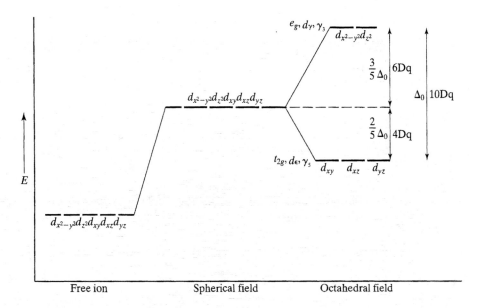

FIGURE 5.6   The presence of ligands near the transition metal ion causes the degeneracy of the d orbitals to be lifted. This figure shows the general splitting pattern for d orbitals in an octahedral field.

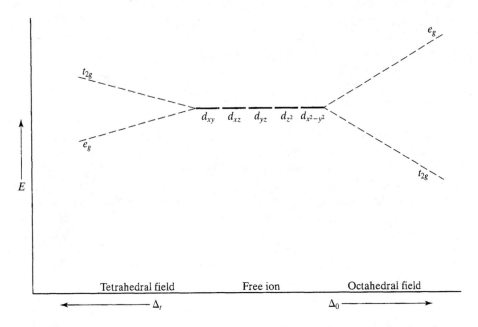

FIGURE 5.7 The general splitting pattern for an octahedral complex compared to that of a tetrahedral complex.

shown in Figure 5.7. The relative magnitudes of the tetrahedral crystal field splitting (Δ_t) and the octahedral splitting (Δ_0) are given by Eq. (5.16) for the case where all factors—that is, the kind of ligand and metal and the metal-ligand distance—are the same:

$$\Delta_t = 4/9 \; \Delta_0 \tag{5.16}$$

Of course it is difficult to conceive of a series of complex ions with different coordination numbers where these factors are exactly the same; however, Eq. (5.16) indicates that, in general, crystal field effects in the tetrahedral case should be about half those observed for octahedral coordination.

As the field of the ligands increases, the energy levels in the transition metal ion split to a greater and greater extent (Figure 5.8). Ligands are classified in terms of the relative strengths of the fields they generate. The common ligands have been arranged in the following order of increasing crystal field splitting: $I^- < Br^- < Cl^- \approx SCN^- < F^- < OH^- < H_2O \approx \; < NH_3 < CN^-$. The absorption of visible light involves the excitation of an electron from a lower energy level to a higher level; the electron that is excited exists in the lower set of the split energy levels of a complexed transition metal ion and it gets excited to one of the upper set of orbitals. Qualitatively, it is easy to see that light of different energy will be absorbed by different complexed transition metal ions because the different ligands will cause a difference in splitting between the lower and upper set of orbitals (Figure 5.7). The difference in the energy of the orbitals corresponds to the energy of the light absorbed by the compound, which is manifested visually as a complementary color. Thus, a compound that absorbs yellow light because of the energy-level difference will appear blue to most people.

The characteristic color of transition metal ions in the presence of certain ligands is a useful property in qualitative analysis; we shall have several occasions to use such observations.

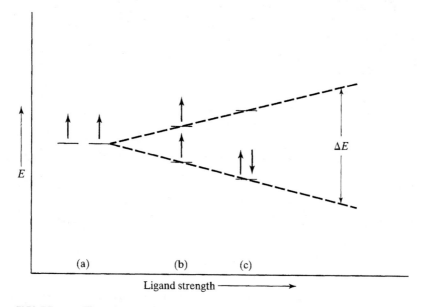

FIGURE 5.8 The splitting of d orbitals is dependent on the strength of the ligand field. In this case, we concentrate on two d orbitals each with an electron; (a) represents the situation for the free gaseous ion. The two orbitals are of equal energy and the electrons are unpaired. If the splitting is small, as compared to the pairing energy for electrons, the electrons remain unpaired as in (b). When the splitting energy is larger than the pairing energy, the electrons are paired as in (c).

STABILITY OF COMPLEX IONS

All complex ions are unstable to a greater or lesser degree and dissociate in characteristic fashion to give—usually—the ions or molecules from which they were formed. Table 5.4 gives the dissociation equations for a number of complex ions.

Since the equations in Table 5.4 represent true equilibrium reactions, their equilibrium constants can be represented in the conventional manner. For example, for the equation

$$Ag(NH_3)_2^+ \rightleftharpoons Ag^+ + 2NH_3 \tag{5.17}$$

the equilibrium formulation yields the familiar expression

$$K_{inst.} = \frac{[Ag] + [NH_3]_2}{[Ag(NH_3)_2^+]} \tag{5.18}$$

For a complex ion that is very stable and, hence, only slightly dissociated, the quantities in the numerator will be very small, the quantity in the denominator will be relatively large, and the value of K will be very small. On the other hand, a complex ion that is relatively unstable and, hence, dissociates to a fairly high degree will have a relatively large value of K. The value of K is therefore a measure of the instability of the ion. If K is very small, the ion is very stable; if K is large, the ion is relatively unstable. For that reason K is called the *instability constant* and is designated by $K_{inst.}$. The instability constants for a number of important ions are given in Table 5.4. Certain of these

TABLE 5.4 SOME COMPLEX ION EQUILIBRIA

Equation[a]	Instability Constant
$Ag(NH_3)_2^+ \rightleftharpoons Ag^+ + 2NH_3$	5.9×10^{-8}
$Ag(CN)_2^- \rightleftharpoons Ag^+ + 2CN^-$	1.8×10^{-19}
$Cu(NH_3)_4^{2+} \rightleftharpoons Cu^{2+} + 4NH_3$	4.7×10^{-15}
$Cu(CN)_2^- \rightleftharpoons Cu^+ + 2CN^-$	5×10^{-28}
$Cd(NH_3)_4^{2+} \rightleftharpoons Cd^{2+} + 4NH_3$	7.5×10^{-8}
$Cd(CD)_4^{2-} \rightleftharpoons Cd^{2+} + 4CN^-$	1.4×10^{-19}
$Co(NH_3)_6^{3+} \rightleftharpoons Co^{3+} + 6NH_3$	2.2×10^{-34}
$Fe(CN)_6^{4-} \rightleftharpoons Fe^{2+} + 6CN^-$	1×10^{-35}
$FeF_6^{3-} \rightleftharpoons Fe^{3+} + 6F^-$	1×10^{-16}
$Ni(NH_3)_6^{2+} \rightleftharpoons Ni^{2+} + 6NH_3$	1.8×10^{-9}
$Ni(CN)_4^{2-} \rightleftharpoons Ni^{2+} + 4CN^-$	1×10^{-22}
$Zn(NH_3)_4^{2+} \rightleftharpoons Zn^{2+} + 4NH_3$	3.4×10^{-10}
$Zn(CN)_4^{2-} \rightleftharpoons Zn^{2+} + 4CN^-$	1.2×10^{-18}

[a] It is understood that the free metal ion shown at the right-hand side of these equations is hydrated, that is, complexed by water molecules. Since the number of such water molecules is not often known with certainty, the metal ions are usually written as the "free" species. Such equilibria represent, in effect, the competition of two Lewis bases, the ligand indicated and water molecules, for a Lewis acid.

ions, it will be noted, are extremely stable, while others are quite unstable. All of these complex ions can be classed as weak electrolytes with respect to the dissociation into their component parts.

The formation of complex ions offers a method for controlling the concentration of a metal ion in solution. Thus, a metal ion can be prevented from precipitating, or taking part in a reaction, by forming an appropriate complex ion and decreasing the concentration of the metal ion in question.

EXAMPLE 5.1

A solution contains $0.010\ M$ each of Ni^{2+} and Zn^{2+}. If KCN is added to this solution until the $[CN^-]$ is $1.0\ M$ and $[S^{2+}]$ is maintained at $0.50\ M$, how much Ni^{2+} and Zn^{2+} will remain in solution?

1. First assume that no NiS or ZnS precipitates.
2. Addition of CN^- leads to the formation of the metal-cyano complexes

$$Ni^{2+} + 4CN^- \rightleftharpoons Ni(CN)_4^{2-}$$
$$Zn^{2+} + 4CN^- \rightleftharpoons Zn(CN)_4^{2-}$$

for which the corresponding *instability* constants are (Table 5.4)

$$K_{inst.} = 1.0 \times 10^{-22} = \frac{[Ni^{2+}][CN^-]^4}{[Ni(CN)_4^{2-}]}$$

$$K_{inst.} = 1.2 \times 10^{-18} = \frac{[Zn^{2+}][CN^-]^4}{[Zn(CN)_4^{2-}]}$$

Note that the instability constants are written for the *reverse* of the equations for the formation of the complex ions.

3. From the original information we see that

$$[CN^-] = 1.0 \ M$$

The mass balance for the metal ions gives

$$[Ni(CN)_4^{2-}] + [Ni^{2+}] = 0.010 \ M$$
$$[Zn(CN)_4^{2-}] + [Zn^{2+}] = 0.010 \ M$$

4. Since the instability constants are so small, it is reasonable to assume that the concentration of the uncomplexed metal ions will be small compared to that of the complexed metal ions. This leads to the approximation

$$[Ni(CN)_4^{2-}] \cong 0.010 \ M$$
$$[Zn(CN)_4^{2-}] \cong 0.010 \ M$$

5. The uncomplexed metal ion concentration can be estimated from the instability constants in step (2) and the concentrations established in step (3), as modified by the assumption in step (4).

$$[Ni^{2+}] = 1.0 \times 10^{-22} \frac{[Ni(CN)_4^{2-}]}{[CN^-]^4}$$

$$\cong 1.0 \times 10^{-22} \times \frac{(0.010)}{(1.0)^4}$$

$$\cong 1.0 \times 10^{-24} \ M$$

$$[Zn^{2+}] = 1.2 \times 10^{-18} \frac{[Zn(CN)_4^{2-}]}{[CN^-]^4}$$

$$\cong 1.2 \times 10^{-18} \times \frac{(0.010)}{(1.0)^4}$$

$$\cong 1.2 \times 10^{-20} \ M$$

6. Now we are in a position to decide whether NiS or ZnS will precipitate. The K_{sp} expressions for these substances are (Table 4.2)

$$K_{sp} = 1.0 \times 10^{-22} = [Ni^{2+}][S^{2-}]$$
$$K_{sp} = 3 \times 10^{-23} = [Zn^{2+}][S^{2-}]$$

7. Since the solution contains sulfide at $[S^{2-}] = 0.50 \ M$, we see that the ion product for NiS,

$$[Ni^{2+}][S^{2-}] = 1.0 \times 10^{-24} \ (0.50)$$

$$= 5 \times 10^{-25}$$

is less than its K_{sp},

$$5 \times 10^{-25} < 1.0 \times 10^{-22}$$

which means that NiS does not precipitate. On the other hand, the ion product for ZnS in this solution,

$$[Zn^{2+}][S^{2-}] = 1.2 \times 10^{-20} \ (0.5)$$
$$= 6 \times 10^{-21}$$

is greater than its K_{sp},

$$6 \times 10^{-21} > 3 \times 10^{-23}$$

and ZnS must precipitate from this solution.

8. The $[Zn^{2+}]$ remaining in solution can now be calculated from the K_{sp} of ZnS.

$$K_{sp} = 3 \times 10^{-23} = [Zn^{2+}][S^{2-}]$$

Since the solution contains $[S^{2-}] = 0.50$, it follows that

$$[Zn^{2+}] = \frac{(3 \times 10^{-23})}{(0.50)} = 6 \times 10^{-23} \ M$$

Just as it is possible to dissolve the insoluble salt of a weak acid by adjusting the $[H^+]$ in a solution, many precipitates containing transition metal cations can be dissolved through complex-ion formation.

EXAMPLE 5.2

What concentration of NH_3 is required to dissolve completely 0.010 moles of AgBr in 1 L of solution?

1. The overall equation for the dissolution of AgBr is given by

$$AgBr(s) + 2NH_3 \rightleftharpoons Ag(NH_3)_2^+ + Br^-$$

2. The pertinent complex-ion equilibrium is

$$Ag^+ + 2NH_3 \rightleftharpoons Ag(NH_3)_2^+$$

The instability constant for this complex ion is given by (Table 5.4)

$$K_{inst.} = \frac{[Ag^+][NH_3]^2}{[Ag(NH_3)_2^+]} = 5.9 \times 10^{-8}$$

Note that the chemical equilibrium for which $K_{inst.}$ is valid is the *reverse* of the equilibrium expression given above.

3. The solubility of AgBr is given by

$$AgBr \rightleftharpoons Ag^+ + Br^-$$

for which K_{sp} is (Table 4.2)

$$K_{sp} = 5.0 \times 10^{-13} = [Ag^+][Br^-]$$

4. If all the AgBr dissolves in 1 L of solution, the concentration of all silver-containing species must be 0.010 M:

$$[Ag^+] + [Ag(NH_3)_2^+] = 0.010 \ M$$

and

$$[Br^-] = 0.010 \ M$$

Assuming that $[Ag^+]$ is very small compared to $[Ag(NH_3)_2^+]$ because $K_{inst.}$ and K_{sp} are so small,

$$[Ag(NH_3)_2^+] \cong 0.10 \ M$$

5. Now $[Ag^+]$ can be obtained from steps 3 and 4.

$$[Ag^+] = \frac{K_{sp}}{[Br^-]} = \frac{5.0 \times 10^{-13}}{0.010} = 5.0 \times 10^{-11} \ M$$

6. Next, $[NH_3]$ can be obtained from $K_{inst.}$ using steps 2 and 5.

$$[NH_3]^2 = \frac{K_{inst.}[Ag(NH_3)_2^+]}{Ag^+}$$

$$[NH_3]^2 = \frac{5.9 \times 10^{-8}(0.010)}{(5.0 \times 10^{-11})} = 11.8$$

$$[NH_3] \cong 3.4 \ M$$

THE ROLE OF COMPLEX IONS IN QUALITATIVE ANALYSIS

The ability of metal ions to form complex compounds is used in qualitative analysis in two fundamental ways. Complex ions often have different solubility characteristics, which is to our advantage. For example, silver chloride (AgCl), which is insoluble, is separated from the other insoluble chlorides by treating the mixture with aqueous NH_3 (see Procedure 4, Chapter 9). The NH_3 present in the latter reagent converts the insoluble AgCl in the precipitate into a soluble complex [Eq. (5.19)].

$$AgCl + 2NH_3(aq) \rightleftharpoons Ag(NH_3)_2^+ + Cl^- \qquad (5.19)$$

Other examples are discussed in the appropriate sections of Part III.

The other major role of complex ions is in the formation of characteristically colored species. Thus, Cu^{2+} is identified by the deep blue color of the $Cu(NH_3)_4^{2+}$ ion (see Procedure 9, Chapter 10), Fe^{3+} by the characteristic red-brown color of the complex ion $Fe(H_2O)_5(SCN)^{2+}$ (see Procedure 17, Chapter 11), nickel by the formation of the red complex with dimethylglyoxime (see Procedure 19, Chapter 11), and cobalt by the blue color developed in the formation of the $Co(SCN)_4^{2-}$ complex ion (Procedure 18, Chapter 11).

PROBLEMS

5.1 Write Lewis formulations for each of the following species.

a. $CaCl_2$

b. CH_3OH

c. $NH_3 \cdot BH_3$

d. PF_5

e. H_2SO_4

f. H_3O^+

g. H_2S

h. CrO_4^{2-}

5.2 Give the effective atomic number of the transition metal ions in the following species.

a. $Na_3[TiCl_6]$

b. $K_2[CoF_4]$

c. $Na_2[NiCl_4]$

d. $K_2[CuCl_4]$

e. $Na_4[Ru(CN)_6]$

f. $Rb_2[CdCl_4]$

5.3 Give the hybridization and the geometry about the transition metal ion in the following species.

a. $ZnCl_4^{2-}$

b. CoF_6^{3-}

c. $Cr(CO)_6$

d. $MnCl_4^{2-}$

5.4 Give the instability constant expression for the following complex species.

a. $CuCl_4^{2-}$

b. $Zn(NH_3)_4^{2+}$

c. $Cr(CN)_6^{3-}$

d. $Co(NH_3)_6^{3+}$

5.5 Give the formal oxidation state of the central metal ion in the following species.

a. $Fe(CN)_6^{3-}$

b. $MnCl_6^{2-}$

c. $[CrCl_2(H_2O)_4]^+$

d. $[V(H_2O)_4Cl_2]^{2+}$

5.6 Name the complex ions given in Problem 5.5

Oxidation– Reduction Reactions

The close relationship between electricity and matter is apparent from the study of atomic structure. A very large number of reactions of interest to us involve reactants that undergo a change in electron configurations of the species involved. Therefore, it is not surprising that the passage of an electric current through matter under certain conditions can cause chemical reactions to occur. In other situations, such as the common lead storage cell or the dry cell, chemical reactions can be arranged to produce an electric current.

All spontaneous chemical reactions occur because the chemical potential energy of the products is less than that of the reactants. That is, when a spontaneous reaction occurs, energy is liberated. When methane burns [Eq. (6.1)], sodium metal reacts with chlorine gas [Eq. (6.2)], metallic aluminum reacts with hydrochloric acid [Eq. (6.3)], or metallic zinc reacts with chlorine gas [Eq. (6.4)], energy is liberated in the form of heat, light, or both.

$$CH_4 + 2O_2 \rightarrow CO_2 + 2H_2O \tag{6.1}$$

$$2Na + Cl_2 \rightarrow 2NaCl \tag{6.2}$$

$$2Al + 6HCl \rightarrow 2AlCl_3 + 3H_2 \tag{6.3}$$

$$Zn + Cl_2 \rightarrow ZnCl_2 \tag{6.4}$$

These are examples of the class of reactions called *oxidation–reduction reactions* (or *redox reactions*). Each of these reactions can be considered to occur in two steps, one involving the loss of electrons *(oxidation)* and the other involving the gain of electrons *(reduction)*. As an example, let us consider the reaction of sodium and chlorine to form sodium chloride. Recall that NaCl is an ionic substance consisting of equal numbers of Na^+ and Cl^- ions. This means that during the reaction [Eq. (6.2)], neutral sodium atoms become positively charged—they lose an electron [Eq. (6.5)]—whereas neutral chlorine molecules acquire a negative charge to become chloride ions [Eq. (6.6)].

$$Na \rightarrow Na^+ + e^- \tag{6.5}$$

$$Cl_2 + 2e^- \rightarrow 2Cl^- \tag{6.6}$$

If we assume that the electrons lost by the sodium atoms in Eq. (6.5) are the electrons gained by the chlorine molecules in Eq. (6.6), it is apparent that the two sodium atoms must lose their electrons for every chlorine molecule involved [Eqs. (6.7) and (6.8)], which is fortunate because the overall reaction [Eq. (6.9)] requires these quantities of products to yield a balanced equation.

$$2Na \rightarrow 2Na^+ + 2e^- \tag{6.7}$$

$$Cl_2 + 2e^- \rightarrow 2Cl^- \tag{6.8}$$

$$\overline{2Na + Cl_2 \rightarrow 2NaCl} \tag{6.9}$$

The overall reaction [Eq. (6.9)] is obtained by adding the two *half-reactions* [Eqs. (6.7) and (6.8)]; the electrons that are lost in the first step are gained in the second step. Both reactions occur simultaneously, and one cannot occur without the other.

OXIDATION NUMBERS

As in all chemical reactions, there are two important quantities in the redox equations: the identities of the species that constitute the reactants and products and the mass balance involved. Thus, it is first important to be able to recognize that a reaction belongs to the redox class. To help with this problem, chemists have invented the concept of an *oxidation number* as the basis for book-keeping the gain and loss of electrons among species. Each atom in a compound is assigned an oxidation number according to the following rather rigid and, in some senses, artificial set of rules:

1. For an element in a binary *ionic* compound, the oxidation number is equal to the charge of the ion. Thus, in ionic compounds containing Sn^{4+} and S^{2-} the oxidation numbers of Sn and S are 4+ and 2−, respectively.
2. The oxidation number of an atom in any free element, such as Fe, O_2, and S, is zero.
3. In its compounds or ions, H has an oxidation number of 1+. (The exceptions to this rule are the metal hydrides, such as NaH and CaH_2, where the oxidation number of H is 1−.)
4. In its compounds or ions, O has an oxidation number of 2−. The exceptions to this rule are OF_2, in which the oxidation number of O is 2+; the peroxides (such as H_2O_2 and Na_2O_2), in which its oxidation number is 1−; and the superoxides (such as NaO_2) in which the oxidation number is $\frac{1}{2}-$. The "exceptions" illustrate the fact that oxygen was chosen as a "reference" oxidation number because it forms compounds with every other element, which makes it a useful choice for a "reference." However, oxygen forms some "nonstandard" binary compounds in which an O_2-unit (rather than an O atom) is involved. In such compounds the "reference oxidation state" shifts to the other element in the binary compound, that is, Na in NaO_2 and Na_2O_2 and F in OF_2.
5. In any neutral molecule the sum of the positive oxidation numbers equals the sum of the negative oxidation numbers. The charge on a complex ion equals the difference between the positive and negative oxidation numbers. For example, in the compound H_3AsO_4, the total oxidation number for the four atoms of oxygen is 8−; since the sum of the oxidation numbers of the three hydrogen atoms is 3+, the oxidation number of arsenic is 5+. In the ion $Cr_2O_7^{2-}$, the total oxidation number for the seven atoms of oxygen is 14−; since the charge on the ion is 2−, the total positive oxidation number is 12+, or 6+ for each of the two atoms of chromium. Note that the definition of oxidation number is made coincidental with the charge on an ion by Rule 1. This does not imply that the oxidation number of sulfur (6+) in H_2SO_4 means that sulfur bears a 6+ charge; rather sulfur has a "tendency" to be positively charged in comparison with oxygen, which has a "tendency" to be negatively charged (as in the ionic compound Na_2O).

HALF-CELL POTENTIALS

Chemical reactions of the oxidation–reduction type can be made to liberate electrical energy, rather than energy in the form of heat or light, using devices known as *galvanic* or *voltaic cells*. A galvanic cell is constructed so that the oxidation and reduction reactions occur in two different parts of the apparatus. The species that gives up electrons is physically isolated from the species that gains electrons.

The reaction between zinc and copper sulfate [Eq. (6.10)] serves as a good illustration of the operation of a galvanic cell because of the ease with which the reaction occurs and the simplicity of the apparatus.

$$Zn + Cu^{2+} + SO_4^{2-} \rightleftharpoons Zn^{2+} + SO_4^{2-} + Cu \tag{6.10}$$

This reaction can be carried out by adding metallic zinc to a solution of copper sulfate; zinc metal goes into solution as zinc ions, metallic copper deposits on the excess zinc, and the energy liberated in this reaction appears as heat. The overall reaction can be divided into an oxidation step [Eq. (6.11)] and a reduction step [Eq. (6.12)].

$$Zn \rightarrow Zn^{2+} + 2e^- \qquad \text{(oxidation)} \tag{6.11}$$
$$Cu^{2+} + 2e^- \rightarrow Cu \qquad \text{(reduction)} \tag{6.12}$$

It is apparent that the sulfate ion does not take part in the reaction other than to maintain electrical neutrality.

A diagrammatic representation of a galvanic cell in which this reaction can occur is shown in Figure 6.1. Zinc and copper electrodes are immersed in a zinc sulfate solution and a copper

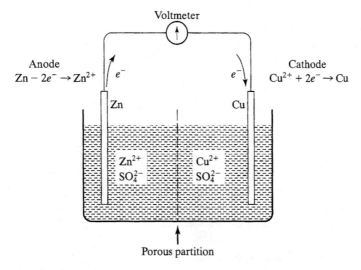

FIGURE 6.1   In a galvanic cell the reduction half-reaction is physically separated from the oxidation half-reaction. A porous partition permits the passage of ions from one side of the cell to the other to maintain electrical neutrality. The electrons generated in the oxidation half-cell pass through a wire and eventually reach the reduction half-cell where they are consumed in a chemical reaction.

sulfate solution, respectively. The two solutions are separated by a porous partition that prevents mechanical mixing, but does not obstruct the movement of ions from one compartment to the other; if mechanical mixing occurred, we would observe directly the reaction described by Eq. (6.10). The combination of an electrode and an ion derived from the electrode material (such as Zn and Zn^{2+} or Cu and Cu^{2+}) is called a *half-cell*, and the reaction that occurs at that electrode is called a *half-reaction*. A galvanic cell, just like the redox process on which it is based, is composed of two half-cells, one involving an oxidation reaction and the other a reduction reaction. By separating the oxidation and reduction processes in a galvanic cell, the chemical energy of the reaction is converted into electrical energy, which is often a more easily controlled form of energy.

When the cell is in operation, zinc atoms at the surface of the electrode go into solution as zinc ions, leaving electrons behind on the electrode. At the same time, copper ions acquire electrons at the copper electrode surface and become copper atoms. The zinc electrode is negatively charged with respect to the copper electrode, and the flow of electrons in the external circuit is from the zinc electrode to the copper electrode. The current produced by a galvanic cell always flows in one direction and is called a *direct current* (dc). If electrons are permitted to flow from the zinc electrode to the copper electrode, the latter increases in weight, while the former decreases. In other words, zinc is consumed and copper is produced, which is exactly the same result as that observed in the chemical reaction between zinc and copper sulfate (Eq. 6.10).

The *voltage*, or potential difference, between the two electrodes is dependent on the concentration and temperature of the solutions in the two compartments; with 1 M solutions of zinc sulfate and copper sulfate, the voltage of the cell is 1.10 volts (v).

Galvanic cells can be constructed to incorporate most oxidation–reduction reactions, although in some cases the experimental details are more complex and hence perhaps less convenient for practical uses than those in the preceding example. Note in Figure 6.1 that a given half-cell consists of the oxidized and reduced forms of the substance of interest in contact with each other. Thus, in principle, we would construct a series of half-cells, connect them in pairs using every possible combination, and measure the potentials of the pairs of half-cells. Although the potential of a given half-cell can never be measured alone, we could assign half-cell potentials relative to some arbitrary standard; the standard established is that the potential of the hydrogen half-cell [Eq. (6.13)] is zero when hydrogen gas is present at 1 atm pressure and hydrogen ions are present at a concentration of 1 M.

$$\tfrac{1}{2}H_2 \leftrightharpoons H^+ + e^- \qquad E^0 = 0.0 \text{ v} \qquad (6.13)$$

Tables 6.1 and 6.2 list the standard half-cell potentials for several common species; the potentials given are those observed for the oxidized and reduced forms of the species present in their standard state, which is defined as 1 M for ionic substances, gaseous substances at 1 atm pressure, and pure metals in their most stable form at 25°C.

An inspection of Tables 6.1 and 6.2 reveals several conventions that have been employed in their construction. Understanding these conventions will make it easier for us to use the information in the tables. The half-reactions are listed so that the substance on the left of the arrow gains one or more electrons and is thereby converted to the substance on the right; that is, the reactions are written as reduction processes. A gain of electrons means that a substance has been reduced and, since the substance that is reduced must thereby be functioning as an oxidizing agent, it follows that each substance on the left of the arrow is functioning as an oxidizing agent. The fact that any reaction reaches a state of equilibrium means that in each redox reaction, the

TABLE 6.1   STANDARD REDUCTION POTENTIALS IN ACID SOLUTION

Reaction	Potential, E^0 (in volts)
1. $K^+ + e^- \rightleftharpoons K$	-2.925
2. $Ba^{2+} + 2e^- \rightleftharpoons Ba$	-2.90
3. $Ca^{2+} + 2e^- \rightleftharpoons Ca$	-2.87
4. $Na^+ + e^- \rightleftharpoons Na$	-2.714
5. $Mg^{2+} + 2e^- \rightleftharpoons Mg$	-2.37
6. $Al^{3+} + 3e^- \rightleftharpoons Al$	-1.66
7. $Mn^{2+} + 2e^- \rightleftharpoons Mn$	-1.18
8. $Zn^{2+} + 2e^- \rightleftharpoons Zn$	-0.763
9. $Cr^{3+} + 3e^- \rightleftharpoons Cr$	-0.74
10. $Te + 2H^+ + 2e^- \rightleftharpoons H_2Te$	-0.72
11. $As + 3H^+ + 3e^- \rightleftharpoons AsH_3$	-0.60
12. $H_3PO_3 + 2H^+ + 2e^- \rightleftharpoons H_3PO_2 + H_2O$	-0.50
13. $Fe^{2+} + 2e^- \rightleftharpoons Fe$	-0.440
14. $Cr^{3+} + e^- \rightleftharpoons Cr^{2+}$	-0.41
15. $Cd^{2+} + 2e^- \rightleftharpoons Cd$	-0.403
16. $Se + 2H^+ + 2e^- \rightleftharpoons H_2Se$	-0.40
17. $PbSO_4 + 2e^- \rightleftharpoons Pb + SO_4^{2-}$	-0.356
18. $Co^{2+} + 2e^- \rightleftharpoons Co$	-0.277
19. $H_3PO_4 + 2H^+ + 2e^- \rightleftharpoons H_3PO_3 + H_2O$	-0.276
20. $Ni^{2+} + 2e^- \rightleftharpoons Ni$	-0.250
21. $Sn^{2+} + 2e^- \rightleftharpoons Sn$	-0.136
22. $Pb^{2+} + 2e^- \rightleftharpoons Pb$	-0.126
23. $2H_2SO_3 + H^+ + 2e^- \rightleftharpoons HS_2O_4^- + 2H_2O$	-0.08
24. $2H^+ + 2e^- \rightleftharpoons H_2$	0.000
25. $HCOOH + 2H^+ + 2e^- \rightleftharpoons HCHO + H_2O$	0.056
26. $P + 3H^+ + 3e^- \rightleftharpoons PH_3$	0.06
27. $S_4O_6^{2-} + 2e^- \rightleftharpoons 2S_2O_3^{2-}$	0.08
28. $2H^+ + S + 2e^- \rightleftharpoons H_2S$	0.141
29. $Sn^{4+} + 2e^- \rightleftharpoons Sn^{2+}$	0.15
30. $SO_4^{2-} + 4H^+ + 2e^- \rightleftharpoons H_2SO_3 + H_2O$	0.17
31. $Hg_2Cl_2 + 2e^- \rightleftharpoons 2Hg + 2Cl^-$	0.2676
32. $Cu^{2+} + 2e^- \rightleftharpoons Cu$	0.337
33. $H_2SO_3 + 4H^+ + 4e^- \rightleftharpoons S + 3H_2O$	0.45
34. $I_2(aq) + 2e^- \rightleftharpoons 2I^-$	0.5355
35. $H_3AsO_4 + 2H^+ + 2e^- \rightleftharpoons H_3AsO_3 + H_2O$	0.559
36. $MnO_4^- + e^- \rightleftharpoons MnO_4^{2-}$	0.564
37. $O_2 + 2H^+ + 2e^- \rightleftharpoons H_2O_2$	0.682

Continued

TABLE 6.1   *Continued*

38.	$Fe^{3+} + e^- \rightleftharpoons Fe^{2+}$	0.771
39.	$Ag^+ + e^- \rightleftharpoons Ag$	0.7991
40.	$NO_3^- + 2H^+ + e^- \rightleftharpoons NO_2 + H_2O$	0.80
41.	$Hg^{2+} + 2e^- \rightleftharpoons Hg$	0.854
42.	$NO_3^- + 4H^+ + 3e^- \rightleftharpoons NO + 2H_2O$	0.96
43.	$HNO_2 + H^+ + e^- \rightleftharpoons NO + H_2O$	1.00
44.	$HOI + H^+ + 2e^- \rightleftharpoons I^- + H_2O$	1.00
45.	$Br_2(aq) + 2e^- \rightleftharpoons 2Br^-$	1.065
46.	$SeO_4^{2-} + 4H^+ + 2e^- \rightleftharpoons H_2SeO_3 + H_2O$	1.15
47.	$O_2 + 4H^+ + 4e^- \rightleftharpoons 2H_2O$	1.229
48.	$MnO_2 + 4H^+ + 2e^- \rightleftharpoons Mn^{2+} + 2H_2O$	1.23
49.	$Cr_2O_7^{2-} + 14H^+ + 6e^- \rightleftharpoons 2Cr^{3+} + 7H_2O$	1.33
50.	$Cl_2(aq) + 2e^- \rightleftharpoons 2Cl^-$	1.3595
51.	$ClO_3^- + 6H^+ + 6e^- \rightleftharpoons Cl^- + 3H_2O$	1.45
52.	$PbO_2 + 4H^+ + 2e^- \rightleftharpoons Pb^{2+} + 2H_2O$	1.455
53.	$HOCl + H^+ + 2e^- \rightleftharpoons Cl^- + H_2O$	1.49
54.	$MnO_4^- + 8H^+ + 5e^- \rightleftharpoons Mn^{2+} + 4H_2O$	1.51
55.	$Ce^{4+} + e^- \rightleftharpoons Ce^{3+}$	1.61
56.	$PbO_2 + SO_4^{2-} + 4H^+ + 2e^- \rightleftharpoons PbSO_4 + 2H_2O$	1.685
57.	$MnO_4^- + 4H^+ + 3e^- \rightleftharpoons MnO_2 + 2H_2O$	1.695
58.	$HBiO_3 + 5H^+ + 2e^- \rightleftharpoons Bi^{3+} + 3H_2O$	1.70
59.	$H_2O_2 + 2H^+ + 2e^- \rightleftharpoons 2H_2O$	1.77
60.	$Co^{3+} + e^- \rightleftharpoons Co^{2+}$	1.82
61.	$S_2O_8^{2-} + 2e^- \rightleftharpoons 2SO_4^{2-}$	1.91
62.	$O_3 + 2H^+ + 2e^- \rightleftharpoons O_2 + H_2O$	2.07
63.	$F_2 + 2e^- \rightleftharpoons 2F^-$	2.87
64.	$F_2 + 2H^+ + 2e^- \rightleftharpoons 2HF(aq)$	3.06

substance on the right of the arrow will react to form the substance on the left. To do so it must lose one or more electrons. When a substance loses electrons, it is oxidized, and during this process of being oxidized it is functioning as a reducing agent. That means all the substances on the right are reducing agents.

On the extreme right in the tables opposite each equation is a number (expressed in volts) that represents the driving force, or potential, with which the reduction reaction as written occurs. The potential is a measure of the intensity with which the substance on the left reacts to form the substance on the right. Note that the potentials at the top of the tables are large negative numbers, which change regularly through zero (at hydrogen) to large positive numbers. Going from a large negative number to a positive number is an *increase* in the value of the potential. Thus in Table 6.1 the potential gradually increases from -2.925 at the top to $+3.06$ at the bottom. That means that the substance at the top and the left of the arrow is the weakest oxidizing agent.

TABLE 6.2 STANDARD REDUCTION POTENTIALS IN BASIC SOLUTION

Reaction	Potential, E^0 (in volts)
1. $K^+ + e^- \leftrightharpoons K$	-2.925
2. $Al(OH)_4^- + 3e^- \leftrightharpoons Al + 4OH^-$	-2.35
3. $PO_4^{3-} + 2H_2O + e^- \leftrightharpoons HPO_3^- + 3OH^-$	-1.12
4. $2SO_3^{2-} + 2H_2O + 2e^- \leftrightharpoons S_2O_4^{2-} + 4OH^-$	-1.12
5. $CNO^- + H_2O + 2e^- \leftrightharpoons CN^- + 2OH^-$	-0.97
6. $SO_4^{2-} + H_2O + 2e^- \leftrightharpoons SO_3^{2-} + 2OH^-$	-0.93
7. $Sn(OH)_6^{2-} + 2e^- \leftrightharpoons Sn(OH)_4^{2-} + 2OH^-$	-0.90
8. $Sn(OH)_4^{2-} + 2e^- \leftrightharpoons Sn + 4OH^-$	-0.76
9. $CrO_4^{2-} + 4H_2O + 3e^- \leftrightharpoons Cr(OH)_4^- + 4OH^-$	-0.48
10. $O_2 + 2H_2O + 2e^- \leftrightharpoons H_2O_2 + 2OH^-$	-0.076
11. $MnO_2 + 2H_2O + 2e^- \leftrightharpoons Mn(OH)_2 + 2OH^-$	-0.05
12. $Cu(NH_3)_4^{2+} + e^- \leftrightharpoons Cu(NH_3)_2^+ + 2NH_3$	0.0
13. $Co(NH_3)_6^{3+} + e^- \leftrightharpoons Co(NH_3)_6^{2+}$	0.1
14. $Co(OH)_3 + e^- \leftrightharpoons Co(OH)_2 + OH^-$	0.17
15. $2H_2O + 4e^- \leftrightharpoons 4OH^-$	0.401
16. $IO^- + H_2O + 2e^- \leftrightharpoons I^- + 2OH^-$	0.49
17. $NiO_2 + 2H_2O + 2e^- \leftrightharpoons Ni(OH)_2 + 2OH^-$	0.49
18. $MnO_4^- + e^- \leftrightharpoons MnO_4^{2-}$	0.54
19. $MnO_4^- + 2H_2O + 3e^- \leftrightharpoons MnO_2 + 4OH^-$	0.57
20. $MnO_4^{2-} + 2H_2O + 2e^- \leftrightharpoons MnO_2 + 4OH^-$	0.60
21. $OBr^- + H_2O + 2e^- \leftrightharpoons Br^- + 2OH^-$	0.76
22. $H_2O_2 + 2e^- \leftrightharpoons 2OH^-$	0.88
23. $O_3 + H_2O + 2e^- \leftrightharpoons O_2 + 2OH^-$	1.24

It is possible to use the information in Tables 6.1 and 6.2 to establish whether a redox reaction will occur. For example, it is a fact that Fe^{3+} will oxidize I^-; how is this information contained in Table 6.1 and how can it be extracted? The equation for this process [Eq. (6.16)] is the sum of two half-reactions [Eq. (6.14) and (6.15)] with their corresponding half-cell potentials.

$$Fe^{3+} + e^- \leftrightharpoons Fe^{2+} \qquad E^0 = 0.771 \qquad (6.14)$$

$$\underline{2I^- \leftrightharpoons I_2 + 2e^- \qquad E^0 = -0.536} \qquad (6.15)$$

$$2Fe^{3+} + 2I^- \leftrightharpoons 2Fe^{2+} + I_2 \qquad E^0 = 0.235 \qquad (6.16)$$

Note that the oxidation of I^- [Eq. (6.14)] is the reverse of the process listed in Table 6.1 and its potential is given with a reversal of sign, and that, even though the coefficient must be doubled for balancing purposes, the potential is *not* doubled; the reduction of Fe^{3+} corresponds to the process listed in Table 6.1, and the sign of the potential is unchanged. Since the sum of the half-cell processes gives the overall process, the sum of the half-cell potentials gives the potential of the overall process. When the potential for a process is positive [as for Eq. (6.16)], the process will occur spontaneously; negative potentials correspond to nonspontaneous processes.

In Chapter 2 we discussed the relationship between the standard free energy of a system, ΔG^0—which is a measure of the *maximum* work that a system can do—and the equilibrium constant [see Eq. (2.41)]. The standard free-energy change, ΔG, for a redox process is the maximum electrical work that can be obtained from that process and is related by Eq. (6.17) to the standard potential for the cell, E^0_{cell}, which describes that process:

$$\Delta G^0 = -nFE^0_{cell} \tag{6.17}$$

where n is the number of electrons in the overall process and F is a universal constant called the *Faraday*. In effect, the product nF is the charge passed when the cell is discharged; nF units of charge moving across a potential of E^0_{cell} corresponds to electrical work. If, as we have said, a positive potential for a cell corresponds to a spontaneous process, Eq. (6.17) indicates that the free energy change for that process is a negative value. A complementary argument is that nonspontaneous redox processes exhibit a negative E^0_{cell}, leading to a positive ΔG^0.

We have just discussed how the half-reaction potentials in Tables 6.1 and 6.2 can be used to establish whether a redox process will occur; if the sum of the half-reaction potentials is a positive number, the reaction as written is spontaneous. We do not really have to know quantitatively how much positive the potential is—only that it is positive.

From this point of view, the relative location of the reduction process in the list of reduction potentials (Table 6.1) with respect to the oxidation process gives us all the information we need. To illustrate, consider a redox process we know occurs; for example, consider the reaction of zinc with copper sulfate (Cu^{2+}) as described by Eq. (6.10). The zinc half-reaction as a reduction process is shown as Eq. (8) in Table 6.1; its reduction potential is a negative number (-0.763). However, zinc as a reducing agent, which it is in Eq. (6.10), involves the reverse of Eq. (8) in Table 6.1; reversing the equation reverses the potential, which now makes it a positive potential ($+0.763$). The other half-reaction involving copper is shown as Eq. (32) in Table 6.1. This process, the reduction of copper from copper ion, is exactly what happens in Eq. (6.10), so that the potential, which is positive, is used as listed. Thus, to obtain the standard potential of the process in Eq. (6.10), we must add these two half-reactions and their corresponding potential. Because each half-reaction potential is positive, we shall get a positive potential for the overall reaction Eq. (6.10), which, recall, means that the reaction is spontaneous.

Applying these ideas, we see from the tables that Fe^{3+} ions [Eq. (38) in Table 6.1] will oxidize I^- ions [Eq. (34)], but will not oxidize Br^- [Eq. (45)]; $Cr_2O_7^{2-}$ ions [Eq. (49)] will oxidize Sn^{2+} [Eq. (29)]; H_2SO_3 [Eq. (30)] will reduce NO_3^- [Eq. (42)].

Table 6.2 lists the standard reduction potentials as they occur in basic solution that should be contrasted with the information presented in Table 6.1, which lists potentials in acid solution. Since the nature of the species is often different in acid and basic solution, we would expect the potentials for the processes to be different. Thus, for example, Al^{3+} in basic solution exists as the ion $Al(OH)_4^-$, Eq. (2) in Table 6.2, and the reduction potential for this species is different from the reduction of Al^{3+} (-2.35 in Table 6.2 *versus* -1.66 in Table 6.1). Note also that the acidity or basicity of the solution does not affect the reduction potential; compare the reduction potential for K^+, for example.

Not only will the tables of potentials tell us whether two substances will react with each other but they will also indicate what products will be formed. Also, they provide a simple basis for balancing the resulting redox equation. In a balanced oxidation–reduction equation, the total number of electrons lost by the element or elements oxidized equals the total number of electrons gained by the element or elements reduced. Thus, Eqs. (49) and (35) in Table 6.1 show that H_3AsO_3 will reduce $Cr_2O_7^{2-}$ to Cr^{3+} and will itself be oxidized to H_3AsO_4. The two half-reactions are

$$Cr_2O_7^{2-} + 14H^+ + 6e^- \leftrightarrows 2Cr^{3+} + 7H_2O \tag{6.18}$$

$$H_3AsO_3 + H_2O \leftrightarrows H_3AsO_4 + 2H^+ + 2e^- \tag{6.19}$$

Since six electrons are added in the first reaction while only two are given off in the second, it is necessary to triple the coefficients in the second reaction so that the number of electrons in the two reactions will be equal. When this is done and the two equations are summed by adding the substances on each of the two sides of the equality signs, the following balanced equation is obtained:

$$3H_3AsO_3 + Cr_2O_7^{2-} + 9H^+ \leftrightarrows 3H_3AsO_4 + 2Cr^{3+} + 4H_2O \tag{6.20}$$

Note that the number of electrons gained, 6, is the same as the number lost; since 6 electrons appear on each side of the equation, they cancel out.

No half-reaction can occur by itself; it must be coupled with another half-reaction. Every redox reaction consists of a pair of half-reactions.

It should be emphasized that the potentials given in the tables are for 1 M solutions at 25°C. Since the reactions represent true equilibria, it follows from the mass law that increasing the concentration of a reactant will increase its potential. An increase in temperature also generally increases the potential. For these reasons, substances that, according to the tables, do not react at 25°C in 1 M solution are found to do so at higher temperatures and concentrations.

NERNST EQUATION

Redox processes, of course, can occur between species that are not present at a concentration of 1 M. The *Nernst equation* relates the potential for a redox process under any conditions to the standard potential. Thus, for any general redox process [Eq. (6.21)],

$$aA + bB = cC + dD \tag{6.21}$$

The Nernst equation is given by Eq. (6.22),

$$E = E^0 - \frac{RT}{nF} \ln \frac{[C]^c[D]^d}{[A]^a[B]^b} \tag{6.22}$$

where E is the potential for the process under nonstandard conditions, E^0 is the standard potential of the cell, R is the ideal gas constant, T is the absolute temperature, n is the number of electrons transferred, and F is the value of a physical constant called the Faraday. At 25°C and with base 10 logarithm, the Nernst equation becomes

$$E = E^0 - \frac{0.059}{n} \log \frac{[C]^c[D]^d}{[A]^a[B]^b} \tag{6.23}$$

Redox Equilibria   Like all chemical reactions, redox processes can reach a condition of equilibrium. As a general example, take the process in Eq. (6.21); when equilibrium is achieved, the equilibrium constant, expressed in the usual manner, is

$$K_{eq} = \frac{[C]^c[D]^d}{[A]^a[B]^b} \tag{6.24}$$

The Nernst equation [Eq. (6.23)] must be valid for all possible combinations of reactant and product concentration; thus, at equilibrium the Nernst equation becomes

$$E = E^0 - \frac{0.059}{n} \log K_{eq} \tag{6.25}$$

At equilibrium, however, the potential for the overall reaction is zero; in effect, equilibrium is attained when the two half-cells that represent the overall reaction are connected, the electrodes are short-circuited, and the resulting voltaic cell is allowed to "run down." If the potential of this cell is zero, the Nernst equation becomes

$$0 = E = E^0 - \frac{0.059}{n} \log K_{eq} \tag{6.26}$$

Rearranging Eq. (6.26) yields

$$\log K_{eq} = \frac{nE^0}{0.059} \tag{6.27}$$

Remember that the factor 0.059 is valid for processes that occur at 25°C; the equivalent factor under any other conditions can be calculated from Eq. (6.22). Thus, the equilibrium constant for any redox process can be obtained from Eq. (6.27) and a table of standard potentials.

EXAMPLE 6.1

Compute the equilibrium constant for the reaction

$$H_3AsO_3 + I_2 + H_2O \leftrightharpoons H_3AsO_4 + 2H^+ + 2I^-$$

1. The two half-reactions for this process are given by

$$H_3AsO_3 + H_2O \leftrightharpoons H_3AsO_4 + 2H^+ + 2e^- \qquad E^0 = -0.559$$

$$I_2 + 2e^- \leftrightharpoons 2I^- \qquad E^0 = +0.536$$

2. From this, E^0 for the overall process is given by

$$E^0 = -0.559 + 0.536 = -0.023 \text{ v}$$

3. Two electrons are transferred in the overall process so that $n = 2$.

4. The equilibrium constant is calculated from Eq. (6.27).

$$\log K_{eq} = \frac{(-0.023)(2)}{0.059} = -0.780$$

$$K_{eq} = 0.166$$

Some very simple redox equations—such as Eqs. (6.28) and (6.29)—can be balanced readily by inspection; however, the more involved processes require a systematic procedure for the sake of efficiency.

$$2Na + Cl_2 \leftrightarrows 2NaCl \tag{6.28}$$

$$2H_2 + O_2 \leftrightarrows 2H_2O \tag{6.29}$$

Two such procedures have been developed; both give the correct answer—a balanced equation—and neither possesses a fundamental advantage from the standpoint of being the most correct. Both are presented for your consideration. Each has its advantages. Learn the one that is the most comfortable for your mental processes.

BALANCING REDOX EQUATIONS BY THE METHOD OF HALF-REACTIONS

In the discussion of oxidation potentials we observed that every redox reaction consists of two half-reactions. In one of these half-reactions, an oxidizing agent (oxidant) is reduced; in the other, a reducing agent (reductant) is oxidized. Also, in the balanced net equation for the redox reaction, the total number of electrons lost by the element or elements oxidized equals the total number of electrons gained by the element or elements reduced.

These two observations are the basis for balancing redox equations by the method of half-reactions (also called the *ion-electron method*). In this method, the two half-reactions are built up by using water to balance the oxygen content (step 2), hydrogen ions to balance the hydrogen content (step 3), and electrons to balance the charge (step 4). Finally, the number of electrons in the reduction process is made equal to the number of electrons in the oxidation process (step 5), and the resulting two half-reactions are added to give the overall reaction (step 6). To illustrate the method we shall balance the equation for the reaction that occurs when H_2SO_3 and MnO_4 react in acidic solution to form SO_4^{2-} and Mn^{2+}. The steps in the balancing process are as follows.

EXAMPLE 6.2

Step 1: Write the skeleton equations for the two half-reactions.

$$H_2SO_3 \leftrightarrows SO_4^{2-} \qquad \text{(The reductant } H_2SO_3 \text{ is oxidized to } SO_4^{2-}.)$$

$$MnO_4^- \leftrightarrows Mn^{2+} \qquad \text{(The oxidant } MnO_4^- \text{ is reduced to } Mn^{2+}.)$$

Step 2: For each half-reaction, balance the oxygen by adding H_2O on the appropriate side of the equation.

$$H_2SO_3 + H_2O \leftrightarrows SO_4^{2-}$$

$$MnO_4^- \leftrightarrows Mn^{2+} + 4H_2O$$

Step 3: Balance the hydrogen by adding H^+ on the appropriate side of the equation.

$$H_2SO_3 + H_2O \rightleftharpoons SO_4^{2-} + 4H^+$$

$$MnO_4^- + 8H^+ \rightleftharpoons Mn^{2+} + 4H_2O$$

Step 4: Balance the charges on each side of the equality sign by adding electrons on the appropriate side of the equation.

$$H_2SO_3 + H_2O \rightleftharpoons SO_4^{2-} + 4H^+ + 2e^-$$

$$MnO_4^- + 8H^+ + 5e^- \rightleftharpoons Mn^{2+} + 4H_2O$$

Step 5: Balance the electrons in the two half-reactions by multiplying each by an appropriate number to yield a common factor. Since the first equation loses $2e^-$ and the second gains $5e^-$, by multiplying the first equation by 5 and the second by 2, a total of 10 electrons will be involved in each half-reaction.

$$5H_2SO_3 + 5H_2O \rightleftharpoons 5SO_4^{2-} + 20H^+ + 10e^-$$

$$2MnO_4^- + 16H^+ + 10e^- \rightleftharpoons 2Mn^{2+} + 8H_2O$$

Step 6: Add the two balanced half-reactions to give the final net equation; cancel quantities that appear on both sides of the equation.

$$5H_2SO_3 + 2MnO_4^- \rightleftharpoons 5SO_4^{2-} + 2Mn^{2+} + 4H^+ + 3H_2O$$

EXAMPLE 6.3

As another example of this method, consider the reaction in which solid P_4S_3 reacts with a solution of HNO_3 to yield H_3PO_4, SO_4^{2-}, and NO. The basic reaction is:

$$P_4S_3 + HNO_3 \rightleftharpoons H_3PO_4 + SO_4^{2-} + NO$$

Step 1: The two half-reactions are:

$$P_4S_3 \rightleftharpoons 4H_3PO_4 + 3SO_4^{2-}$$

$$NO_3^- \rightleftharpoons NO$$

Note that each half-reaction contains the minimum number of moles of products per mole of reactant.

Step 2: Balance the oxygen by adding H_2O on the appropriate side.

$$28H_2O + P_4S_3 \rightleftharpoons 4H_3PO_4 + 3SO_4^{2-}$$

$$NO_3^- \rightleftharpoons NO + 2H_2O$$

Step 3: Balance the hydrogen by adding H^+.

$$28H_2O + P_4S_3 \rightleftharpoons 4H_3PO_4 + 3SO_4^{2-} + 44H^+$$

$$4H^+ + NO_3^- \rightleftharpoons NO + 2H_2O$$

Step 4: Balance the charges by adding electrons.

$$28H_2O + P_4S_3 \rightleftharpoons 4H_3PO_4 + 3SO_4^{2-} + 44H^+ + 38e^-$$

$$3e^- + 4H^+ + NO_3^- \rightleftharpoons NO + 2H_2O$$

Step 5: Balance the electrons in the two half-reactions by multiplying the first by 3 and the second by 38.

$$84H_2O + 3P_4S_3 \rightleftharpoons 12H_3PO_4 + 9SO_4^{2-} + 132H^+ + 114e^-$$

$$114e^- + 152H^+ + 38NO_3^- \rightleftharpoons 38NO + 76H_2O$$

Step 6: Add the two half-reactions to give the final net equation.

$$3P_4S_3 + 38NO_3^- + 20H^+ + 8H_2O \rightleftharpoons 12H_3PO_4 + 9SO_4^{2-} + 38NO$$

Finally, we consider the balancing process using the method of half-reactions if the redox reaction occurs in basic solution. The H^+ ions added to balance the hydrogen (step 3) will immediately react with the OH^- ions to form H_2O. Otherwise, the balancing routine is the same as already outlined for acidic solutions. Consider the reaction in which CN^- reacts with CrO_4^{2-} in basic solution to form CNO^- and $Cr(OH)_4^-$.

EXAMPLE 6.4

The basic reaction is:

$$CN^- + CrO_4^{2-} \rightleftharpoons CNO^- + Cr(OH)_4^-$$

Step 1: Write the half-reactions.

$$CN^- \rightleftharpoons CNO^-$$

$$CrO_4^{2-} \rightleftharpoons Cr(OH)_4^-$$

Step 2: Balance the oxygen by adding water.

$$H_2O + CN^- \rightleftharpoons CNO^-$$

$$CrO_4^{2-} \rightleftharpoons Cr(OH)_4^- \qquad \text{(Oxygen is already balanced.)}$$

Step 3: Balance the hydrogen by adding H^+; remember that where a hydrogen ion appears in basic solution, it will be converted into water.

$$H_2O + CN^- \leftrightharpoons CNO^- + 2H^+$$
$$\underline{2OH^- + 2H^+ \leftrightharpoons 2H_2O}$$
$$CN^- + 2OH^- \leftrightharpoons CNO^- + H_2O$$

$$CrO_4^{2-} + 4H^+ \leftrightharpoons Cr(OH)_4^-$$
$$\underline{4H_2O \leftrightharpoons 4H^+ + 4OH^-}$$
$$CrO_4^{2-} + 4H_2O \leftrightharpoons Cr(OH)_4^- + 4OH^-$$

Step 4: Balance charges by adding electrons.

$$CN^- + 2OH^- \leftrightharpoons CNO^- + H_2O + 2e^-$$
$$CrO_4^{2-} + 4H_2O + 3e^- \leftrightharpoons Cr(OH)_4^- + 4OH^-$$

Step 5: Find the lowest multiplier for each step to give the same number of electrons.

$$3CN^- + 6OH^- \leftrightharpoons 3CNO^- + 3H_2O + 6e^-$$
$$2CrO_4^{2-} + 8H_2O + 6e^- \leftrightharpoons 2Cr(OH)_4^- + 8OH^-$$

Step 6: Add the two half-reactions and cancel species that appear on both sides.

$$3CN^- + 2CrO_4^{2-} + 5H_2O \leftrightharpoons 3CNO^- + 2CR(OH)_4^- + 2OH^-$$

EXAMPLE 6.5

For our final example, consider the reaction in basic solution for which the skeleton equation is

$$OBr^- + HPO_3^- \leftrightharpoons Br^- + PO_4^{3-}$$

Step 1: Separate reduction and oxidation processes.

$$HPO_3^- \leftrightharpoons PO_4^{3-}$$
$$OBr^- \leftrightharpoons Br^-$$

Step 2: Balance oxygen.

$$HPO_3^- + H_2O \leftrightharpoons PO_4^{3-}$$
$$OBr^- \leftrightharpoons Br^- + H_2O$$

Step 3: Balance hydrogen.

$$HPO_3^- + H_2O \rightleftharpoons PO_4^{3-} + 3H^+$$

$$\underline{3H^+ + 3OH^- \rightleftharpoons 3H_2O}$$

$$HPO_3^- + 3OH^- \rightleftharpoons 2H_2O + PO_4^{3-}$$

$$OBr^- + 2H^+ \rightleftharpoons Br^- + H_2O$$

$$\underline{2H_2O \rightleftharpoons 2H^+ + 2OH^-}$$

$$OBr^- + H_2O \rightleftharpoons Br^- + 2OH^-$$

Step 4: Balance charge.

$$HPO_3^- + 3OH^- \rightleftharpoons 2H_2O + PO_4^{3-} + e^-$$

$$OBr^- + H_2O + 2e^- \rightleftharpoons Br^- + 2OH^-$$

Step 5: Make the number of electrons in each process the same.

$$2HPO_3^- + 6OH^- \rightleftharpoons 4H_2O + 2PO_4^{3-} + 2e^-$$

$$OBr^- + H_2O + 2e^- \rightleftharpoons Br^- + 2OH^-$$

Step 6: Add half-reactions.

$$2HPO_3^- + OBr^- + 4OH^- \rightleftharpoons 2PO_4^{3-} + Br^- + 3H_2O$$

The obvious advantage of the ion-electron method is that you do not have to establish the oxidation number of the species involved; indeed, you do not even have to know ahead of time which species is oxidized and which is reduced. This information comes out of the process of balancing.

BALANCING REDOX EQUATIONS USING THE CHANGE IN OXIDATION NUMBER METHOD

In any redox process, the oxidation number of at least one element increases, and the oxidation number of at least one element decreases. We can use this observation as the basis for balancing redox equations. As an example, consider the reactions that occur when an acidified solution (H_2SO_4) of $FeSO_4$ is treated with a solution of $KMnO_4$. The steps in the balancing process follow.

Step 1: Write the formulas for the reactants and products.

$$Fe^{2+} + MnO_4^- \rightleftharpoons Fe^{3+} + Mn^{2+}$$

Step 2: Identify the element or elements oxidized and the element or elements reduced. This is best accomplished by assigning oxidation numbers to all the elements in the species present using the previously established rules (see Appendix I). Note the initial and final oxidation number of each of these elements. Note the change in the oxidation number of each of these elements.

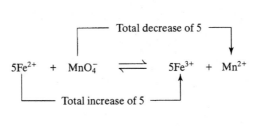

Here Fe is oxidized; Mn is reduced.

Step 3:   Adjust the number of moles of each reactant (find the lowest common factor) so that the total increase in oxidation number equals the total decrease.

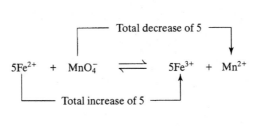

Step 4:   Balance the charges on each side of the equation by adding the necessary H^+ ions. If the solution is alkaline, the charges can be balanced by adding OH^- ions. If the solution is neutral, either H^+ or OH^- ions may be added; H_2O will provide these ions. As the equation is written in step 3, the net charge on the left (from the five Fe^{2+} ions and the one MnO_4^-) is +9 and the net charge on the right (from the five Fe^{3+} ions and the one Mn^{2+} ion) is +17. By adding eight H^+ ions to the left, the charge on each side will be +17.

$$5Fe^{2+} + MnO_4^- + 8H^+ \leftrightharpoons 5Fe^{3+} + Mn^{2+}$$

Step 5:   Balance the hydrogen by adding H_2O to the appropriate side. If the work has been correct up to this point, balancing the hydrogen will also balance the oxygen and, thus, balance the equation.

$$5Fe^{2+} + MnO_4^- + 8H^+ \leftrightharpoons 5Fe^{3+} + Mn^{2+} + 4H_2O$$

If more than one element is oxidized (and/or reduced), the total increase in oxidation number is the sum of the increases for each element. The oxidation of FeS by HNO_3 illustrates such a reaction.

Step 1:   Write formulas for reactants and products.

$$FeS + NO_3^- \leftrightharpoons NO + SO_4^{2-} + Fe^{3+}$$

Step 2: Assign oxidation states.

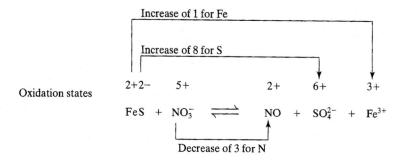

Step 3: Note total increases and decreases.

Step 4: Make total increase equal to total decrease using appropriate multiplier.

$$FeS + 3NO_3^- \rightleftharpoons 3NO + SO_4^{2-} + Fe^{3+}$$

Step 5: Add H^+ to balance charges and H_2O to balance hydrogen; the oxygen content should be balanced.

$$FeS + 3NO_3^- + 4H^+ \rightleftharpoons 3NO + SO_4^{2-} + Fe^{3+} + 2H_2O$$

To become more comfortable with the balancing processes and to establish your personal preference for one, it might be useful if you went back and balanced these two examples using the other process.

THE ROLE OF REDOX PROCESSES IN QUALITATIVE ANALYSIS

Redox processes often are used to produce unambiguous products that can help identify species that elude identification using the other reaction types described thus far.

Thus, mercurous ion, Hg_2^{2+}—which forms a hot-water insoluble white precipitate with Cl^-, just like Ag^+—can be identified as such because it undergoes a redox reaction in the presence of aqueous ammonia.

$$Hg_2Cl_2 + 2NH_4OH \rightleftharpoons Hg + HgNH_2Cl + NH_4Cl + 2H_2O \qquad (6.30)$$

The reaction product is black because the liquid elemental mercury formed is finely divided.

In this process, mercurous ion (oxidation state of $1+$) undergoes both oxidation and reduction to yield mercury in a higher oxidation state (Hg^{2+}), as well as elemental mercury (oxidation state of 0). The process in which one element undergoes both oxidation and reduction is fairly uncommon; such processes are called *disproportionation reactions*.

The formation of elemental mercury is also used to establish the presence of Sn^{2+}, which acts as a reducing agent toward Hg^{2+} [Eq. (6.31)].

$$3HgCl_2 + Sn^{2+} \rightleftharpoons Hg + Hg_2Cl_2 + Sn^{4+} + 4Cl^- \tag{6.31}$$

The mixture of Hg and Hg_2Cl_2 is gray.

There are only a few good reactions with which Sb^{3+} can be identified; the redox process shown in Eq. (6.32) in which black elemental antimony is formed is useful.

$$Sb^{3+} + Al \rightleftharpoons Al^{3+} + Sb(s) \tag{6.32}$$

The analysis of the aluminum-nickel group involves the formation of an aqueous solution possibly containing Fe^{3+} (yellow), Co^{2+} (pink), Ni^{2+} (green), and Mn^{2+} (pink). The first three members of this group are identified by the colors of the coordination compounds formed with different ligands. Manganese is identified by the color of the oxidation product formed when Mn^{2+} is treated with $NaBiO_3$ [Eq. (6.33)].

$$2Mn^{2+} + 5BiO_3^- + 4H^+ \rightleftharpoons 5BiO^+ + 2MnO_4^- + 2H_2O \quad \text{(purple)} \tag{6.33}$$

Redox processes are sometimes used to dissolve precipitates that cannot be dissolved through the formation of weakly ionized substances or very stable complex ions. Thus, in the analysis of the copper-arsenic group, the sulfides of lead (PbS), bismuth (Bi_2S_3), copper (CuS), cadmium (CdS), and mercury (HgS) cannot be dissolved in a strong nonoxidizing acid such as HCl because these substances are so insoluble that the H^+ concentration cannot be made high enough to form a sufficient amount of the weak acid H_2S to cause these sulfides to dissolve (as in Example 4.8). Instead, the low concentration of S^{2-} in these solutions is attacked with an oxidizing agent (HNO_3), which converts it to sulfur [Eq. (6.34)].

$$3S^{2-} + 2NO_3^- + 8H^+ \rightleftharpoons 3S + 2NO + 4H_2O \tag{6.34}$$

The sulfates of these cations are more soluble than the sulfides. Dilute nitric acid will separate PbS, Bi_2S_3, CuS, and CdS from HgS; the first four sulfides dissolve, while HgS does not. Although HgS is not soluble in either HNO_3 or HCl, a mixture of these acids (called *aqua regia*) is a more powerful oxidizing agent than HNO_3 and will dissolve HgS [Eq. (6.35)].

$$3HgS + 2NO_3^- + 8H^+ + 12Cl^- \rightleftharpoons 3HgCl_4^{2-} + 2NO + 4H_2O + 3S \tag{6.35}$$

Note that in Eq. (6.35), a redox process *and* complex-ion formation (see Chapter 5) assist in the dissolution process.

PROBLEMS

6.1 Give the oxidation number of the atom indicated in bold lettering.

    a. H_3PO_4

    b. $NaSbF_6$

    c. K_2O_2

    d. $S_2O_3^{2-}$

    e. PH_4^+

    f. $AsCl_4^+$

6.2 Predict whether the following substances will react with each other in a redox process and, if they do, what products will be formed.

    a. $Cr_2O_7^{2-} + Fe^{3+}$

    b. $MnO_4^- + Sn^{2+}$

    c. $H_2SO_3 + Fe^{2+}$

    d. $Cl^- + I^-$

    e. $H_3AsO_3 + I_2$

    f. $H_3PO_3 + Sn^{4+}$

    g. $Hg + Pb^{2+}$

6.3 For the reactions in Problem 6.2 that do occur, write balanced equations.

6.4 For the reactions in Problem 6.2 that occur spontaneously, estimate the potential for the overall process assuming that the species are present in their standard states.

6.5 Estimate the potential of the cell shown in Figure 6.1 when $[Zn^{2+}] = 0.6\ M$ and $[Cu^{2+}] = 0.01\ M$.

6.6 Calculate the equilibrium constants for the following redox systems.

    a. $Br_2 + 2I^- \rightleftharpoons 2Br^- + I_2$

    b. $Zn + Cu^{2+} \rightleftharpoons Zn^{2+} + Cu$

    c. $CrO_7^{2-} + 3Sn^{2+} + 14H^+ \rightleftharpoons Cr^{3+} + 3Sn^{4+} + 7H_2O$

    d. $2Fe^{3+} + Sn^{2+} \rightleftharpoons 2Fe^{2+} + Sn^{4+}$

6.7 Balance the following redox reactions.

    a. $H_2O_2 + MnO_4^- + H^+ \rightleftharpoons O_2 + Mn^{2+} + H_2O$

    b. $Ag + NO_3^- + H^+ \rightleftharpoons Ag^+ + NO + H_2O$

    c. $SO_3^{2-} + H^+ + MnO_4^- \rightleftharpoons SO_4^{2-} + Mn^{2+} + H_2O$

    d. $Cr_2O_7^{2-} + H^+ + Cl^- \rightleftharpoons Cr^{3+} + H_2O + Cl_2$

    e. $PbO + NH_3 \rightleftharpoons Pb + N_2 + H_2O$

    f. $NO_2 + H_2 \rightleftharpoons NH_3 + H_2O$

PART

THE PRACTICE
OF QUALITATIVE
ANALYSIS

The Practical Aspects of Qualitative Analysis

In this chapter we organize the principles discussed in Chapters 2 to 6 on a practical basis as they apply to laboratory work in qualitative analysis. This organization should provide a useful basis for translating chemical principles described in Chapters 2 to 6 into laboratory practice.

THE PERIODIC TABLE

The study of qualitative analysis involves the study of reactions that occur in aqueous solutions. We know (Chapter 3) that aqueous solutions of substances can contain molecules, cations, or anions in several possible combinations. Thus, the study of qualitative analysis involves the study of how molecules, cations, and anions react in aqueous solutions. Since the periodic table provides the best available basis for the systematic organization of the properties of the elements and their compounds, we should expect the periodic table to serve as a useful guide in many aspects of qualitative analysis. Specific examples of this usefulness will be indicated as the various analytical procedures are carried out.

The close similarity in the chemical behavior of chlorides, bromides, and iodides, which is to be expected from the location of the three halogens in Group 17, causes them to interfere with each other in their identifications. The lower reactivity of iodine and the greater reducing power of the iodide ion, however, enable one to liberate iodine from an iodide with an oxidizing agent that will not be reduced by either bromide ion or chloride ion. The list of redox potentials (Table 6.1) gives us an indication of the species that can accomplish this result in acid solution.

The marked similarity in the solubilities and colors of phosphates and arsenates is to be expected from the position of arsenic and phosphorus in Group 15. This similarity, however, also becomes a complicating factor in the identification of these two anions.

Because the barium-magnesium group is composed entirely of metals from periodic Groups 1 and 2, and the silver, copper-arsenic, and aluminum-nickel groups are made up exclusively of metals from other groups, there are very marked differences in the character of the reactions observed in the barium-magnesium group as compared to those taking place in the other three groups. The metals in the periodic table Groups 1 and 2 exist, in their compounds or ions, in only one oxidation state. Therefore, oxidation–reduction reactions are not important among the barium-magnesium qualitative analysis group of ions. A number of transition metals and metals in the right half of the periodic table have several oxidation states; therefore, we find numerous oxidation–reduction processes among the reactions used to separate and identify the elements in the silver, copper-arsenic, and aluminum-nickel qualitative analysis groups. The cations of the strongly electropositive metals in the periodic table Groups 1 and 2 form few stable complex ions; by contrast, the transition metals are strongly inclined to form complex ions. As a result no complex ions are encountered in the analysis of the barium-magnesium qualitative analysis group, while complex ions are found in the analysis of the other three qualitative analysis groups. Certain of these complex ions play a prominent role in separation and identification.

The aluminum-nickel qualitative analysis group includes some of the metals that fall along the borderline between metals and nonmetals. We should, therefore, not be surprised to find that the amphoteric character of the hydroxides of these borderline elements is utilized to effect their separation.

The nonmetallic character of arsenic, antimony, and tin, as indicated by their positions in the periodic table, accounts for the acid character of their sulfides; this acid character is the basis for the separation of the sulfides of arsenic, antimony, and tin from the more basic sulfides of mercury, lead, bismuth, copper, and cadmium.

The close similarity of the oxides and sulfides of the various cations is noted at several points during the analysis. This close similarity is to be expected from the fact that oxygen and sulfur both fall at the top of periodic table Group 16.

CHEMICAL REACTIONS

We discussed the nature of chemical reactions in Chapter 2 and elaborated on several types of reactions in Chapters 5 and 6. There are several generalizations concerning chemical reactions that are useful in qualitative analysis laboratory practice.

Direct Combination

The large majority of reactions in qualitative analysis involve simple, direct combinations of an anion with a cation. As a general rule, whenever two electrolytes are brought together in solution or under such conditions that a solution can form, the positive ion of one electrolyte can combine with the negative ion of the other electrolyte; a compound is produced from the combination if the product is insoluble or if it is weakly ionized. The chemical formula of the compound will be determined by the charges on the combining ions. The simple combination of oppositely charged ions to form a compound will actually take place only if the following occur:

1. The resulting compound is insoluble (Eq. 7.1) or decomposes to give an insoluble substance.

$$Ag^+ + Cl^- \rightarrow AgCl \tag{7.1}$$

$$Ag^+ + OH^- \rightarrow [AgOH] \rightarrow Ag_2O + H_2O \tag{7.2}$$

2. The resulting compound is a weak electrolyte and, hence, is only slightly ionized (Eq. 7.3).

$$Cu(OH)_2(s) + 2H^+ \rightarrow Cu^{2+} + H_2O \tag{7.3}$$

3. Neither of the ions becomes involved in an oxidation–reduction reaction.

The solubility rules given in Chapter 4 identify most of the insoluble compounds; the schemes of analysis indicate any insoluble compounds not covered by the solubility rules.

The weak electrolytes with which we shall be concerned are H_2O, the weak base aqueous NH_3, and the weak acids H_2S, $HC_2H_3O_2$, H_3BO_3, H_3AsO_4, and H_3PO_4; H_2CO_3 and H_2SO_3 are unstable and decompose to give the sparingly soluble gases CO_2 and SO_2, which can be detected by their properties.

$$H_2CO_3(aq) \rightarrow H_2O + CO_2(g) \tag{7.4}$$

$$H_2SO_3 \rightarrow H_2O + SO_2(g) \tag{7.5}$$

Complex-Ion Formation

Although it is not possible to state simple rules that will enable us to know when a complex ion (Chapter 5) will be formed and what its formula will be, the following generalizations provide some clues:

1. The cations of the B-group elements (old designations) in the periodic table have a strong tendency to form complex ions. The silver, copper-arsenic, and aluminum-nickel group elements are made up, predominantly, of B-group cations and, accordingly, form many stable complex ions.

2. The periodic table Group 1 and 2 metals form few complex ions. For this reason, no complex ions are encountered in the barium-magnesium qualitative analysis group.

3. All cations except those in the barium-magnesium group form stable complex chloro ions of the type $HgCl_4^{2-}$ in solutions of high Cl^- ion concentration, as, for example, when a metal or its sulfide is dissolved in aqua regia.

4. When a precipitate redissolves in an excess of the precipitating ions or molecules, a complex ion often forms. Thus, when $Cu(OH)_2$, $Cd(OH)_2$, $Zn(OH)_2$, $Ni(OH)_2$, and $Co(OH)_2$, which are precipitated when aqueous ammonia (NH_4OH) is added to solutions of the respective cations, redissolve in an excess of aqueous NH_3 (which contains NH_4^+ and OH^-), we can be sure that complex ions of the type $Cu(NH_3)_4^{2+}$ and $Ni(NH_3)_6^{2+}$ are formed. There is a possibility with aqueous NH_3 that the cations might form soluble hydroxy complexes, especially if the metal is amphoteric, but since NH_3 is a weak base, the concentration of OH^- is not sufficiently high to produce this type of complex ion. Likewise, when $Al(OH)_3$ dissolves in excess strong base and $Cd(CN)_2$ dissolves in excess CN^-, complex ions of the types $Al(OH)_4^-$ and $Cd(CN)_4^{2-}$, respectively, are formed.

Oxidation–Reduction Reactions

The following facts and generalizations will help determine whether or not an oxidation–reduction reaction (Chapter 6) will occur:

1. Before a particular ion or molecule can function as an oxidizing agent, it must contain an atom that can exist in more than one oxidation state; furthermore, in the particular ion or molecule the atom must be present in the higher of two oxidation states. Chromium has an oxidation number of 6+ in $Cr_2O_7^{2-}$ and 3+ in Cr^{3+}. Therefore, $Cr_2O_7^{2-}$ is able to function as an oxidizing agent. Chromium can exhibit no higher oxidation state than 6+ in $Cr_2O_7^{2-}$, so this species cannot act as a reducing agent.

2. In order for a particular ion or molecule to function as a reducing agent, it must contain an atom that can exist in more than one oxidation state; furthermore, in this particular ion or

molecule the atom must be present in the lower of two oxidation states. Sulfur has an oxidation number of $2-$ in H_2S and 0 in elemental sulfur, S. Therefore H_2S is able to function as a reducing agent. Since the oxidation state of sulfur cannot be lower than $2-$, as in H_2S, the latter cannot act as an oxidizing agent.

3. When an oxidizing agent and a reducing agent are brought together in solution, an oxidation–reduction reaction will take place, provided the former is a strong enough oxidizing agent and the latter is a strong enough reducing agent. Whether they are strong enough to react can be decided by noting their relative positions in a table of reduction potentials, such as Table 6.1.

Equations and Chemical Reactions

Practically every separation and identification in qualitative analysis involves one or more reactions in solution. The discussion of these reactions requires writing equations, so we need to say something about the manner in which the equation for a chemical reaction in solution may be presented. Three kinds of equations may be written: *molecular equations, detailed equations,* and *net equations*. The differences among these three kinds of equations can best be shown by using them to represent typical reactions.

The molecular equation for the reaction that takes place when a solution of the strong acid HCl is neutralized by a solution of the strong base NaOH is given by Eq. (7.6).

$$NaOH + HCl \rightleftharpoons NaCl + H_2O \qquad (7.6)$$

Equation (7.6) gives the formulas of the solid NaOH that was used in preparing the solution of sodium hydroxide, the HCl gas that was used in preparing the solution of hydrochloric acid, the solid sodium chloride that would be recovered if the resulting solution were evaporated to dryness, and the water molecules that are formed as a product. Equation (7.6) tells us that 1 mole of NaOH will neutralize exactly 1 mole of HCl to form exactly 1 mole of NaCl and 1 mole of H_2O; in other words, it represents the exact stoichiometry of the reaction. It enables us to calculate how many grams of solid NaOH would have to be dissolved in water in order to give a solution that would exactly neutralize a solution formed by dissolving a definite number of grams of HCl in water, and it allows us to calculate exactly how many grams of NaCl would be recovered if the resulting solution were evaporated to dryness. The molecular equation makes no pretense whatever of telling anything about the details of the reaction that takes place in solution; it simply gives the molecular formulas of the substances used in preparing the solutions and represents the molecular ratio in which they react.

From Chapter 3, we know that NaOH, HCl, and NaCl are strong electrolytes; in dilute solutions of strong electrolytes, the solute is completely ionized (NaOH and NaCl, being ionic compounds, are completely ionized in the solid state). Therefore, a solution of NaOH contains Na^+ and OH^- ions, but no NaOH molecules; a solution of HCl contains H^+ and Cl^- ions, but no HCl molecules. Water is such an extremely weak electrolyte that it exists almost exclusively in the form of H_2O molecules. When a solution of NaOH reacts with a solution of HCl to form a solution of NaCl and H_2O, it actually is the reaction of the ions in solution; the solution of Na^+ and OH^- ions reacts with a solution of H^+ and Cl^- ions to form a solution of Na^+ and Cl^- ions also containing H_2O molecules. The detailed equation for the reaction would then be

$$Na^+ + OH^- + H^+ + Cl^- \rightleftharpoons Na^+ + Cl^- + H_2O \qquad (7.7)$$

We note that all the sodium exists as Na^+ ions on the right side of the equation, as well as on the left side. Likewise, all the chlorine appears as Cl^- ions on both right and left. Evidently, then, nothing happens to either Na^+ or Cl^- ions. Therefore, we can eliminate Na^+ and Cl^- from both sides of the equation; in a sense Na^+ and Cl^- are spectators. That means that the only reaction that actually happens involves H^+ ions combining with OH^- ions to form H_2O molecules. Note that the reaction occurs because a cation (H^+) and an anion (OH^-) combine to form a non-electrolyte (H_2O). We can represent the simple overall picture of what actually takes place by the following net equation:

$$H^+ + OH^- \rightleftharpoons H_2O \qquad (7.8)$$

The reaction that takes place when a solution of the weak base aqueous ammonia (NH_3) is neutralized by the strong acid HCl can be represented by the following molecular equation:

$$NH_3 + HCl \rightleftharpoons NH_4Cl + H_2O \qquad (7.9)$$

All that has been said about the molecular equation in Eq. (7.6) applies exactly to Eq. (7.9), which gives the molecular species of the substances used in preparing the two solutions and the molecular species of the products and represents the exact stoichiometry of the reaction; Eq. (7.9) makes no pretense of describing the details associated with the reaction that takes place. Thus, there is considerably more information required about the nature of the species in solution before we can establish the details in this chemical reaction. It should be noted at this point that there is some question about the molecular species that are present in the solution formed when NH_3 gas is dissolved in water. The assumption is made in this discussion that NH_3 reacts with H_2O to form NH_4^+ and OH^- ions. Since not all the NH_3 in solution is converted into the corresponding ions (aqueous NH_3 is a weak base), the equilibrium shown in Eq. (7.10) is established.

$$NH_3 + H_2O \rightleftharpoons NH_4^+ + OH^- \qquad (7.10)$$

As already noted, HCl is a strong acid; a dilute solution of HCl, therefore, contains H^+ and Cl^- ions but essentially no un-ionized HCl molecules. A solution of aqueous ammonia contains a large number of NH_3 molecules and relatively few NH_4^+ and OH^- ions but no NH_4OH molecules. Thus, a solution of H^+ and Cl^- ions reacts with a solution containing a few NH_4^+ and OH^- ions in equilibrium with a great many NH_3 molecules to form a solution containing NH_4^+ and Cl^- ions and H_2O molecules. We could represent the actual situation by the equation

$$H^+ + Cl^- + NH_4^+ + OH^- \rightleftharpoons NH_4^+ + Cl^- + H_2O \qquad (7.11)$$

When electrolytes react in solution, the detailed reaction is between the appropriate ions. Therefore, the reaction in Eq. (7.11) indicates that H^+ ions from HCl react with the OH^- ions from aqueous NH_3 to form H_2O molecules. This reduces the concentration of OH^- ions. Molecules of NH_3 continue to form NH_4^+ and OH^-, as predicted from Le Chatelier's principle; as fast as the OH^- ions are consumed, additional molecules of NH_3 give more OH^- ions. This process continues until all the NH_3 has been converted to NH_4^+ and OH^- ions and all the OH^-

ions produced have been combined with H^+ ions to form H_2O molecules. We could represent the detailed reaction as follows:

$$NH_3 + H_2O$$

$$\updownarrow$$

$$H^+ + Cl^- + NH_4^+ + OH^- \leftrightharpoons NH_4^+ + Cl^- + H_2O \qquad (7.12)$$

A more realistic representation of the process is given by Eqs. (7.13) through (7.15).

$$NH_3 + H_2O \leftrightharpoons NH_4^+ + OH^- \qquad (7.13)$$

$$H^+ + OH^- \leftrightharpoons H_2O \qquad (7.14)$$

$$NH_4^+ + OH^- + H^+ \leftrightharpoons NH_4^+ + H_2O \qquad (7.15)$$

Equations (7.13) and (7.14) are the two detailed equations. When these two equations are added, H_2O and OH^- being canceled in the process, the result is the net equation for the reaction, Eq. (7.16).

$$NH_3 + H^+ \leftrightharpoons NH_4^+ \qquad (7.16)$$

Equation (7.16) tells us that the predominant species involved in the reaction are H^+ ions and NH_3 molecules. It correctly represents the stoichiometry of the reaction between the predominant species. It does not, however, pretend to represent the actual mechanism of the reaction.

Summarizing, we can say that the molecular equation gives the formula of the molecular species of the substances used in making up the reacting solutions and the formulas of the molecular species that can be recovered as products. It represents the correct stoichiometric ratio in which these substances react; it does not represent the mechanism of the reaction and does not imply that the molecules as such either take part in the reaction or exist in solution. The detailed equation attempts to give a more accurate picture of the reaction process. The net equation gives the predominant species involved in the reaction and also represents the correct stoichiometric ratio in which they react.

Because the net equation shows the species that predominate in a reacting system and because, in qualitative analysis, we are more interested in the species that are present than how they come and go, as a general rule we use net equations rather than detailed equations. When a more exact description of a reaction process is under discussion, the detailed equation will be presented. In using the net equation, it must be recognized that the identities of the predominant species may be in question. Likewise, in presenting a detailed equation it must be understood that we cannot be absolutely certain about the detailed mechanism of any reaction.

The next examples emphasize further the relationship among the three kinds of equations.

1. The reaction of hydrogen sulfide gas with a solution of lead chloride:

$$\text{Molecular:} \qquad PbCl_2 + H_2S \leftrightharpoons PbS + 2HCl \qquad (7.17)$$

$$\text{Detailed:} \qquad H_2S \leftrightharpoons 2H^+ + S^{2-} \qquad (7.18)$$

$$Pb^{2+} + S^{2-} \leftrightharpoons PbS \qquad (7.19)$$

$$\text{Net:} \qquad Pb^{2+} + H_2S \leftrightharpoons PbS + 2H^+ \qquad (7.20)$$

2. The reaction of hydrochloric acid with calcium carbonate:

$$\text{Molecular:} \qquad CaCO_3(s) + 2HCl \rightleftharpoons CaCl_2 + H_2O + CO_2(g) \qquad (7.21)$$

$$\text{Detailed:} \qquad\qquad CaCO_3(s) \rightleftharpoons Ca^{2+} + CO_3^{2-} \qquad (7.22)$$

$$CO_3^{2-} + 2H^+ \rightleftharpoons H_2O + CO_2(g) \qquad (7.23)$$

$$\text{Net:} \qquad CaCO_3(s) + 2H^+ \rightleftharpoons H_2O + CO_2(g) + Ca^{2+} \qquad (7.24)$$

It is obvious that to write a correct net equation a decision must be made about whether a substance exists predominantly as an ion or as a neutral unit (a molecule or a neutral ionic crystal, such as solid $CaCO_3$). The following summary will aid in making this decision.

1. The following soluble compounds, which are strong electrolytes, exist in solutions as ions.
   a. All soluble salts (see solubility rules, Chapter 4).
   b. All soluble metal hydroxides (see solubility rules, Chapter 4).
   c. The common acids H_2SO_4, HNO_3, HCl, HBr, HI, and $HClO_4$.
2. The following soluble compounds, which are weak electrolytes, exist in solution predominantly as molecules.
   a. All soluble acids not listed in part 1c above.
   b. Ammonia.
   c. Water.
3. All insoluble substances (solids, liquids, gases) exist as molecules or neutral ionic units.
4. Complex ions are soluble weak electrolytes and exist predominantly as undissociated ions.

The following net equations will illustrate applications of the rules above:

1. The reaction of 5 M aqueous NH_3 with a solution of $FeCl_3$:

$$Fe^{3+} + 3NH_3 + 3H_2O \rightleftharpoons Fe(OH)_3 + 3NH_4^+$$

2. Dissolution of solid ZnS by excess 3 M HCl:

$$ZnS + 2H^+ \rightleftharpoons Zn^{2+} + H_2S$$

3. Reaction of aqueous NH_3 with acetic acid:

$$NH_3 + HC_2H_3O_2 \rightleftharpoons NH_4^+ + C_2H_3O_2^-$$

4. Dissolution of solid Ag_3AsO_4 by excess 3 M HNO_3:

$$Ag_3AsO_4 + 3H^+ \rightleftharpoons 3Ag^+ + H_3AsO_4$$

5. Dissolution of insoluble $AgCl$ by excess 5 M aqueous NH_3:

$$AgCl + 2NH_3 \rightleftharpoons Ag(NH_3)_2^+ + Cl^-$$

6. Addition of 2 M $(NH_4)_2C_2O_4$ to 0.2 M $CaCl_2$:

$$Ca^{2+} + C_2O_4^{2-} \rightleftharpoons CaC_2O_4$$

EQUILIBRIUM CONSIDERATIONS

Since the practice of qualitative analysis is concerned to a very large degree with forming and dissolving precipitates, it follows that, from both a practical and theoretical point of view, the most significant equilibria are those involving solids (precipitates) and solutions. Consideration of these equilibria immediately brings up certain practical questions, the answers to which are fundamental to an understanding of qualitative analysis. Why are some substances soluble in water while others are insoluble? Why will a certain combination of ions yield a precipitate under one set of conditions but not under another? What are the best conditions for removing the maximum quantity of an ion from solution? For example, the separation of the copper-arsenic qualitative analysis group from the aluminum-nickel qualitative analysis group is governed by adjusting the sulfide ion concentration, S^{2-}, derived from the ionization of H_2S, the *overall* processes for which are given by Eq. (7.25).

$$H_2S \rightleftharpoons 2H^+ + S^{2-} \tag{7.25}$$

If the weak base, aqueous ammonia, is added to a solution containing H_2S, the sulfide ion concentration, S^{2-}, *increases* to accommodate to the reaction of OH^- with H^+ (Figure 7.1).

$$H_2S \rightleftharpoons 2H^+ + S^{2-}$$
$$+$$
$$NH_3 + H_2O \rightleftharpoons OH^- + NH_4^+$$
$$\downarrow\uparrow$$
$$H_2O$$

FIGURE 7.1   The equilibria present in a solution of H_2S containing aqueous ammonia.

This increase in sulfide ion concentration is sufficient to cause the sulfides of some of the cations of the aluminum-nickel qualitative analysis (namely, MsS, NiS, CoS, and FeS) to precipitate. However, in the presence of a strong acid like HCl, the ionization of H_2S is repressed because the excess H^+ shifts the H_2S equilibrium to the left, which *decreases* the S^{2-} concentration in such solutions. The S^{2-} concentration can be made sufficiently small (Figure 7.2) by this technique as to allow the precipitation of only the most insoluble sulfides that make up the copper-arsenic group (Bi_2S_3, CuS, CdS, As_2S_5, Sb_2S_3, SnS_2).

$$H_2S \rightleftharpoons 2H^+ + S^{2-}$$
$$\Uparrow$$
$$HCl \rightarrow H^+ + Cl^-$$

FIGURE 7.2   The repression of the ionization of H_2S in the presence of excess H^+.

In other words, by adjusting the pH of the solution containing H_2S, the sulfide ion can be adjusted to cause the separation of one group of cations from another group.

The theoretical bases on which the experimental details of such questions are answered are discussed in Chapters 2 and 4.

FORMATION OF PRECIPITATES

As implied in the discussion of solubility products in Chapter 4, a specific ionic compound will precipitate from solution when, and only when, the product of the concentrations of its ions (momentarily) exceeds its solubility product. Since the solid ionic compound that is precipitated is in equilibrium with its ions, it follows that to remove the maximum number of cations, for example A^+, from solution as the precipitate AB, an excess of the precipitating reagent B^- should also be added to the solution. However, we must exercise caution because B^- might also be a ligand that complexes strongly with A^+. In such a case, a soluble complex ion such as $AB_n^{(n-1)-}$ can be formed if too large an excess of B^- is added.

A case in point is the precipitation of $PbCl_2$ as a part of the qualitative analysis silver group. If lead is present in the unknown as Pb^{2+}, the addition of the group reagent HCl will cause the precipitation of $PbCl_2$ [Eq. (7.26)], but, if an excess of the group reagent is added, lead can form the soluble complex ion $PbCl_4^{2-}$ [Eq. (7.27)].

$$Pb^{2+} + Cl^- \rightarrow PbCl_2(s) \tag{7.26}$$

$$PbCl_2(s) + 2Cl^- \rightarrow PbCl_4^{2-} \tag{7.27}$$

Under these conditions, lead will appear in the copper-arsenic group. Since it is generally difficult to add "the right amount" of HCl as the group reagent, you should always check for the possibility that the copper-arsenic qualitative analysis group does not contain lead, especially if lead appears in the silver qualitative analysis group.

DISSOLVING PRECIPITATES

A precipitate will go into solution if the product of the concentrations of its ions in solution is less than its solubility product. It will continue to dissolve until the product of the concentrations of its ions in solution equals the solubility product. How a precipitate is dissolved in actual practice, what happens when a precipitate dissolves, and why a certain solvent will dissolve one precipitate but will not dissolve another can best be shown by considering some specific examples.

Formation of Weak Electrolytes

Silver phosphate, Ag_3PO_4, is soluble in dilute HNO_3, which means when HNO_3 is added to solid Ag_3PO_4 a condition is created in which the product of $[PO_4^{3-}]$ and $[Ag^+]^3$ is less than the solubility product for Ag_3PO_4. The actual solution process probably proceeds somewhat as follows. When Ag_3PO_4 is brought in contact with water or a water solution, the following equilibrium is established:

$$Ag_3PO_4 \leftrightarrows 3Ag^+ + PO_4^{3-} \tag{7.28}$$

This is a true equilibrium condition because the rate at which silver ions and phosphate ions combine to form solid Ag_3PO_4 is equal to the rate at which Ag_3PO_4 dissolves and dissociates into Ag^+ and PO_4^{3-} ions. The rate at which Ag^+ and PO_4^{3-} ions combine is directly proportional to the concentration of each ion. The rate at which Ag_3PO_4 dissolves from a unit surface of solid Ag_3PO_4 is constant at a given temperature; that is, Ag_3PO_4 dissolves at constant rate, and this rate is not affected by the concentration of the Ag^+ and PO_4^{3-} ions in solution. If an excess of the strong acid HNO_3 is added to the system, the H^+ ions that it provides will combine with the PO_4^{3-} ions and form molecules of the weak acid H_3PO_4.

$$PO_4^{3-} + 3H^+ \leftrightarrows H_3PO_4 \tag{7.29}$$

This process reduces the concentration of PO_4^{3-} ions in solution, which in turn reduces the rate at which Ag^+ ions and PO_4^{3-} ions combine to form solid Ag_3PO_4. Since Ag_3PO_4 continues to dissolve at the same constant rate, we now have a situation in which solid Ag_3PO_4 is being formed at a slower rate than it is dissolving. If the rate at which Ag^+ and PO_4^{3-} ions combine is kept low by adding sufficient HNO_3 to give enough H^+ ions to combine with an equivalent number of PO_4^{3-} ions—thus keeping their concentration low—eventually the Ag_3PO_4 will all go into solution.

In terms of the equilibrium represented by Eq. (7.28), reduction of the concentration of PO_4^{3-} ions by tying them up as the weak electrolyte H_3PO_4 reduces the product of $[Ag^+]^3$ and $[PO_4^{3-}]$ below the solubility product for Ag_3PO_4. If $[PO_4^{3-}]$ is made very low by having a high concentration of H^+ in Eq. (7.29), all the Ag_3PO_4 will dissolve. Any solid electrolyte will dissolve if the product of the concentrations of its ions in solution is less than its solubility product.

The net equation for the solution of Ag_3PO_4 in HNO_3 is

$$Ag_3PO_4 + 3H^+ \leftrightarrows 3Ag^+ + H_3PO_4 \qquad (7.30)$$

In a similar fashion, Ag_3AsO_4 dissolves in HNO_3. The net equation is

$$Ag_3AsO_4 + 3H^+ \leftrightarrows 3Ag^+ + H_3AsO_4 \qquad (7.31)$$

The detailed equations for the dissolution of Ag_3AsO_4 are

$$Ag_3AsO_4 \leftrightarrows 3Ag^+ + AsO_4^{3-} \qquad (7.32)$$

$$AsO_4^{3-} + 3H^+ \leftrightarrows H_3AsO_4 \qquad (7.33)$$

Silver chloride will not dissolve in HNO_3. When HNO_3 is added to the system represented by the equilibrium in Eq. (7.34), there is no tendency for either the Ag^+ or Cl^- to react with H^+ or NO_3^- ions because both $AgNO_3$ and HCl are soluble strong electrolytes.

$$AgCl \leftrightarrows Ag^+ + Cl^- \qquad (7.34)$$

Therefore, there is no change in the concentrations of either Ag^+ or Cl^-; hence, no additional $AgCl$ goes into solution when HNO_3 is added.

If we reexamine the three cases we have considered and also note a number of additional cases, we arrive at the following general interpretation of what happens when any solid electrolyte exists in equilibrium with its ions, as represented by Eq. (7.35).

$$Ag_3PO_4 \leftrightarrows 3Ag^+ + PO_4^{3-} \qquad (7.35)$$

$$PO_4^{3-} + 3H^+ \leftrightarrows H_3PO_4 \qquad (7.36)$$

The solvent ion or molecule (H^+, in this case) then reacts with one of these ions (PO_4^{3-}, in this case) and forms a more stable substance (H_3PO_4) in accordance with Eq. (7.36). This process reduces the concentration of the ion (PO_4^{3-}) with which the solvent has combined below the solubility product concentration required for Eq. (7.35). In an effort to raise the concentration of the ion (PO_4^{3-}) to the value required to satisfy the solubility product, some of the solid electrolyte (Ag_3PO_4) goes into solution; eventually it will be completely dissolved.

The more stable substance formed in Eq. (7.36) may be either a soluble weak electrolyte, a complex ion, or a highly insoluble substance. In most instances, Eq. (7.36) represents an incomplete reaction; in a few cases the reaction is complete. In any case, Eqs. (7.35) and (7.36) have an ion in common, and it is imperative that the concentration of this common ion always be kept lower than the equilibrium solubility product concentration needed for Eq. (7.35). As long as this requirement is met, the precipitate will keep on going into solution faster than it precipitates and will eventually completely dissolve. In effect, lowering the concentration of the common ion causes the equilibrium in Eq. (7.35) to be shifted to the right; this shift leads eventually to the complete dissolution of the precipitate. If the concentration of the common ion in equilibrium is made (and kept) less than the concentration required to satisfy the equilibrium in Eq. (7.35), the total reaction will run "downhill" from the precipitate in Eq. (7.35) to the stable substance in Eq. (7.36).

Both Ag_3PO_4 and Ag_3AsO_4 dissolve in HNO_3 because the anions PO_4^{3-} and AsO_4^{3-} combine to form the weak acids H_3PO_4 and H_3AsO_4 (see Table 3.2). The weak acids were formed because Ag_3PO_4 and Ag_3AsO_4 are salts of weak acids. We can state that, as a general rule, the insoluble salt of a weak acid will dissolve in a strong acid. In agreement with this rule, the phosphates, arsenates, carbonates, sulfites, borates, and chromates of all metals are soluble in strong acids.

Since AgCl is the salt of a strong acid, it will not dissolve in HNO_3 because its anion cannot react with H^+ ions to form a weak electrolyte. Likewise AgBr, AgI, $PbSO_4$, and $BaSO_4$ are not soluble in strong acids such as HNO_3, HCl, or H_2SO_4 because they are, themselves, salts of strong acids. We can conclude from these examples that the insoluble salt of a strong acid will not dissolve in a strong acid; since weak acids provide fewer H^+ ions, the insoluble salts of strong acids obviously will not dissolve in weak acids either.

Mercury(II) sulfide is the salt of the weak acid H_2S. In accordance with the rule stated above, it should be soluble in the strong acid HCl. Actually, it is not dissolved by HCl because HgS is so highly insoluble that the concentration of sulfide ions in equilibrium with solid HgS in Eq. (7.37) is so small that it is much lower than the concentration of sulfide ions in Eq. (7.38) even when the equilibrium in the latter equation is shifted far to the right by a high concentration of H^+ ions.

$$HgS \leftrightharpoons Hg^{2+} + S^{2-} \tag{7.37}$$

$$S^{2-} + 2H^+ \leftrightharpoons H_2S \tag{7.38}$$

The failure of HgS to be dissolved by HCl emphasizes the fact that if a solid electrolyte is to dissolve, not only must one of its ions be tied up in the form of a weak electrolyte, a complex ion, or an insoluble substance but also the concentration of this ion in equilibrium with the stable substance must always be less than its concentration in equilibrium with the solid electrolyte.

Formation of Complex Ions

Although AgCl is insoluble in all acids, it will dissolve in ammonium hydroxide. As usual, insoluble AgCl establishes an equilibrium with its constituent ions [Eq. (7.39)].

$$AgCl \leftrightharpoons Ag^+ + Cl^- \tag{7.39}$$

$$Ag^+ + 2NH_3 \leftrightharpoons Ag(NH_3)_2^+ \tag{7.40}$$

Ammonia molecules (NH_3) combine with Ag^+ to form the stable complex ion $Ag(NH_3)_2^+$, in accordance with Eq. (7.40). This process reduces the concentration of Ag^+ below the equilibrium value required for Eq. (7.39).

Silver iodide, AgI, will not dissolve in aqueous NH_3. It is so insoluble that the concentration of Ag^+ ions in the equilibrium

$$AgI \rightleftharpoons Ag^+ + I^- \qquad (7.41)$$

is always less than the concentration of Ag^+ ions in the equilibrium

$$Ag^+ + 2NH_3 \rightleftharpoons Ag(NH_3)_2^+ \qquad (7.42)$$

Silver iodide can be dissolved, however, in a solution containing CN^- ions. The complex $Ag(CN)_2^-$ ion that is involved in the process represented by Eqs. (7.43) and (7.44) is very stable and, as a result, the Ag^+ ion in equilibrium is maintained at a concentration that is even lower than the concentration of Ag^+ ions in equilibrium with AgI.

$$AgI \rightleftharpoons Ag^+ + I^- \qquad (7.43)$$

$$Ag^+ + 2CN^- \rightleftharpoons Ag(CN)_2^- \qquad (7.44)$$

The fact that the instability constant for $Ag(CN)_2^-$ (Table 5.4) is 1.8×10^{-19} and the constant for $Ag(NH_3)_2^+$ is 5.9×10^{-8} shows that $Ag(CN)_2^-$ is much more stable than $Ag(NH_3)_2^+$. In this instance, as in all the other cases that we have discussed, a high concentration of the reacting ion or molecule (CN^-, in this case) shifts equilibrium [Eq. (7.44)] to the right and keeps the concentration of the common ion at a very low level. If this process is attempted in acid solution, there is competition for CN^- by H^+ because HCN is a weak acid. Thus, in acid solution we might not be able to increase the free CN^- concentration to a sufficient level to allow the process in Eq. (7.44) to occur to a significant extent.

Amphoterism

When the hydroxide of a metal—for example, $Zn(OH)_2$—is dissolved by an acid in the manner already discussed under amphoterism (Chapter 5), the conditions that we have set up for dissolving a precipitate are realized.

$$Zn(OH)_2 \rightleftharpoons Zn^{2+} + 2OH^- \qquad (7.45)$$

$$OH^- + H^+ \rightleftharpoons H_2O \qquad (7.46)$$

The common ion is OH^-, the solvent ion is H^+, and the stable, soluble molecule is H_2O.

When an amphoteric hydroxide dissolves in a strong base, the requirements that we have set up are again fulfilled.

$$Zn(OH)_2 \rightleftharpoons Zn^{2+} + 2OH^- \qquad (7.47)$$

$$Zn^{2+} + 4OH^- \rightleftharpoons Zn(OH)_4^{2-} \qquad (7.48)$$

The common ion is Zn^{2+}, and the stable, soluble substance is the complex ion $Zn(OH)_4^{2-}$.

Although $Mg(OH)_2$ is very sparingly soluble in water, it is readily dissolved by a solution of NH_4Cl, NH_4NO_3, or $(NH_4)_2SO_4$. The following reactions are involved.

$$Mg(OH)_2 \rightleftharpoons 2OH^- + Mg^{2+} \qquad (7.49)$$

$$OH^- + NH_4^- \rightleftharpoons NH_3 + H_2O \qquad (7.50)$$

The NH_4^+ ions from the ammonium salt combine with the OH^- ions formed in Eq. (7.49) to form molecules of NH_3 and H_2O in the manner represented by Eq. (7.50).

Redox Process

In certain cases it is not possible to find any solvent that will form a soluble ion or molecule of sufficient stability to cause the solid to dissolve. In some such cases one can resort to the technique of destroying an ion by having it take part in a reaction that goes essentially to completion. This is the process that occurs when the sulfides of copper, lead, bismuth, and cadmium are dissolved in nitric acid in the analysis of the copper group (Chapter 10).

$$CuS \rightleftharpoons Cu^{2+} + S^{2-} \tag{7.51}$$

$$3S^{2-} + 2NO_3^- + 8H^+ \rightleftharpoons 3S + 2NO + 4H_2O \tag{7.52}$$

The common ion in this case is S^{2-}. Notice that the sulfide ion, S^{2-}, is oxidized to elemental sulfur, thus removing it from the equilibrium expressed as Eq. (7.51); as the S^{2-} is removed, more CuS must dissolve (to maintain K_{sp} for CuS). Eventually, all the CuS dissolves.

A very insoluble substance that, accordingly, is in equilibrium with a very low concentration of ions will be not only very difficult to dissolve at all, but will also dissolve very slowly. In such a case, it may be expedient to work on both of its ions at the same time. This is the process occurring when HgS dissolves in a mixture of HNO_3 and HCl.

$$HgS \rightleftharpoons Hg^{2+} + S^{2-} \tag{7.53}$$

$$Hg^{2+} + 4Cl^- \rightleftharpoons HgCl_4^{2-} \tag{7.54}$$

$$3S^{2-} + 2NO_3^- + 8H^+ \rightleftharpoons 3S + 2NO + 4H_2O \tag{7.55}$$

Formation of Gaseous Products

Dissolving carbonates and sulfites by a strong acid illustrates the special case where the soluble substance formed is not only sparingly ionized but is also present in very low concentration, due to the fact that it is partially decomposed into an insoluble gaseous product.

$$CaCO_3 \rightleftharpoons Ca^{2+} + CO_3^{2-} \tag{7.56}$$

$$CO_3^{2-} + 2H^+ \rightleftharpoons H_2CO_3 \rightleftharpoons H_2O + CO_2 \tag{7.57}$$

Formation of Less Soluble Substances

In certain instances the dissolution of one solid electrolyte is attended by the precipitation of another electrolyte. Thus, if a solution of $0.20\ M$ NaCl is added to some solid Ag_3PO_4, the Ag_3PO_4 dissolves, but a precipitate of AgCl forms.

$$Ag_3PO_4 \rightleftharpoons 3Ag^+ + PO_4^{3-} \tag{7.58}$$

$$Ag^+ + Cl^- \rightleftharpoons AgCl \tag{7.59}$$

In a solution containing a high concentration of Cl^-, the concentration of Ag^+ ions required to exceed the solubility product for AgCl in Eq. (7.59) is less than the solubility product concentration of Ag^+ in Eq. (7.58).

When $PbSO_4$ is dissolved by Na_2CO_3 solution, the following equilibria are established:

$$PbSO_4 \leftrightharpoons Pb^{2+} + SO_4^{2-} \tag{7.60}$$

$$Pb^{2+} + CO_3^{2-} \leftrightharpoons PbCO_3 \tag{7.61}$$

If the concentration of carbonate ions is kept very high by using a saturated solution of Na_2CO_3, the concentration of Pb^{2+} ions in equilibrium in Eq. (7.61) will be kept so low that it will be below the concentration of Pb^{2+} required for the equilibrium [Eq. (7.60)]. As a result $PbSO_4$ will go into solution. Any water-insoluble salt will behave as $PbSO_4$ does; in each instance the cation (metal) of which the solid is composed will be precipitated as the carbonate, but the anion will go into solution. Therefore, this provides a convenient method for preparing a solution of the anions of water-insoluble salts [see Procedure 30(C) in Chapter 14]. Since the metal carbonates are all readily soluble in dilute acids, digestion of an acid-insoluble salt with sodium carbonate solution, followed by treatment of the metal carbonate with dilute acid, is a convenient method for preparing a solution containing the cations of such a solid.

The common ion involved in the solution equilibria just discussed may be involved in other equilibria that have no direct connection with the actual dissolving process. For example, solid As_2S_5 is readily dissolved by a solution of $(NH_4)_2S$, the stable soluble complex ion AsS_4^{3-} being formed in the process (see Procedure 6 in Chapter 10). The two reactions directly involved can be considered to be

$$As_2S_5 \leftrightharpoons 2As^{5+} + 5S^{2-} \tag{7.62}$$

$$As^{5+} + 4S^{2-} \leftrightharpoons AsS_4{}^{3-} \tag{7.63}$$

The As^{5+} ions are very strongly hydrolyzed, however, according to the equilibrium reaction

$$As^{5+} + 4H_2O \leftrightharpoons H_3AsO_4 + 5H^+ \tag{7.64}$$

Equation (7.64) is not directly involved in the solution process. It does play a very important role, however, in determining the concentration of As^{5+} ions that are in equilibrium with the solid As_2S_5.

A substance that will dissolve a precipitate will obviously keep that precipitate from forming in the first place. Thus, Ag_3PO_4 and Ag_3AsO_4 will not precipitate from a solution when the hydrogen ion concentration (acidity) is too high (see Procedure 13 in Chapter 11), while $Mg(OH)_2$ will not precipitate from a solution that contains a high concentration of NH_4^+ ions (see Procedure 15 in Chapter 11). The latter case is of particular interest since, in the precipitation of the cations of the aluminum-nickel qualitative analysis group, ammonium chloride is deliberately added to prevent $Mg(OH)_2$ from being precipitated when aqueous NH_3 is added as a group reagent.

LABORATORY SAFETY

One of the more difficult, yet very important, aspects of laboratory work that needs to be learned is laboratory safety, which demands that you develop a personal outlook and philosophy over an extended period of time. Laboratory safety depends upon individual diligence, recognition of hazards, and a knowledge of how to deal with such hazards. For most people, information about recognizing and dealing with hazards can be readily acquired by reading or observing; diligence and discipline usually come with constant practice.

We can conveniently divide the discussion of working safely in a laboratory environment into two categories: (1) recognizing potential hazards in anticipation of dealing with them in a safe and acceptable manner, and (2) dealing with emergencies if they should arise. The details of emergency procedures—what to do in case of fires, personal injuries, and so on—are highly dependent upon local conditions. For example, the locations of fire extinguishers (and how they work), eyewash fountains (or their equivalents), and safety showers are not necessarily standardized in all laboratories, so it would not be useful to discuss these aspects of laboratory safety here. You should, however, inquire into such details during your first visit to the laboratory room to which you will be assigned. Your teacher will probably mention such details to your class soon after you meet for the first time.

On the other hand, it is possible to discuss, in a general way, some ideas that will help you foresee potential hazards and thus take the appropriate action to keep them from happening. We can start with the recognition that there are a number of other students in the laboratory with you and that you can do several things to keep yourself and your fellow students safe. If each of you does the maximum to be safe, the entire class will have safer working conditions.

Personal safety in the laboratory begins with maintaining a clean, orderly immediate working environment. Think systematically at the start of each laboratory session about how to arrange your equipment and reagents to encourage good technique and good housekeeping with a minimum of lost effort. It may take you several sessions to find the optimum arrangement, but, if you persist, you should find the best one for you. Try to maintain the discipline of keeping everything orderly and systematic; your effort will be rewarded with better results and, indirectly, a safer working environment.

Personal safety also involves always using equipment to protect yourself from accidents. Thus, safety glasses should *always* be worn in a laboratory environment; indeed, your teacher may have a rule to this effect. It is also wise to wear a laboratory coat or apron when working with chemicals. These two items of equipment will help protect your person in case you, or a classmate, have an unforeseen accident.

Many potentially hazardous situations can be anticipated by being aware of the properties of chemicals. Volatile substances (or gases) should be treated with caution because they can be inhaled; they may also be flammable. Volatile compounds (and noxious gases) should be handled in a well-ventilated area (a chemical fume hood), and volatile substances that are also flammable should be kept away from open flames or electrical equipment that may produce sparks. This makes it apparent why the general prohibition against smoking is enforced in most laboratories.

Since no one would ever go around a laboratory tasting chemicals, the usual rule of not eating or smoking in a laboratory makes sense. If the possibility of accidentally bringing a chemical into your mouth (or inhaling it) is decreased, it is apparent that the frequency of concomitant hazards is lessened.

There are, of course, a myriad of possible hazards that could develop because of specific properties of the compounds you may use in any laboratory environment. We have attempted in the remainder of this book to give details of such hazards in the explanatory notes. You should read and understand these before you do the tests to which the notes refer. If you have any questions on such notes, talk to your instructor about them.

Laboratory Practices

SUMMARY OF LABORATORY WORK

The separation and identification of soluble species follow an orderly plan of action, and the execution of this plan calls for careful and extensive laboratory work. Understanding the plan requires knowledge of many principles and facts. To provide you with a sound basis to learn these principles and facts, acquire the techniques required for the various procedures, and learn to recognize the colors and other characteristics of the solutions and precipitates that make identification of the various cations possible, the following program of work is suggested.

- A practice solution (a "known" solution) containing all the cations of a particular group is analyzed first. The analysis of a known should allow you to correlate your perception of written descriptions ("AgCl precipitates as a white, curdy solid") with the reality of what you see happen in a test tube.
- This analysis of the known solution is then followed by the analysis of an unknown solution (or solid) for that group, for which you go through the same steps taken with the known.
- The series of group knowns and unknowns for the four groups should be followed by an unknown that contains combinations of all four cation groups in the form of an alloy or a mixture of cations, either as a mixture of solid salts or as a solution.

Thus, bit-by-bit, separation and identification of the cations can be studied effectively and efficiently. You are then ready to extend your efforts to a study of the identification of anions, which is accomplished by analyzing salts and salt mixtures.

In keeping with our objective of studying the qualitative procedures of the separate groups before undertaking a complete analysis, it is suggested that the following substances be analyzed in the order indicated.

1. One silver group known. This solution contains *all* the cations of the silver group. It will not contain cations of other groups. Obtain it and all subsequent knowns from your laboratory instructor. Follow the directions given in Procedures 1 to 4, inclusive, in Chapter 9.

2. One silver group unknown. This solution contains some of the cations of the silver group. It will not contain cations of other groups. Obtain this and all subsequent unknowns from the instructor. Follow the directions given in Procedures 1 to 4, inclusive, in Chapter 9.

3. One copper-arsenic group known. This solution contains *all* the cations of the copper-arsenic group. It will not contain cations of other groups. Follow the directions given in Procedures 5 to 14, inclusive, in Chapter 10.

4. One copper-arsenic group unknown. This solution contains some of the cations of the copper-arsenic group. It will not contain cations of other groups. Follow the directions given in Procedures 5 to 14, inclusive, in Chapter 10.

5. One aluminum-nickel group known. This solution contains *all* the cations of the aluminum-nickel group and will not contain cations of other groups. Follow the directions given in Procedures 15 to 22, inclusive, in Chapter 11.

6. One aluminum-nickel group unknown. This solution contains some of the cations of the aluminum-nickel group. It will not contain cations of other groups. Analyze according to the directions given in Procedures 15 to 22, inclusive, in Chapter 11.[1]

7. One barium-magnesium group known. This solution contains *all* the cations of the barium-magnesium group. It will not contain cations of other groups. Follow the directions given in Procedures 23 to 27, inclusive, in Chapter 12.

8. One barium-magnesium group unknown. This solution contains some of the cations of the barium-magnesium group. It will not contain cations of other groups. Analyze according to the directions given in Procedures 23 to 27, inclusive, in Chapter 12.

9. One alloy or one solid mixture of salts. This unknown may contain any or all of the metals or cations in the four groups. If it is an alloy, follow Procedure 28 in Chapter 13. If it is a mixture of cations, run through the complete analysis of cations starting with five drops of solution in Procedure 1 in Chapter 9 (if time is limited, this unknown may be omitted).

10. Preliminary tests and specific tests for anions. Follow the directions given in Experiments 1, 2, 3, and 4 in Chapter 14.

11. One anion unknown. This is a solution of the sodium or potassium salts of two or more anions. Follow the instructions given in Chapter 14 (if time is limited, this unknown may be omitted).

12. Complete analysis of solids for both cations and anions. The remainder of the laboratory work will consist of the complete analysis of solid unknowns for cations and anions. Follow the directions given in Chapter 14.

RECORD OF LABORATORY WORK AND LABORATORY REPORTS

A record of all laboratory work is to be kept in a bound notebook with blank pages. Keeping a *clear* record of your laboratory activities is an important skill to be developed. A clean record will help you to create, or permit a review of, accurate deductions from your laboratory activities. Even the keenest of observers can become confused when trying to remember critical details a day or two after experiments were performed; "was that a tan precipitate that formed on the solution that led me to believe that zinc was present, or was it that the tan precipitate dissolved in excess NaOH?"

---

1. In the presence of phosphates or borates, a modified plan of analysis must be used for the aluminum-nickel group; see, for example, L. J. Curtman and T. B. Greenslade, *J. Chem. Ed.*, 13, (1936):238. Unknowns ordinarily issued in the course will not contain phosphates or borates that necessitate using this or other modified procedures.

Thus, it is important that you develop the habit of recording your observations in a clear and logical manner. Several methods are acceptable in the sense that they produce an accurate record of your observations and the conclusions derived therefrom. We describe below a method of taking notes that produces an accurate and acceptable record of your work. When studying chemical reactions, it is important to note:

- The reagent and any special experimental circumstances that were applied when the test was performed.
- The changes observed.
- An equation that represents the reaction occurring.
- A conclusion derived from the test.

You might be tempted to copy parts of this text in keeping a notebook, but that process would not help you develop your own "style" of keeping notes. A useful way of keeping notes in this course is to consider the two blank pages of an open notebook as the focus for your note taking. The left page is divided vertically into two equal sections. The left column of the left page is headed *TEST*, and it carries a brief description of the test performed; the description includes the experimental conditions. The right-hand column of the left page is headed *OBSERVATIONS*; it should contain a description of the visible change (if any) that occurred during the test. The entire right-hand page is reserved for an *EXPLANATION* of the observations, which should include a chemical equation that represents the observations made. An example of this technique of making notes is shown (compressed) in Figure 8.1 for the reactions of silver ion, Ag^+.

TEST	OBSERVATIONS	EXPLANATION
Group I, Ag^+ 1. add HCl	White ppt	$Ag^+ + Cl^- \rightarrow AgCl(s)$ insoluble silver chloride
+ NH_3	White ppt dissolves	$AgCl(s) + 2NH_3 \rightarrow Ag(NH_3)_2^+ + Cl^-$ soluble amine complex forms
2. add NH_3	Black ppt	$Ag^+ + OH^- \rightarrow AgOH(s)$ insoluble silver hydroxide
+ excess NH_3	Black ppt dissolves	$AgOH(s) + 2NH_3 \rightarrow Ag(NH_3)_2^+ + OH^-$ soluble amine complex forms

FIGURE 8.1   An example of the entries in a laboratory notebook. Two pages of an open notebook are depicted. The left page is divided into two columns, while the right page consists of only one column.

Your instructor will inform you of how you will keep your records. The following information should be a part of your record.

1. As each group, *known* or *unknown*, is being analyzed, the outline for that group must be developed as a part of your laboratory notebook. The object of making such an outline is to

enable you to visualize the steps of the analysis and record your own findings. Each step in the outline must be recorded as that step is performed in the laboratory.

2. All confirmatory tests obtained in the analysis of a known mixture must be approved by the instructor. Confirmatory tests obtained in the analysis of an unknown need not be approved.

3. The known mixture must be finished and the completed outline for this known must be approved by your instructor before an unknown will be issued. You must demonstrate to your instructor that you are familiar with the analysis of the group before you are permitted to start the analysis of the unknown.

4. When an unknown is reported to the instructor, your notebook should contain these points.
   a. The approved outline of the analysis of the group known.
   b. All equations for the reactions that took place in the analysis of the known. The information in Figure 8.1 will suffice for this requirement.
   c. The outline for the analysis of the unknown.

5. An alloy will not be issued until you are able to demonstrate a proper understanding of the analysis of the alloy to your instructor.

6. When the alloy is reported, your notebook should contain these points.
   a. A composite outline showing how it was analyzed.
   b. Equations for dissolving the metals eventually proved to be present in the alloy.

7. The report of the analysis of an all-group mixture of cations should show, schematically, the procedures that were carried out.

8. Salts and salt mixtures will be reported in accordance with the instructions given in Chapter 14.

9. The first salt will not be issued until you can demonstrate a satisfactory understanding of its plan of analysis. Experiments 1, 2, 3, and 4 in Chapter 14 should be performed before the first salt mixture is analyzed.

10. The analysis of special solids will be reported in the same manner as salts.

LABORATORY PROCEDURES

Good laboratory technique in qualitative analysis is important if results like those described in the procedures are to be obtained. Careless work will produce precipitates of the wrong color, or precipitates where you should get solutions, and in general it will make you less successful than you might otherwise have been. Good laboratory practice not only involves the careful manipulation of equipment and compounds, but also involves doing these in a safe, nonhazardous manner. As we have indicated, this includes wearing appropriate protective clothing—an apron or a laboratory jacket—and eye protection in the form of approved safety glasses. Although you might feel that you have little to fear from your own actions in the laboratory, you must remember that the average teaching laboratory environment involves people other than yourself. Thus, you should take the position that you are protecting yourself as well as others when you engage in safe practices.

The analysis described here will be carried out on the semimicro level. The amounts of chemicals used are neither large (grams) nor very small (micrograms). The sample solutions are typically about 0.1 M, and the volume of those solutions will be about 1 mL, which means that there will be about 10 mg of solute in the sample. The identification tests will be ineffective if you do not have at least 1 mg of solute present. The experiments must be carried out carefully to avoid losing the major portion of the components during the analysis.

Measuring Liquids.   Typically, procedures call for adding less than 1 mL of a reagent to a solution. This does not mean that you should use a small graduated cylinder to measure the reagent. The dropwise addition of water, solutions, or liquid reagents is done with a medicine dropper or with an appropriate dropping bottle. All liquid-reagent bottles should be fitted with droppers. Each student's desk will be provided with several medicine droppers. Droppers should be tested to see that they deliver drops of volume such that from 20 to 23 are required to give 1 mL of liquid. With a given liquid, the smaller the tip of the dropper, the smaller the drop delivered.

Measuring Solids.   Dispensing an approximate amount of solid can be accomplished in much the same way as for liquids. A small amount of solid is usually added from the tip of a spatula, the amount being estimated by the volume it occupies (the assumption is that most solid substances have nearly the same densities so that equal volumes will give approximately equal weights). To gain experience in judging the amount of solid from its volume, use the following procedure: Weigh a small empty beaker on a sensitive balance and add small portions of a solid substance from the tip of a spatula. You should be able to reach the point where you can deliver reproducibly 0.1 ± 0.02 g of solid from a spatula.

Precipitation.   One of the most common reactions in qualitative analysis is precipitation from solution. The indicated amount of precipitating reagent is added to the sample solution and stirred well with a glass stirring rod. Heat the solution in the water bath, if so directed. Since some precipitates form slowly, they must be given enough time to form completely. Frequently, solutions are not mixed thoroughly enough. When you think the precipitation is complete, centrifuge the mixture and, before decanting the liquid, add a drop or two of the precipitating reagent just to be sure that you added enough reagent the first time. The directions usually specify enough reagent to furnish an excess, but it is good practice to check that the desired reaction is complete.

Separation of Solids from Liquids.   In the present system of semimicro analytical procedures, solid precipitates are separated from liquids by centrifuging and decantation rather than by filtration. The centrifugal force imparted by the whirling centrifuge causes the heavy precipitate to be thrown to the bottom of the tube, the lighter liquid remaining on top. Any precipitate that happens to be clinging to the sides of the tube above the level of the liquid will remain there and will not be thrown down with the rest of the precipitate. Its presence on the walls of the tube will interfere with subsequent decantation. Therefore, any precipitate adhering to the inside walls of the test tube should always be washed down with a few drops of water or other appropriate reagent before the solution is centrifuged.

The test tube containing the material to be centrifuged must always be balanced in the centrifuge by a test tube containing an equal amount of water or other solution.

In all future procedures in which precipitation is used to accomplish a separation of substances, you should understand that you must make a test for complete precipitation. If the test shows that precipitation is not complete, add some more of the precipitating reagent, mix thoroughly by stirring the contents of the test tube with a glass stirring rod, and then centrifuge. The length of centrifuging time required will depend upon the nature of the precipitate. Most precipitates require only 15 to 30 s of centrifuging; however, some very finely divided precipitates may require several minutes. Proper centrifuging should give a clear supernatant liquid, with the precipitate packed into the bottom of the tube. If the supernatant liquid is not clear, more centrifuging is required. As a result of settling by centrifuging, most precipitates are so well packed into the bottom of the test tube that the supernatant liquid can be decanted (poured off) without much danger of disturbing the precipitate. The last drop of decantate can even be removed from the lip of the test tube by gently tapping. Some precipitates, however, are so light and fluffy that decantation of the supernatant liquid can be accomplished only with great care, and even then a part of the supernatant liquid must be allowed to remain behind with the precipitate. In a few instances, noted in the procedures, the precipitate is so light that decantation cannot be accomplished even with the exercise of great care; in such instances the top 75 to 80 percent of the supernatant liquid is drawn off with a medicine dropper.

If the decantate is to be discarded, as in a washing operation, the loss of bits of precipitate in the decantate is of no consequence. If, however, the decantate is to be submitted to further analysis, it must not contain any precipitate; if precipitate does get into the decantate, the process of centrifuging and decantation must be repeated.

Because the 3-in. test tubes are of such small diameter ($\frac{3}{8}$ in.), surface tension may cause the failure of decantate to flow out over the lip of the tilted tube. In such a case, the decantate will flow out if the lip of another test tube or the end of a stirring rod is touched to the solution at the lip of the tube.

Washing Solids.   The liquid decanted from the test tube after centrifuging out a solid does not, in general, contain any of the solid and may be used directly in a following analysis step. The solid remaining in the tube, however, has residual liquid around it. Since this liquid contains ions that may interfere with further tests on the solid, they must be removed by diluting the liquid with a wash liquid (often water), which does not interfere with the analysis. To wash the solid, add the indicated amount of wash liquid to the test tube and mix well with a glass stirring rod to disperse the solid well in the wash liquid. After mixing thoroughly, centrifuge out the solid and decant the wash liquid, which usually may be discarded. In key separations, it is best to perform the washing operation twice because an uncontaminated precipitate tends to give much better results than one mixed with even a trace of contamination. Failure to wash precipitates thoroughly is one of the main sources of error in qualitative analysis. On the other hand, care must be taken not to overwash a precipitate because it may undergo *peptization,* which is the reverse of the coagulation process—that is, breaking large, well-formed particles into smaller particles.

If a precipitate or solution is to be preserved from one laboratory period to the next, the test tube in which it is kept should be stoppered. Stoppering prevents contamination and also keeps solutions from evaporating and precipitates from drying out. Test tubes should be labeled so that their contents can be correctly identified.

Heating.   Many reactions are best carried out when the reagents are hot. You should never heat a test tube containing a reaction mixture directly on a Bunsen flame; invariably the contents become overheated and they splatter out of the test tube. It is more convenient, safer, and much more considerate of your colleagues to heat the test tube in a water bath. A 100-mL beaker about three-fourths full of boiling water serves as an effective bath for heating a 3-in. test tube and its contents. Simply set the tube in the boiling water. If the beaker is fitted with a flanged aluminum cover into which three or four holes—each large enough to hold a small test tube—have been punched, the tube will be held in a more nearly vertical position and three or four tubes can be heated conveniently at one time.

When you are following a procedure in which you need to heat a solution, keep the water bath hot or boiling by using a small flame to heat the beaker on a piece of asbestos-covered wire gauze. This way, the bath is ready whenever required. A mixture under study can be brought to approximately 100°C without boiling it. If the sample is heated in an open flame, it inevitably bumps out of the tube and onto the laboratory bench or onto you or someone else. If necessary, the contents of a test tube can be boiled by direct heating in the open flame; however, great care must be taken—otherwise the liquid may be thrown completely out of the tube. Hold the tube well above the flame with a test-tube holder and move it back and forth so that the top, as well as the bottom, of the liquid in the tube is heated. As a general rule, if a solution is to be boiled or evaporated, it should be transferred to a casserole and treated as described in the following section.

Evaporating a Liquid.   In some cases, it will be necessary to decrease the volume of a liquid to concentrate a species or to remove a volatile reagent. Sometimes, this will be necessary because the reaction of interest will proceed readily only in a boiling solution. When you must decrease the volume of a liquid, perform the evaporation in a small 30-mL beaker on a square of asbestos-covered wire gauze and use a small Bunsen flame, judiciously applied to maintain controlled, gentle boiling. Since the volumes involved usually are of the order of 3 or 4 mL, they can easily be overheated.

If you are supposed to stop the evaporation at a volume of about 1 mL, do not heat to dryness or "bake" it; you may decompose the sample or render it inert. Since concentrated solutions, or slurries, tend to bump, it is often helpful to encourage smooth boiling by scratching the bottom of the beaker with a stirring rod as the boiling proceeds. Good judgment in decreasing the volume of a sample by boiling it is important.

Occasionally the liquid that is evaporated is highly acidic (HCl or HNO_3) and the boiling causes evolution of noticeable amounts of noxious gases into the laboratory. In some cases the amounts are small enough to be ignored. If you notice that bothersome vapors are escaping from a boiling mixture, transfer the operation to the hood.

Transferring a Solid.   Sometimes it is necessary to transfer a solid from the beaker in which it was prepared to a test tube for centrifuging, or from the test tube to the beaker. The amount of solid involved is never large, about 50 mg at most. Such transfers can be made in the presence of a liquid, which serves as a carrier. When you are ready to make the transfer, stir the solid well into the liquid to form a slurry and then, without delay, pour the slurry into the other container. The transfer is not quantitative but, if done properly, you can move 90 percent of the sample to the other container. Forget about the rest. In general, do not attempt to transfer wet solids with a spatula because it is not easy, and all too often the spatula reacts with the liquid present, contaminating it.

Adjusting the pH of a Solution.   One of the most important experimental variables controlling chemical reactions is the pH of the solution. Frequently it is necessary to make a basic solution acidic, or *vice versa*, in order to make a desired reaction take place. For example, if you are directed to add 6 *M* HCl to a mixture until it is acidic, you should proceed as follows: knowing how the solution was prepared, make a quick mental calculation of about how much acid is needed—1 drop, 1 mL, or perhaps more. Then add the acid drop by drop, until you think the pH is about right. Mix well with a stirring rod and then touch the end of the rod to a piece of blue litmus paper on a piece of paper towel or filter paper. If the color does not change, add another drop or two of acid, mix, and test again. Frequently the system changes its character at the neutral point; a precipitate may dissolve or form or the color may change. In any event, add enough acid so that, after mixing, the litmus paper turns red when touched with the stirring rod.

Similarly, if you are told to make a solution basic with 6 M NH_3 or 6 M NaOH, add the reagent, drop by drop, until the solution, after being well mixed, turns red litmus blue. Adjustment of pH is not difficult, but it must be done properly if the desired reaction is to occur.

 General Work Environment. A beaker or flask of hot distilled water should always be available on the desk. A medicine dropper to be used exclusively for hot water should be kept in this beaker or flask. A second beaker or flask and medicine dropper should be provided for cold distilled water. Additional medicine droppers should be available for transferring solutions. Use of one dropper for all operations is liable to result in contamination. Since avoidance of contamination is imperative if good results are to be obtained, it is important that all pieces of equipment be cleaned with tap water and rinsed with distilled water *directly after they have been used* and before they are put on the desk. If this rule is followed, everything will be clean and available for immediate use when you need it. The pieces are most conveniently rinsed by squirting them with distilled water from a medicine dropper. There is no need for special wash bottles or other equipment for dispensing distilled water. The practice of placing a clean towel, folded to approximately a 12-in. square, on the desk top during the laboratory period and keeping clean test tubes, stirring rods, extra medicine droppers, test-tube brush, and casseroles on this towel is highly recommended.

The Silver Group

The first qualitative analysis group of cations, called the *silver group*, consists of Ag^+, Pb^+, and Hg_2^{2+}. The silver group of cations all form insoluble chlorides, which is the basis for the separation of the silver group from all other cations.

THE CHEMISTRY OF THE SILVER-GROUP IONS

Silver Ion, Ag^+

The most common oxidation state of silver is $1+$. Most silver compounds are insoluble in water; the common exceptions are $AgNO_3$ and AgF, which are very soluble, and Ag_2SO_4, which is slightly soluble. Most insoluble silver salts dissolve in cold $6\,M$ HNO_3, the main exception being the silver halides, $AgSCN$, and Ag_2S.

Most silver compounds are white; the colored compounds of silver that we are likely to encounter in qualitative analyses include:

Ag_2S	Black	Ag_3PO_4	Yellow
Ag_2O	Dark brown	AgI	Yellow
Ag_2CrO_4	Dark red	$AgBr$	Light yellow

Silver ions form many stable complexes in which the coordination number of the silver is 2, AgL_2. Of these, the best known is probably the $Ag(NH_3)_2^+$ ion. This complex is sufficiently stable that it forms when $AgCl$ or $AgSCN$ is treated with $6\,M$ NH_3; the reaction, Eq. (9.1), that occurs is useful for dissolving those solids.

$$AgX(s) + NH_3 \rightarrow Ag(NH_3)_2^+ + X^- \tag{9.1}$$
$$X = Cl \text{ or } SCN$$

Both $AgBr$ and AgI are less soluble than $AgCl$; $AgBr$ will dissolve in $15\,M$ NH_3, but AgI is so insoluble that it will not dissolve in aqueous NH_3. The silver thiosulfate complex ion, $Ag(S_2O_3)_2^{3-}$, is extremely stable and is an important species in photographic processes, where it is formed during the "fixing" reaction in which unreacted AgBr is removed from the developed negative.

$$AgBr(s) + 2S_2O_3^{2-} \rightarrow Ag(S_2O_3)_2^{3-} + Br^- \tag{9.2}$$

Lead Ion, Pb^{2+}

Of the two oxidation states of lead, Pb^{2+} and Pb^{4+}, the most commonly encountered in qualitative analysis is Pb^{2+}. Most compounds of Pb^{2+} are insoluble in water; lead nitrate and acetate are the only well-known soluble lead salts. Lead chloride is not nearly as insoluble in water as silver chloride and becomes moderately soluble if the water is heated; $PbSO_4$ is one of the few insoluble sulfates. Lead forms a stable hydroxide complex ion, $Pb(OH)_4^{2-}$, and a weak chloride complex, $PbCl_4^{2-}$.

$$Pb^{2+} + 2OH^- \xrightarrow{} Pb(OH)_2(s) \xrightarrow{2OH^-} Pb(OH)_4^{2-} \tag{9.3}$$

$$Pb^{2+} + 2Cl^- \xrightarrow{} PbCl_2(s) \xrightarrow{2Cl^-} PbCl_4^{2-} \tag{9.4}$$

Although lead ordinarily forms compounds in which it exhibits a 2+ oxidation state, some lead compounds containing the 4+ state exist. The most common is PbO_2 (brown); it is insoluble in most reagents but will dissolve in 6 M HNO_3 if H_2O_2 is present.

Mercury(I) Ion, Hg_2^{2+}

Mercury exhibits two oxidation states, Hg_2^{2+} and Hg^{2+}; both oxidation states are reasonably common. Mercury(I) appears as a member of the silver group, but mercury(II) occurs in the copper-arsenic group (Chapter 10). Except for $Hg_2(NO_3)_2$, most mercury(I) salts are insoluble. The mercury(I) species is very susceptible to a simultaneous oxidation–reduction process, called *disproportionation* [Eq. (9.5)].

$$Hg_2^{2+} \leftrightharpoons Hg^{2+} + Hg(l) \tag{9.5}$$

Disproportionation of Hg_2^{2+} occurs readily in the presence of species that form very insoluble precipitates or stable complex ions with mercury(II). When disproportionation of Hg_2^{2+} occurs, metallic mercury is formed [Eq. (9.5)] as very small droplets, which gives the mixture a gray color; the intensity of the color depends upon the amount of mercury formed and could range from an off-white to a charcoal gray.

PRECIPITATION AND ANALYSIS OF THE SILVER GROUP

The chlorides of Ag^+, Hg_2^{2+}, and Pb^{2+} are all insoluble (see Note 1) in cold water and in cold, dilute hydrochloric acid; the chlorides of other cations are soluble (see solubility rules, Chapter 4). These facts form the basis for the separation of the silver-group cations from all other metallic cations (Figure 9.1). Note that in Figure 9.1 we make the assumption that all the ionic species covered by this qualitative analysis scheme are present in the solution. Recall (from Chapter 1) that by making this "worst-case" assumption, we are assured of performing *all* the tests that will produce sufficient data to decide for each species whether it is present in our solutions.

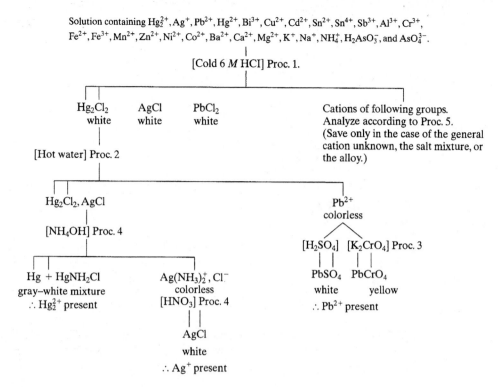

FIGURE 9.1   Analysis of the silver group.

When HCl or some other soluble chloride (see Note 2) is added to a cold solution containing all the common cations, AgCl, Hg_2Cl_2, and $PbCl_2$ are precipitated. All other metals remain in solution as soluble chlorides (see Figure 9.1). The ions of silver, mercury(I), and lead constitute the silver group. Since HCl is used to separate this group of ions from other cations by precipitating them as insoluble chlorides, it is called the *group reagent* for the silver group.

The object of the silver-group precipitation is to remove the Ag^+, Hg_2^{2+}, and Pb^{2+} ions from solution as completely as possible by precipitating them as AgCl, Hg_2Cl_2, and $PbCl_2$, respectively. A precipitate will form when the product of the concentrations of the ions that react to form the precipitate just exceeds the solubility product (Chapter 4). Furthermore, in any precipitation process, the insoluble compound will keep precipitating until the concentrations of its ions remaining in solution reach values at which their product just equals the solubility product. When this point is reached, the precipitate will be in equilibrium with its ions, which means that the rate at which the precipitate is forming is equal to the rate at which it is dissolving and dissociating into its ions. This is a true equilibrium, and the equilibrium point will be shifted by changes in the concentrations of the reacting ions in accordance with the mass law. It follows, therefore, that an excess of the group reagent ion—in this case Cl^-—is desirable, since, if the concentration of Cl^- is high, the concentrations of Ag^+, Hg_2^{2+}, and Pb^{2+} remaining in solution will be low. An inordinately large excess of the reagent may, however, form soluble complex ions, thus nullifying the usual advantage resulting from having excess reagent. A very large excess of HCl will actually increase the amount of lead remaining in solution. The excess chloride ions react with lead ions to form the soluble stable complex ion $PbCl_4^{2-}$ [Eq. (9.7)]. This will reduce the concentration of lead ions in solution below the value required for the solubility product of $PbCl_2$. As a result, some $PbCl_2$ will go back into solution, in accordance with the principles discussed in Chapter 2. The equations for the reactions are

$$PbCl_2(s) \rightleftharpoons Pb^{2+} + 2Cl^- \tag{9.6}$$

$$Pb^{2+} + 4Cl^- \rightleftharpoons PbCl_4^{2-} \tag{9.7}$$

Silver has a slight tendency to form the stable complex ion $AgCl_2^-$ with excess HCl.

$$AgCl_2(s) \rightleftharpoons Ag^+ + Cl^- \tag{9.8}$$

$$Ag^+ + 2Cl^- \rightleftharpoons AgCl_2^- \tag{9.9}$$

Since $PbCl_2$ is appreciably soluble in cold dilute HCl, lead will not be completely precipitated in the silver group. The lead ions remaining in solution will be completely precipitated as lead sulfide in the next group, the copper-arsenic group (see Chapter 10).

Notes

1. For all practical purposes, broad statements relating to solubilities—such as the solubility rules—are true, and they are useful for pointing out general facts that form the basis for analytical separations.
2. Other soluble chlorides, such as NH_4Cl, can be substituted for HCl, since the reaction is one between ions.

$$Pb^{2+} + 2Cl^- \rightleftharpoons PbCl_2(s)$$

PROCEDURE 1

Precipitation of the Silver Group

Place 5 drops of the solution to be analyzed in a 3-in. test tube and add 5 drops of water. Add 2 drops of 6 M HCl and mix the contents thoroughly with a glass stirring rod. With a few drops of cold water, wash down into the solution any precipitate (see Note 1) that adheres to the inside of the test tube above the level of the solution; centrifuge. Test for complete precipitation by adding another drop of 6 M HCl to the clear supernatant solution in the test tube. When precipitation is complete, centrifuge and then decant. Save the decantate, which contains the cations of the groups that follow in the analysis scheme (Figure 1.1) for Procedure 5 in Chapter 10 (see Note 2). Wash the precipitate once with 5 drops of cold water (see Note 3), adding the washings to the decantate being saved for Procedure 5. Allow the precipitate to remain in the test tube (see Note 4) and analyze it according to Procedure 2 (see Note 5).

Notes

1. If no precipitate is formed with cold HCl, the absence of the ions of silver and mercury(I) is definitely proved. Lead may be present in small quantities, however, since $PbCl_2$ is appreciably soluble even in cold water.

2. If the solution being analyzed is known to contain only cations of the silver group (see Summary of Laboratory Work, 1 and 2, in Chapter 8), this decantate may be discarded.

3. Wash a precipitate as follows: Add the water or other wash liquid to the precipitate in the test tube, mix thoroughly by stirring the contents of the tube with a glass stirring rod, centrifuge, and decant. *Failure to wash precipitates thoroughly is one of the main sources of error in qualitative analysis.*

4. If a precipitate or solution is to be preserved from one laboratory period to the next, the test tube in which it is kept should be stoppered. Stoppering prevents contamination and also keeps solutions from evaporating and precipitates from drying out. Test tubes should be labeled so that their contents can be correctly identified.

5. A scheme of analysis of the kind shown by the flowchart in Figure 9.1 should be started on two facing blank pages in your notebook just as soon as the analysis of a silver-group solution is started. This scheme should be developed step-by-step and procedure-by-procedure as the analysis is actually carried out. Notes should be made indicating the colors of precipitates or solutions formed. The scheme will then be a progress chart showing exactly what was done and what was observed. A scheme of this sort should be developed for every group solution that is analyzed.

SEPARATION OF LEAD FROM SILVER AND MERCURY(I)

The silver-group precipitate is a mixture of $PbCl_2$, Hg_2Cl_2 (see Note 1), and AgCl. It is necessary to separate individual members of the group so that the presence of each may be confirmed. Lead is separated first. $PbCl_2$ is soluble in hot water (see Note 2); AgCl and Hg_2Cl_2 are insoluble. This difference is the basis for the separation of lead ions from mercury(I) and silver ions.

Notes

1. Mercury(I) exists in solution as the stable diatomic ion Hg_2^{2+}. Therefore, the precipitate of mercury(I) chloride is Hg_2Cl_2 rather than HgCl.

2. 100 mL of water at 0°C dissolves 0.673 g of $PbCl_2$. 100 mL of water at 100°C dissolves 3.34 g of $PbCl_2$.

PROCEDURE 2

Separation of Lead from Silver and Mercury(I)

Add 15 to 20 drops of hot water to the test tube containing the precipitate from Procedure 1, stir well until all the precipitate is in suspension, and then heat the tube by placing it in a 100-mL beaker of boiling water for about 1 min. Stir frequently. Centrifuge at once and decant into another test tube immediately after centrifuging; save the decantate, which contains Pb^{2+}, for Procedure 3. Wash the precipitate twice with 10-drop portions of hot water (see Note 1); save the precipitate in the test tube for Procedure 4.

Notes

1. Unless specified otherwise, washings are always discarded.

PROCEDURE 3

Detection of Lead

Cool the decantate from Procedure 2 and divide into two parts. To one part, add 1 drop of 0.2 M K_2CrO_4 solution; a yellow precipitate ($PbCrO_4$) proves the presence of lead. To the second part, add 1 drop of 2 M H_2SO_4; a white precipitate ($PbSO_4$), which may form slowly, is further proof of the presence of lead.

Notes

1. The solubility product of $PbSO_4$ is 1.3×10^{-8}, and that of $PbCrO_4$ is 2×10^{-16}, at 20°C. Obviously $PbCrO_4$ provides a much more sensitive test for lead than does $PbSO_4$.
2. One confirmatory test is usually all that is required to establish the presence or absence of a given ion. However, two confirmatory tests for lead ions are carried out to show that one of these tests is much more sensitive than the other. A certain concentration of lead ions will give a much stronger test when treated with chromate ions than when treated with sulfate ions because lead sulfate is about 8000 times as soluble as lead chromate. Just as the two confirmatory tests for lead differ in sensitivity, so may confirmatory tests for other ions differ greatly in sensitivity. The best test known is used for each ion.

SEPARATION AND DETECTION OF SILVER AND MERCURY

Silver chloride, AgCl, is soluble in aqueous NH_3, and Hg_2Cl_2 reacts with aqueous NH_3 to form Hg and $HgNH_2Cl$, both of which are insoluble. These facts are the basis for the separation of silver ions from mercury(I) ions.

Silver chloride dissolves in aqueous ammonia according to Eq. (9.10).

$$AgCl(s) + 2NH_3 \rightleftharpoons Ag(NH_3)_2^+ + Cl^- \qquad (9.10)$$

When a solution containing the complex ion $Ag(NH_3)_2^+$ and Cl^- is acidified with nitric acid, AgCl is precipitated in accordance with Eq. (9.11).

$$Ag(NH_3)_2^+ + Cl^- + 2H^+ \rightleftharpoons AgCl(s) + 2NH_4^+ \qquad (9.11)$$

 Formation of the complex diamminesilver ion, $Ag(NH_3)_2^+$, is an example of a type of reaction that will be encountered in connection with the detection of other ions. Similar complexes are $Cu(NH_3)_4^{2+}$, $Co(NH_3)_6^{3+}$, $Ni(NH_3)_6^{2+}$, $Zn(NH_3)_4^{2+}$, and $Cd(NH_3)_4^{2+}$ (see the discussion in Chapter 5).

Aqueous ammonia, NH_4OH, reacts with Hg_2Cl_2 to produce a mixture of black, finely divided mercury (Hg) and white mercury(II) amido chloride ($HgNH_2Cl$) according to Eq. (9.12). This mixture may appear as some shade of gray, depending on the size of the mercury droplets formed.

$$Hg_2Cl_2(s) + 2NH_3 \rightleftharpoons Hg(l) + HgNH_2Cl(s) + NH_4^+ + Cl^- \qquad (9.12)$$

The compound $HgNH_2Cl$ may be considered to be a derivative of $HgCl_2$ in which the amino group (NH_2) has replaced one atom of chlorine, as in (**1**).

1

The formation of Hg and $HgNH_2Cl$ from Hg_2Cl_2 may be considered to be taking place in two steps. First, the Hg_2Cl_2 disproportionates (1 mole of Hg in Hg_2Cl_2 functions as an oxidizing agent, and 1 mole functions as a reducing agent) according to Eq. (9.13).

$$Hg_2Cl_2(s) \leftrightharpoons Hg(l) + HgCl_2(s) \tag{9.13}$$

The $HgCl_2$ then reacts with aqueous NH_3 to form $HgNH_2Cl$.

$$HgCl_2(s) + 2NH_3 \leftrightharpoons HgNH_2Cl(s) + NH_4^+ + Cl^- \tag{9.14}$$

The dissolving of AgCl by aqueous NH_3 and the failure of AgCl to be dissolved by HNO_3 may be interpreted in terms of the general rules given in Chapter 7. The detailed equations for the reactions are

$$AgCl(s) \leftrightharpoons Ag^+ + Cl^- \tag{9.15}$$

$$Ag^+ + 2NH_3 \leftrightharpoons Ag(NH_3)_2^+ \tag{9.16}$$

PROCEDURE 4

Separation and Detection of Mercury(I) and Silver

Add 4 drops of 15 M aqueous NH_3 to the precipitate from Procedure 2, mix thoroughly, centrifuge, and decant into another test tube, saving the decantate (see Note 1) for testing the silver. A gray-to-black residue (the mixture of Hg + $HgNH_2Cl$) proves the presence of mercury(I). To the decantate, add 16 M HNO_3 drop by drop and mix constantly with a stirring rod until slightly acidic (see Note 2). A white precipitate (AgCl) proves the presence of silver.

Notes

1. If $PbCl_2$ is not completely removed from the precipitate of AgCl and Hg_2Cl_2, it is converted by aqueous ammonia into the finely divided insoluble white basic salt Pb(OH)Cl, which may give a turbid decantate. The basic salt will dissolve in HNO_3 and will not interfere with the confirmatory test for silver.
2. Test for the acidity or alkalinity of a solution as follows: Place a piece of litmus paper on a clean towel or drape it over the edge of a beaker. Withdraw the stirring rod used for stirring the solution and touch the end of it to the piece of litmus paper. Litmus paper turns red in the presence of an acid and blue in the presence of a base.

PROBLEMS

9.1 Using the scheme of analysis as a guide, write net equations for all reactions that take place in the precipitation and analysis of the silver group.

9.2 You are asked to make up a water solution of an unknown that will contain lead, silver, mercury(I), copper, manganese, barium, and sodium ions, all in the same solution. What salt (sulfate, nitrate, carbonate, acetate, chloride) of each of the metals would you use to obtain a clear solution containing all seven of the ions above?

9.3 What reagent could be used in place of HCl as the group reagent for the silver group?

9.4 Cold dilute H_2SO_4 was accidentally used in place of cold concentrated HCl as the group reagent for the silver group in the analysis of an unknown solution. A white precipitate formed. Explain.

9.5 A solution of Na_2CO_3 in water was accidentally used in place of HCl as the group reagent for the silver group in the analysis of an unknown solution. A heavy precipitate formed. Explain.

9.6 State the fact upon which each of the following separations is based.

   a. The separation of silver, lead, and mercury(I) ions from all other cations.
   b. The separation of lead ions from silver and mercury(I) ions.
   c. The separation of silver ions from mercury(I) ions.

9.7 In the analysis of an unknown solution, what difficulty, if any, would arise in each event?

   a. Hot concentrated HCl was used instead of cold HCl in the precipitation of the silver group.
   b. The precipitate of AgCl and Hg_2Cl_2 was not washed free of $PbCl_2$.
   c. The silver-group precipitate (AgCl, Hg_2Cl_2, and $PbCl_2$) was washed too long with cold water.
   d. A large excess of concentrated HCl was used in the silver-group precipitation.
   e. Not enough HNO_3 was added in the confirmatory test for silver.

9.8 Using the scheme of analysis as a guide, give the formula for a chemical substance that will behave as indicated.

   a. Form a precipitate with KCl solution and also with $CuCl_2$.
   b. Form a precipitate with HCl and also with H_2SO_4.
   c. Form a precipitate with NH_4Cl solution and also with K_2CrO_4.
   d. Form a precipitate with HCl and also with HNO_3.
   e. Form a precipitate with HCl, but not with HNO_3.
   f. Form a precipitate with $AgNO_3$ solution, but not with HNO_3.
   g. Form a precipitate with $Hg_2(NO_3)_2$ solution, but not with $Hg(NO_3)_2$.
   h. Form a precipitate with $Ag(NH_3)_2Cl$ solution, but not with $AgNO_3$.
   i. Readily dissolve $PbCl_2$, but not AgCl.
   j. Readily dissolve $ZnCl_2$, but not $PbCl_2$.
   k. Readily dissolve AgCl, but not Hg_2Cl_2.
   l. Readily dissolve $HgCl_2$, but not Hg_2Cl_2.
   m. Readily dissolve $Pb(NO_3)_2$, but not $PbSO_4$.
   n. Form a precipitate with Na_2CrO_4, but not with $NaNO_3$.

9.9 By means of what single reagent could you distinguish between the following? (Tell what happens to each substance.)

   a. *Solutions:*

        $AgNO_2$ and $Zn(NO_3)_2$ H_2SO_4 and HCl

        K_2CrO_4 and KNO_3 H_2SO_4 and HNO_3

        HCl and HNO_3 $Hg(NO_3)_2$ and $Hg_2(NO_3)_2$

        $Pb(NO_3)_2$ and $Hg(NO_3)_2$ $Ag(NH_3)_2Cl$ and $AgNO_3$

   b. *Solids:*

        Hg_2Cl_2 and $HgCl_2$ $AgCl$ and $ZnCl_2$

        $PbCl_2$ and Hg_2Cl_2 $AgCl$ and Hg_2Cl_2

        $PbSO_4$ and $Pb(NO_3)_2$ $PbCrO_4$ and K_2CrO_4

9.10 A solution, which was known to contain only cations of the silver group, gave no precipitate when heated to boiling and treated with hot concentrated HCl, but gave a white precipitate when cooled and treated with cold concentrated HCl. What conclusions can be drawn?

9.11 In the analysis of a solution that was known to contain only cations of the silver group, the white precipitate obtained by adding cold HCl to the unknown solution partly dissolved in hot water. The residue dissolved completely in aqueous ammonia (NH_3). What cations were present? Which ones were absent?

9.12 The white precipitate obtained when cold HCl was added to a solution was completely insoluble in both hot water and aqueous ammonia (NH_3). What conclusions can be drawn?

9.13 The white precipitate obtained when cold HCl was added to an unknown solution was completely insoluble in hot water and completely soluble in aqueous NH_3. What conclusions can be drawn?

9.14 A solution that is known to contain only cations of the silver group forms a white precipitate when treated with 6 M HCl. The white precipitate is washed with cold water. When treated with 15 M aqueous NH_3, the washed precipitate remains completely unchanged, the only indication of reaction being the formation of a slight milky color in the 15 M aqueous NH_3. What cations are shown to be absent? Which are present?

9.15 A student was given a silver-group unknown and was told, correctly, that it contained only one cation. She treated the unknown with 6 M HCl as directed in the scheme of analysis and obtained a white precipitate. She washed this precipitate twice with cold water. She then treated the washed precipitate with one reagent. By observing what happened when this one reagent was added, she was able to decide within a few seconds what cation was present in the unknown. What one reagent did she add to the precipitate? What would she have observed on addition of the one reagent to the washed precipitate if the unknown solution contained, as the one cation, Ag^+, Hg_2^{2+}, or Pb^{2+}?

9.16 A white solid is known to consist of a mixture of equimolar amounts of only two of the five water-soluble salts listed below. When an adequate amount of cold water is added to the solid and the mixture is well stirred, a precipitate remains. The precipitate is separated from the supernatant liquid and is divided into two parts. One part of the precipitate is found to be readily and completely soluble in hot 3 M HCl. The other part of the precipitate is found to be readily and completely soluble in cold 3 M HNO_3. Indicate whether each salt is present or absent, or whether it is impossible to determine.

$$Pb(C_2H_3O_2)_2, \ AgNO_3, \ Na_2SO_4, \ (NH_4)_2CO_3, \ KCl$$

9.17 A white solid is known to consist of one or more of the following water-soluble compounds:

$$Ag_2SO_4, \ Na_2CO_3, \ Pb(NO_3)_2$$

When an adequate amount of cold water is added and the mixture is stirred well for some time, a precipitate remains in the solution in the beaker. The entire precipitate still remains when the contents of the beaker are heated to boiling. The contents of the beaker are divided into two parts. When cold 3 M HNO_3 is added to one part until acidic, a clear solution, with no precipitate, is formed. When hot 3 M HCl is added to the other part until acidic, a precipitate remains in the beaker even when it is heated. Indicate whether each solid is present or absent, or whether it cannot be determined.

9.18 Ag_3AsO_4 is soluble in HNO_3 but insoluble in $HC_3H_3O_2$. Explain why, giving the detailed equations for all reactions involved.

9.19 Is each of the following four observations consistent with each of the other three? Explain.

   a. The solubility product for AgI is less than the solubility product for $AgCl$.

   b. $AgCl$ is more soluble in water than is AgI.

   c. $AgCl$ is soluble in 15 M aqueous ammonia, NH_3; AgI is not.

   d. When solid $AgCl$ is shaken with a solution of 1 M KI, the solid $AgCl$ is dissolved and a precipitate of AgI is formed.

9.20 The reactions that follow in set 1, in which M is a divalent metal, are observed to proceed substantially to completion in the direction written. On the basis of this information alone, decide whether each reaction in set 2 will proceed as written, will fail to proceed, or cannot be predicted.

   Set 1:

$$MCO_3(s) + 2OH^- \rightleftharpoons M(OH)_2(s) + CO_3^{2-}$$
$$M(IO_3)_2(s) + CO_3^{2-} \rightleftharpoons MCO_3(s) + 2IO_3^-$$
$$M(NH_3)_4^{2+} + 2OH^- \rightleftharpoons M(OH)_2(s) + 4NH_3$$
$$M(OH)_2(s) + S^{2-} \rightleftharpoons MS(s) + 2OH^-$$
$$M(OH)_2(s) + Se^{2-} \rightleftharpoons MSe(s) + 2OH^-$$

   Set 2:

$$M(NH_3)_4^{2+} + Se^{2-} \rightleftharpoons MSe(s) + 4NH_3$$
$$MS(s) + Se^{2-} \rightleftharpoons MSe(s) + S^{2-}$$
$$MS(s) + CO_3^{2-} \rightleftharpoons MCO_3(s) + S^{2-}$$
$$M(IO_3)_2(s) + S^{2-} \rightleftharpoons MS(s) + 2IO_3^-$$
$$M(IO_3)_2(s) + 4NH_3 \rightleftharpoons M(NH_3)_4^{2+} + 2IO_3^-$$

9.21 Silver chloride, $AgCl$, and AgI both dissolve in a solution of KCN. Silver chloride, $AgCl$, dissolves in NH_4OH; AgI does not.

   a. How does the magnitude of $K_{inst.}$ of $Ag(CN)_2^-$ compare with the magnitude of $K_{inst.}$ of $Ag(NH_3)_2^+$?

   b. How does the magnitude of K_{sp} of $AgCl$ compare with the magnitude of K_{sp} of AgI?

9.22 Write net equations for the reactions that occur.

   a. 0.002 mole of $AgNO_3$ contained in 10 mL of solution is treated with 0.001 mole of NH_4OH in the form of 5 M NH_4OH.

b. A large excess of 5 M NH_4OH is added to the product formed in part (a). The solubility product of AgOH is 2.0×10^{-9}. The ionization constant for aqueous NH_3 is 1.8×10^{-5}. The solubility product of AgCl is 2.8×10^{-10}.

9.23 Suppose you were to calculate the solubility product for AgCN from its solubility in water on the assumption that only Ag^+ and CN^- ions are present, not realizing that the stable $Ag(CN)_2^-$ complex ion is formed. Would this calculated value be greater than, less than, or equal to the true value of the product $[Ag^+][CN^-]$?

9.24 A student, thinking the formula of mercury(I) chloride was HgCl, calculated its solubility product from the weight of the compound present in a liter of solution. She assumed the reaction occurring when the salt dissolved was $2HgCl \rightarrow Hg_2^{2+} + 2Cl^-$. The correct formula is Hg_2Cl_2, and the correct reaction is $Hg_2Cl_2 \rightarrow Hg_2^{2+} + 2Cl^-$. Was her calculated value of the solubility product equal to, less than, or greater than the true value? (Assume an activity coefficient of 1 for each ion.)

9.25 The solubility product of AgCNS, which is pale amber in color, is 1×10^{-12}. The solubility product of $Al(OH)_3$ is 5×10^{-33}. When a dilute solution of NaCNS is added to a beaker containing AgCl, the precipitate changes in color from white to pale amber. The saturated solution formed when solid $Fe(OH)_3$ is shaken with water is less basic than the saturated solution formed when $Al(OH)_3$ is shaken with water. The water solution in equilibrium with solid Ag_3PO_4 has four times the concentration of Ag^+ present in the water solution in equilibrium with solid AgCl. On the basis of the facts above, list the following in order of decreasing molar solubility in water: $Al(OH)_3$, AgCNS, AgCl, $Fe(OH)_3$, Ag_3PO_4. (Assume an activity coefficient of 1 for each ion.)

9.26 You are given the following facts, established at 25°C.

a. The pH of a saturated solution of $Mg(OH)_2$ in water is 10.

b. The solubility product of AgCl is 2.8×10^{-10}.

c. When a mixture of AgI and AgCNS (both solids) is treated with 5 M NH_4OH, the amber-colored AgCNS dissolves, but the AgI remains as a precipitate.

d. The potential of the half-reaction $Ag \rightleftharpoons Ag^+ + e^-$ is higher when the silver electrode dips into a saturated solution of AgBr prepared by dissolving AgBr in water than when it dips into a saturated solution of AgCl prepared by dissolving AgCl in water.

e. When 0.01 M KCNS is added to a precipitate of AgBr, the color of the precipitate in the beaker changes from cream to amber.

On the basis of the facts above, arrange $Mg(OH)_2$, AgCl, AgI, AgCNS, and AgBr in the order of their increasing molar solubilities in water. (Assume an activity coefficient of 1 for each ion.)

The Copper-Arsenic Group

10

The copper-arsenic group of ions Hg^{2+}, Pb^{2+}, Bi^{3+}, Cu^{2+}, Cd^{2+}, As^{3+}, Sb^{3+}, and Sn^{4+} form sulfides that are insoluble in dilute HCl (Figure 10.1), in contrast with the sulfides of the groups that follow in the analysis scheme (Figure 11.1); this fact forms the basis for the separation of copper-arsenic group cations from those of the following groups in the analysis scheme.

Lead, Pb^{2+}, and mercury, Hg^{2+}, appear in the copper-arsenic group, as well as in the silver group. Lead chloride is somewhat soluble in dilute HCl. For that reason lead ions generally are not completely precipitated in the silver group and are carried over into the copper-arsenic group. Mercury occurs in the monovalent, Hg_2^{2+}, and divalent, Hg^{2+}, states. Although Hg_2Cl_2 is very insoluble in dilute HCl, $HgCl_2$ is very soluble. Therefore, mercury(II) is not precipitated in the silver group.

The sulfides HgS, CuS, Bi_2S_3, SnS_2, CdS, PbS, As_2S_3, and Sb_2S_3 are formed as a result of the combination of the cations with the sulfide ions produced by ionization of the weak acid H_2S.

$$H_2S \rightleftharpoons H^+ + HS^- \qquad (10.1)$$

$$HS^- \rightleftharpoons H^+ + S^{2-} \qquad (10.2)$$

$$S^{2-} + Hg^{2+} \rightleftharpoons HgS(s) \qquad (10.3)$$

The net equation for the reaction is

$$Hg^{2+} + H_2S \rightleftharpoons HgS(s) + 2H^+ \qquad (10.4)$$

A particular metal sulfide will not begin to precipitate until the product of the concentrations of the cation and the sulfide ion equals the solubility product for the sulfide (see Chapter 4).

The sulfide will continue to precipitate until the concentration of the cation has been reduced to a point where the product of its concentration and the sulfide ion concentration is equal to the solubility product of the sulfide. When precipitation ceases, the precipitate will be in equilibrium with its ions in solution, and the rate at which the ions in solution combine to form the precipitate will be exactly equal to the rate at which the precipitate dissolves and dissociates into its ions. Since this is a true equilibrium that obeys the mass law, it follows that the higher the

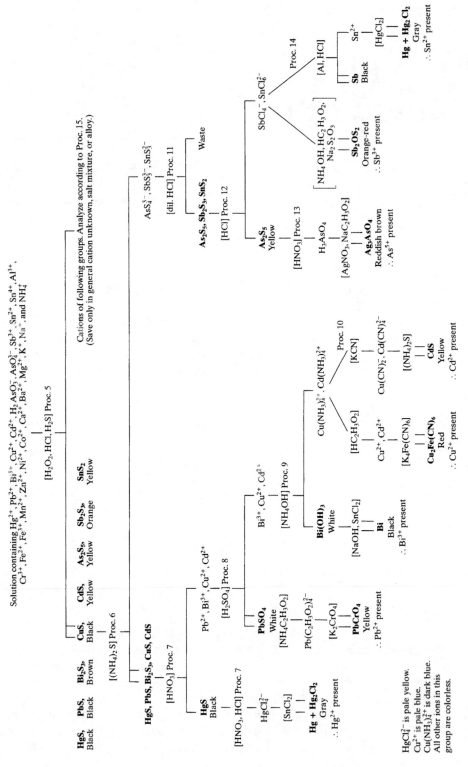

FIGURE 10.1 Flow chart for the analysis of the copper-arsenic group.

concentration of sulfide ions, the lower the concentration of cations left in solution when precipitation is complete.

The sulfide ions required for the precipitation of the copper-arsenic group sulfides, as shown in Figure 10.1, are derived from the ionization of the weak acid H_2S, which is formed when H_2S gas is dissolved in water solution or when thioacetamide hydrolyzes, as we shall see later. The ionization of H_2S takes place in two stages [Eqs. (10.1) and (10.2)]: For the purposes of this discussion, however, the two stages shown as Eqs. (10.1) and (10.2) may be combined into one equation (10.5):

$$H_2S \rightleftharpoons 2H^+ + S^{2-} \tag{10.5}$$

If the hydrogen ion concentration (acidity) of a saturated solution of H_2S is increased by addition of HCl, the point of the equilibrium in Eq. (10.5) is shifted to the left, with a resulting decrease in the S^{2-} ion concentration. If the concentration of the H^+ derived from the HCl is reduced by dilution (H^+ concentration decreases) while the solution is kept saturated with H_2S, the concentration of the S^{2-} ions in the equilibrium above will increase. That is, in a saturated solution of H_2S, the greater the H^+ concentration, the smaller the S^{2-} concentration; the smaller the H^+ concentration, the greater the S^{2-} concentration.

Not only can the concentration of H^+ derived from HCl be reduced by dilution but also it can be more effectively decreased by addition of the salt of a weak acid, such as ammonium acetate, $NH_4C_2H_3O_2$. The $C_2H_3O_2^-$ ions from $NH_4C_2H_3O_2$ react with H^+ in the form of the weak acid $HC_2H_3O_2$, Eq. (10.7) (see "Buffers," Chapter 3).

$$NH_4C_2H_3O_2(s) \rightarrow NH^{4+} + C_2H_3O_2^- \tag{10.6}$$

$$C_2H_3O_2^- + H^+ \rightleftharpoons HC_2H_3O_2 \tag{10.7}$$

Since the concentration of the OH^- ions is already extremely low because the solution is acidic, the added NH_4^+ ions from the $NH_4C_2H_3O_2$ will have negligible effect on the acidity of the solution.

To get complete precipitation of the cations of the copper-arsenic group as sulfides, the concentration of the sulfide ion, which is the group-reagent ion, should be as high as possible; to get a high S^{2-} concentration, the H^+ concentration should be as low as possible. There are two factors in our scheme of analysis that modify the application of this rule. Arsenic, which is usually present as arsenate, is converted into As_2S_5; this reaction requires a high concentration of H^+ (see below). At the other extreme, if the acidity of the solution is too low, the sulfide ion concentration will be so high that the solubility products for the sulfides of iron, zinc, manganese, cobalt, and nickel will be exceeded, and these sulfides will also be precipitated.

As noted in Table 4.2, the solubility products of the sulfides of the divalent cations in the copper-arsenic and aluminum-nickel groups are: HgS, 4×10^{-33}; CuS, 6×10^{-37}; CdS, 8×10^{-28}; PbS, 3×10^{-28}; CoS, 6×10^{-22}; NiS, 1×10^{-22}; ZnS, 3×10^{-23}; FeS, 6×10^{-19}; and MnS, 3×10^{-11}. Since the objective is to precipitate the sulfides of the copper-arsenic group but leave the aluminum-nickel group cations in solution, the sulfide ion concentration must be kept slightly below the value that will exceed the solubility product value for CoS, which is the least soluble of the aluminum-nickel group sulfides.

Since K_{sp} for CoS is 6×10^{-22} and since—at the end of the successive dilutions during the precipitation—the Co^{2+} ion concentration is approximately 0.02 M, the S^{2-} ion concentration must not be allowed to exceed 3×10^{-20} M. As noted from the values in Table 3.2, the value of the overall ionization constant, Eq. (10.5), for H_2S is 1.0×10^{-26}. Since the solubility of H_2S in water at room temperature is about 0.1 M, the value of the ion product, $[H^+]^2 \times [S^{2-}]$, is 1×10^{-27} at room temperature. This means that to keep the S^{2-} ion concentration 3×10^{-20} M, the hydrogen ion concentration must be no lower than about 0.2 M.

Thus, the equilibrium principles outlined in Chapter 4 are transformed into the details described in the procedures to effect the desired separations. The unthinking student can imagine that the experimental procedures are simply empirical recipes, which are to be performed mechanically with little regard to the principles described in Chapter 4. This approach to laboratory work would be a serious intellectual mistake.

THE CHEMISTRY OF THE COPPER-ARSENIC GROUP IONS

Mercury(II), Hg^{2+}

Mercury(II), Hg^{2+}, is the highest common oxidation state of mercury observed in aqueous solution. Mercury(II) halides, acetate, and cyanide are soluble in water. The nitrate, which forms an insoluble basic salt in water, dissolves in dilute acids. Interestingly, many mercury(II) salts are weak electrolytes, and they are only slightly ionized in water. Stable complexes are formed between Hg^{2+} and the halide ions or sulfide ion; NH_3 and OH^- are not good ligands. Aqueous solutions containing soluble Hg^{2+} compounds are poisonous.

Relatively few mercury(II) compounds are colored. Notable exceptions include HgO (yellow), HgS (black), and HgI_2 (red). Black HgS turns red when heated, a form of HgS called cinnabar. Red HgI_2 dissolves in excess I^- to form the colorless complex ion HgI_4^{2-}.

$$HgI_2(s) + 2I^- \leftrightharpoons HgI_4^{2-} \tag{10.8}$$

Lead(II), Pb^{2+}

The general characteristics of lead(II), Pb^{2+}, are discussed in Chapter 9, because this ion is separated in the silver group ions. It is important to include these characteristics in a discussion of the copper-arsenic group ions as well because traces of Pb^{2+} may appear in this group if it is detected in the silver group. The relatively high solubility of $PbCl_2$ can lead to the appearance of Pb^{2+} in the copper-arsenic group.

Bismuth(III), Bi^{3+}

The most common oxidation state of bismuth in aqueous solutions is Bi^{3+}. The free ion is strongly hydrolyzed, forming insoluble basic salts; for instance, Eq. (10.9) represents the hydrolysis of $BiCl_3$.

$$Bi^{3+} + Cl^- + H_2O \leftrightharpoons BiOCl(s) + 2H^+ \tag{10.9}$$

Thus, solutions containing Bi^{3+} must be made highly acidic to keep this cation from precipitating as the oxychloride, $BiOCl_2$.

Bismuth can be oxidized to the 5+ state by very strong oxidizing agents; the most common compound containing Bi(V) is $NaBiO_3$, sodium bismuthate. This substance is, itself, an exceedingly strong oxidizing agent, being sufficiently powerful to oxidize Mn^{2+} to MnO_4^-.

The majority of bismuth compounds are colorless; the notable exception is Bi_2S_3, which is dark brown. Bismuth(III) forms soluble complex ions in concentrated solutions of halides; thus, Bi_2S_3 dissolves in 12 M HCl.

$$Bi_2S_3(s) + 8Cl^- \rightarrow 2BiCl_4^- + 3S^{2-} \tag{10.10}$$

Copper(II), Cu^{2+}

Copper can exist in the 2+, Cu^{2+}, and 1+, Cu^+, oxidation states, the most common being 2+. Cupric chloride, nitrate, and sulfate are common soluble Cu(II) salts; aqueous solutions of Cu(II) salts are blue, the color of the complex ion $Cu(H_2O)_4^{2+}$—which is always present in such systems. Most Cu(II) salts are blue solids, since they are hydrates and invariably contain the $Cu(H_2O)_4^{2+}$ complex ion. Copper(II) readily forms complexes with neutral (e.g., H_2O and NH_3) as well as anionic (e.g., Cl^- and CN^-) ligands.

$$Cu^{2+} + 4L \rightleftharpoons CuL_4^{2+}$$
$$L = NH_3, H_2O \tag{10.11}$$

$$Cu^{2+} + X^- \rightleftharpoons CuX_4^{2-}$$
$$X = Cl^-, CN^- \tag{10.12}$$

All Cu(I) salts are insoluble in water and all are colorless solids except Cu_2O (red) and Cu_2S (black). Copper(I) compounds are often formed as precipitates when soluble Cu(II) salts are reduced. For example, I^- is a sufficiently strong reducing agent to reduce Cu(II).

$$2Cu^{2+} + 4I^- \rightleftharpoons 2CuI(s) + I_2 \tag{10.13}$$

Copper(I) forms stable complexes with halide and cyanide ions and with ammonia; these complexes are readily oxidized in solution by atmospheric oxygen.

Cadmium(II), Cd^{2+}

The only common oxidation state of cadmium in aqueous systems is 2+. The halides, nitrate, sulfate, and acetate of cadmium are soluble in water. Most cadmium salts are colorless, the exceptions being CdS (yellow) and CdO (yellow brown). Cadmium(II) forms complex ions with common ligands (halides, CN^-, and NH_3).

$$Cd^{2+} + 4CN^- \rightleftharpoons Cd(CN)_4^{2-} \tag{10.14}$$

$$Cd^{2+} + 4NH_3 \rightleftharpoons Cd(NH_3)_4^{2+} \tag{10.15}$$

Arsenic(III), As^{3+}

Arsenic in its compounds exists in either the 3+ or 5+ oxidation states. Arsenic is rarely, if ever, present in a solution as As^{3+} or As^{5+}; usually it is present as arsenate (AsO_4^{3-}) or arsenite ($H_2AsO_3^-$). If $AsCl_3$ or $AsCl_5$ is dissolved in water, they immediately hydrolyze to give the weak acids H_3AsO_3 and H_3AsO_4 according to Eqs. (10.16) and (10.17).

$$AsCl_3 + 3H_2O \rightleftharpoons H_3AsO_3 + 3H^+ + 3Cl^- \tag{10.16}$$

$$AsCl_5 + 4H_2O \rightleftharpoons H_3AsO_4 + 5H^+ + 5Cl^- \tag{10.17}$$

When arsenic is present as arsenate or arsenite, the sulfides are precipitated according to the equations

$$2H_3AsO_4 + 5H_2S \rightleftharpoons As_2S_5(s) + 8H_2O \qquad (10.18)$$

$$2H_3AsO_3 + 3H_2S \rightleftharpoons As_2S_3(s) + 6H_2O \qquad (10.19)$$

Some arsenate may be reduced by H_2S to arsenite.

$$H_3AsO_4 + H_2S \rightleftharpoons H_3AsO_3 + H_2O + S \qquad (10.20)$$

The H_3AsO_3 then reacts with H_2S to form As_2S_3 [Eq. (10.11)]. Because the amount of As_2S_3 formed by this series of reactions is very small and because this small amount of As(III) is oxidized to As(V) by $(NH_4)_2S_2$ in Procedure 6 (as explained in Note 4 of Procedure 11), no As(III) species are shown in Figure 10.1 after the original listing of $H_2AsO_3^-$.

Each of these reactions is very slow. Each is speeded up (perhaps by catalysis), however, when the H^+ concentration is increased. To ensure complete precipitation of arsenic, the solution in the first stage of the sulfide precipitation is, therefore, strongly acidic.

Antimony(III), Sb^{3+}

Antimony exists in two oxidation states, Sb^{3+} and Sb^{5+}, in its common compounds. Antimony salts hydrolyze in water to form basic salts [as shown in Eq. (10.21)], which are often insoluble.

$$Sb^{3+} + H_2O + Cl^- \rightleftharpoons SbOCl(s) + 2H^+ \qquad (10.21)$$

Thus, many antimony salts cannot be dissolved in water because of reactions of the type shown in Eq. (10.21). However, antimony(III) tends to form soluble complex ions with halide or hydroxide ion. For example, although $SbCl_3$ is not soluble in water [Eq. (10.23)], it will dissolve in hydrochloric acid solution, forming $SbCl_6^{3-}$, or in bases, forming $Sb(OH)_6^{3-}$. The majority of antimony compounds are colorless; notable exceptions are Sb_2S_3 (orange red) and Sb_2S_5 (yellow).

If a solution containing members of the copper-arsenic group is diluted with water, a milky white suspension may result because of the formation, by hydrolysis, of the insoluble white basic salts of bismuth and antimony according to Eqs. (10.22) and (10.23).

$$Bi^{3+} + Cl^- + H_2O \rightleftharpoons BiOCl(s) + 2H^+ \qquad (10.22)$$

$$Sb^{3+} + Cl^- + H_2O \rightleftharpoons SbOCl(s) + 2H^+ \qquad (10.23)$$

If the milky suspension is treated with HCl, the milkiness disappears, because the point of equilibrium in these two reactions has been shifted to the left. In concentrated HCl solution, antimony exists as the complex tetrachloroantimonate(III) ion, $SbCl_4^-$. This ion is quite stable.

$$SbCl_4^- \rightleftharpoons Sb^{3+} + 4Cl^- \qquad (10.24)$$

The hydrolysis that takes place when a strongly acid solution of antimony chloride is added to water may, therefore, be represented by the equation

$$SbCl_4^- + H_2O \rightleftharpoons SbOCl(s) + 2H^+ + 3Cl^- \qquad (10.25)$$

Tin(II and IV), Sn^{2+} and Sn^{4+}

Pure compounds containing Sn(II)—stannous—and Sn(IV)—stannic—oxidation states are common. In aqueous solutions, both species are stable, but Sn(II) is slowly converted into Sn(IV) by air oxidation. Both tin oxidation states readily undergo hydrolysis to yield stannites [Sn(II)] or stannates [Sn(IV)]. Thus, common tin salts are insoluble in water because the hydrolysis products are insoluble basic salts such as $SnOCl_2$ [Eqs. (10.26) and (10.27)].

$$Sn^{2+} + 2Cl^- + H_2O \leftrightharpoons Sn(OH)Cl(s) + H^+ + Cl^- \tag{10.26}$$

$$Sn^{4+} + 4Cl^- + H_2O \leftrightharpoons SnOCl_2(s) + 2H^+ + 2Cl^- \tag{10.27}$$

The compounds of tin in either oxidation state are usually colorless; the most common exceptions are the sulfides SnS (brown) and SnS_2 (light yellow). Either oxidation state forms complex ions, the most useful ligands being halide and hydroxide ions. Thus, tin(IV) salts, although they undergo hydrolysis to form insoluble products, are soluble in hydrochloric acid and strong bases, where they form the complex ions $SnCl_6^{2-}$ and $Sn(OH)_6^{2-}$, respectively.

$$Sn^{4+} + 6Cl^- \leftrightharpoons SnCl_6^{2-} \tag{10.28}$$

$$Sn^{4+} + OH^- \leftrightharpoons Sn(OH)_6^{2-} \tag{10.29}$$

The Sn(IV) hydroxide complexes are sufficiently stable to cause SnS_2 to dissolve in 6 M NaOH.

$$SnS_2 + 6OH^- \leftrightharpoons Sn(OH)_6^{2-} + 2S^{2-} \tag{10.30}$$

PROCEDURE 5

Precipitation of the Copper-Arsenic Group (See Note 1)

(A) Place 4 drops of the solution to be tested (see Note 2) or the entire decantate from Procedure 1 in Chapter 9 in a casserole and add 2 drops of 3 percent H_2O_2 and 2 drops of 2 M HCl. Carefully boil down to a volume of 1 or 2 drops (see Note 3). Allow to cool. Then add 6 drops of 6 M HCl. Carefully evaporate the contents of the casserole down to a pasty mass, being very particular not to bake the residue (see Note 3). Cool, and then, *if H_2S gas is to be used as the precipitating reagent,* proceed as directed in (B); *if thioacetamide solution is to be used as the precipitating reagent,* proceed as directed in (C).

(B) *Follow this procedure if H_2S gas is to be used as the precipitating reagent.* To the residue in the cool casserole, add exactly 4 drops of 2 M HCl. Swirl the acid around until all the residue is dissolved; if necessary, stir the mixture and warm slightly. Transfer the solution to a 3-in. test tube, heat carefully until it begins to show signs of effervescence, and then treat with H_2S under the hood for 20 to 30 s (See Note 5). Dilute with 10 drops of hot water and treat with H_2S for another 20 to 30 s. Then add 1 drop of 1 M $NH_4C_2H_3O_2$ and treat with H_2S for 20 to 30 s more, being sure that the H_2S gas is bubbled all the way to the bottom of the solution. Finally, add 25 drops of cold water and treat with H_2S for another 20 to 30 s. Centrifuge; then test for complete precipitation by first noting the appearance of the solution and then passing H_2S into the top $\frac{1}{4}$ in. of the supernatant liquid, being careful not to disturb the precipitate. If precipitation is not complete, add 5 drops more of cold water and treat with H_2S for another 20 to 30 s. When tests with H_2S show that precipitation is complete, use 2 or 3 drops of water to wash down the precipitate on the walls of the test tube, centrifuge, and decant into a casserole. Boil the decantate, which contains the cations of the groups that follow in the scheme of analysis, for 1 min and save it for Procedure 15 in Chapter 11 (see Notes 6 and 8). Wash the precipitate twice with 15-drop portions of hot water and analyze according to Procedure 6. If the wash water peptizes the precipitate (causes it to be converted to a colloidal suspension that will not settle when centrifuged), add 10 drops of 1 M $NH_4C_2H_3O_2$, mix well, and heat nearly to boiling before centrifuging.

(C) *Follow this procedure if thioacetamide solution is used as the precipitating reagent.* To the residue in the cool casserole, add exactly 4 drops of 2 M HCl. Swirl the acid around until all the residue is dissolved; if necessary, stir the mixture and warm slightly. Then transfer the solution to a 3-in. test tube, add 4 drops of 1 M thioacetamide solution (see Note 7), mix thoroughly, and heat in a boiling water bath for 4 min (see Note 6). Add 8 drops of hot water, 8 drops of 1 M thioacetamide, and 1 drop of 1 M $NH_4C_2H_3O_2$; mix well and heat in the boiling water bath for 4 min. Centrifuge and decant into a test tube; save both the precipitate and the decantate. Test the decantate for complete precipitation by adding 2 drops of 1 M thioacetamide, mixing well, and allowing to stand for 1 min; if a precipitate forms, indicating that precipitation is not complete, add 2 drops of hot water and 2 drops more of 1 M thioacetamide; mix well and heat in the boiling water bath for 2 min. When a test shows that precipitation is complete, transfer the decantate to a casserole. Boil this decantate, which contains the cations of the following qualitative groups, for 1 min and save it for Procedure 15 in Chapter 11 (see Note 8).

All precipitates obtained in the original precipitation and the successive tests for complete precipitation should be combined by using a few drops of water to flush the latter into the test tube containing the original precipitate. Wash the precipitate three times, once with 10 drops of hot water and twice with 20-drop portions of a hot solution prepared using equal volumes of water and 1 M $NH_4C_2H_3O_2$, and analyze according to Procedure 6. If the wash solution causes the precipitate to go into colloidal suspension, add 10 drops of 1 M $NH_4C_2H_3O_2$ to the suspension, mix well, and heat nearly to boiling; centrifuge, saving the precipitate for Procedure 6. In all the washing operations, be sure that the precipitate and washing liquid are well mixed by vigorous agitation with a stirring rod.

Notes

1. In Procedure 5, the initial H^+ ion concentration is 2 M. In the course of the precipitation, about a tenfold dilution occurs, which should reduce the H^+ ion concentration to 0.2 M. However, H^+ ions are liberated in the precipitation reactions ($Hg^{2+} + H_2S \leftrightharpoons HgS + 2H^+$). As a result, the final H^+ ion concentration is slightly higher than 0.2 M. The H^+ ion concentration range of 2 M to 0.2 M allows for complete precipitation of the copper-arsenic group without precipitation of the cations of the aluminum-nickel group.

    The sulfides of mercury, lead, and copper are black; SnS and Bi_2S_3 are dark brown, but look black when wet. CdS, SnS_2, and As_2S_5 are yellow. Sb_2S_3 is orange. Lead may first form an orange red precipitate of $PbS \cdot PbCl_2$; this changes to black PbS on continued H_2S treatment. Mercury first precipitates as white $HgS \cdot HgCl_2$; this changes through yellow, orange, and brown, to black HgS on continued H_2S treatment, the intermediate colored precipitate being a mixture of $HgS \cdot HgCl_2$ and HgS in varying proportions.

    $CuCl_2$ is green in concentrated solution and blue in dilute solution. The chlorides of all other members of the copper-arsenic group form colorless solutions, except that $BiCl_3$

and $SbCl_3$ hydrolyze as given by Eqs. (10.22) and (10.23). The complete absence of any green or blue color in an unknown solution indicates the absence of copper.

2. If *only* members of the copper-arsenic group are present, the following procedure—which saves considerable time without loss of accuracy of results—may be used.

Place 4 drops of the solution containing the copper-arsenic group cations in a test tube. Add 6 drops of H_2O_2; add 10 drops of ammonium sulfide solution. Then add more ammonium sulfide until the solution turns yellow (about 30 drops). Stir the contents of the tube.

Stir the contents of the tube well and then heat for 3 to 4 min in a boiling water bath while stirring the contents. Avoid heating the tube to the point where excessive foaming of the contents occurs. Centrifuge, and decant into a test tube, and save the decantate, which may contain AsS_4^{3-}, SbS_3^{3-}, and SbS_3^{2-}, for Procedure 11. Repeat the treatment of the precipitate with a second 10-drop portion of ammonium sulfide solution, heating the tube in a boiling water bath for 2 min while stirring. Centrifuge and decant, combining the second decantate with the first; save it in a stoppered test tube for Procedure 11. Wash the precipitate twice with 20-drop portions of a hot solution prepared by mixing equal volumes of water and 1 M $NH_4C_2H_3O_2$ and analyze this precipitate, which may consist of the sulfides of mercury(II), lead, bismuth, copper, and cadmium, according to Procedure 7.

Remember the difference between a true solution and a colloidal solution when you have to determine if the solution is yellow.

3. Evaporate the solution in the casserole using the following procedure: Hold the casserole in your hand and pass it back and forth over the top of the flame. At the end of every two or three back-and-forth passes, tilt the casserole slightly so that the solution will run to its lower edge. In the course of the back-and-forth pass, the solution is swished around over the bottom of the casserole. Be very careful not to overheat to the point where the residue begins to bake. If brown areas develop on the bottom of the casserole, indicating baking, swish the remaining solution around until the brown area is removed. When only 2 or 3 drops of liquid remain, remove from above the flame and let the heat of the casserole complete the evaporation. Baking the residue must be avoided; baking may sublime off the chlorides of arsenic, mercury, and tin.

4. Six drops of 6 M HCl are added before evaporation to dryness so that nitrate ions may be reduced in accordance with the equation

$$2NO_3^- + 6Cl^- + 8H^+ \leftrightharpoons 4H_2O + 2NO + 3Cl_2$$

Nitrate ions, if present, will oxidize the H_2S; S is precipitated.

$$2NO_3^- + 3H_2S + 2H^+ \leftrightharpoons 4H_2O + 2NO + 3S(s)$$

Other oxidizing agents, such as Fe^{3+} or $Cr_2O_7^{2-}$, may be present in a solution to be analyzed. They will react with H_2S in acid solution as follows (see Table 6.1).

$$2Fe^{3+} + H_2S \leftrightharpoons 2Fe^{2+} + 2H^+ + S(s)$$

$$Cr_2O_7^{2-} + 3H_2S + 8H^+ \leftrightharpoons 2Cr^{3+} + 3S(s) + 7H_2O$$

Fe^{3+} will not be affected by evaporation with 6 M HCl. $Cr_2O_7^{2-}$ will be reduced as follows:

$$Cr_2O_7^{2-} + 6Cl^- + 14H^+ \leftrightharpoons 2Cr^{3+} + 3Cl_2 + 7H_2O$$

5. A solution is treated with H_2S gas as follows: Attach a clean glass bubbling tube to the rubber outlet tube at the source of H_2S. This bubbling tube is made by drawing down a glass tube of suitable diameter to a fairly fine constricted end. The overall length of the bubbling tube should be about 5 in. Insert the end of the bubbling tube from which H_2S is escaping into the surface of the solution in the test tube and then gradually bring it down to the bottom of the solution. The constricted tube will deliver a stream of very fine bubbles; large bubbles would tend to throw the solution out of the small test tube. If the tip of the bubbling tube is brought suddenly all the way to the bottom of the solution, the sudden rush of gas may throw the solution out of the test tube. A very rapid rate of H_2S bubbling should be avoided.

An ordinary Kipp generator, placed in a hood where it is accessible to a group of students, is a very satisfactory source of H_2S. A trap, in the form of a bottle or flask, should be placed between the generator valve and the point where the bubbling tube is attached.

One of the most satisfactory sources of H_2S is the commercial mixture of sulfur, hydrocarbon mix, and asbestos, of which "Aitch-Tu-Ess" is an example. This mixture can be purchased in bulk or in small capsules; for class use bulk purchase is recommended. It is used for generation of H_2S as follows. Set up a generator under the hood, as illustrated in Figure 10.2. Fill the 6-in. Pyrex tube, A, one-half- to three-quarters full of H_2S-generating mixture. Insert the rubber stopper, B, to which is attached the delivery tube, C, the rubber connecting tube, D, and the clean bubbling tube, E. When heat is applied to A, H_2S gas will be generated and will issue from the tip of E, which is inserted into the solution to be treated. The heating should be gentle, just strong enough to give a fairly rapid evolution of H_2S bubbles. The burner with which A is being heated can be held in one hand, while the test tube containing the solution being treated can be held in the other hand. Do not heat so strongly that sulfur is distilled over. The evolution of H_2S will cease when heating is stopped.

The tube, E, should be removed from the solution being treated as soon as heating is stopped to avoid liquid being drawn from the test tube into E.

The tube, E, should be removed and cleaned after each series of precipitations. Test tube, A, containing unexpended mixture together with the attachments, B, C, and D, can be

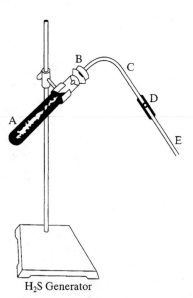

FIGURE 10.2 H_2S generator; see text for detailed description.

H_2S Generator

kept in the student's desk when not in use. When the mixture no longer yields H_2S on heating, tube A can be replaced by new mixture.

The setup described is intended for use by one or two students. If a larger tube is used at A, the generator will serve several students.

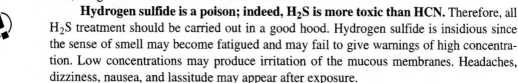

Hydrogen sulfide is a poison; indeed, H_2S is more toxic than HCN. Therefore, all H_2S treatment should be carried out in a good hood. Hydrogen sulfide is insidious since the sense of smell may become fatigued and may fail to give warnings of high concentration. Low concentrations may produce irritation of the mucous membranes. Headaches, dizziness, nausea, and lassitude may appear after exposure.

6. Care should be taken so that the tube is not heated to the point at which excessive foaming results in loss of solution by spillage. If the tube is grasped with a test-tube holder, it can be withdrawn from the boiling water if and when spillage is imminent.

7. The organic compound thioacetamide, CH_3CSNH_2, hydrolyzes in water—particularly at higher temperatures—to yield H_2S according to the equation

$$CH_3CSNH_2 + 2H_2O \leftrightarrows CH_3COOH + NH_3 + H_2S$$

It, therefore, serves as a convenient source of H_2S. The hydrolysis is so slight at room temperature that a 1 M solution of the compound, when preserved in a stoppered bottle, undergoes very little deterioration. At temperatures of about 80°C, the hydrolysis is sufficient so that a 1 M solution yields a solution saturated with H_2S.

8. If the solution being analyzed is a copper-arsenic group known or unknown containing only cations of the copper-arsenic group, this decantate can be discarded.

SEPARATION OF THE COPPER SUBGROUP
FROM THE ARSENIC SUBGROUP

The sulfides of arsenic, antimony, and tin(IV) are soluble in a solution of ammonium sulfide. The sulfides of mercury, lead, bismuth, copper, and cadmium are insoluble. This fact is the basis for the separation of the copper subgroup from the arsenic subgroup.

The statement has been made that, as a general rule, the water-insoluble salt of a weak acid will dissolve in a strong acid (see Chapter 7). By this rule the metal sulfides should dissolve in HCl. The fact that they do not dissolve means that their normal solubility in water is so very small that not even a strong acid can give a concentration of hydrogen ions sufficiently high to cause them to go into solution. Taking CuS as an example, its solubility is very low—its solubility product is 6×10^{-37}; this means that the concentration of sulfide ions in the equilibrium is extremely low.

$$CuS \leftrightarrows Cu^{2+} + S^{2-} \qquad (10.31)$$

$$S^{2-} + 2H^+ \leftrightarrows H_2S \qquad (10.32)$$

The equilibrium reaction that is set up when a strong acid is added to the sulfide is not capable of giving a sulfide ion concentration lower than that in equilibrium with CuS and Cu^{2+}. Consequently, CuS does not dissolve.

On prolonged boiling with HCl, CuS will dissolve very slowly. In this case, the S^{2-} ions are actually removed from solution as H_2S, which is a volatile product. Over time, H_2S

is removed by boiling, shifting the point of equilibrium in (10.32) to the right, which, in turn, causes a shift in the point of equilibrium in (10.31) to the right. The net effect is that CuS dissolves slowly in boiling acid solution.

It should be noted at this point that the solubility product principle tells very little about the *rate* at which an "insoluble" substance will dissolve.

Ammonium sulfide, $(NH_4)_2S$, will dissolve As_2S_3, As_2S_5, Sb_2S_3, and SnS_2, but will not dissolve SnS. Therefore, if tin is to go into solution in $(NH_4)_2S$ along with arsenic and antimony as the separation requires, it must be present as SnS_2. Thus, it must be present as Sn(IV) during treatment with H_2S. To guarantee that tin is present as Sn(IV), the solution is boiled with H_2O_2, which oxidizes Sn^{2+} to Sn^{4+}, before precipitation with H_2S (see Procedure 5, Note 2).

The separation of arsenic, antimony, and tin from mercury(II), lead, bismuth, copper, and cadmium depends on the fact that As_2S_3, As_2S_5, SnS_2, and Sb_2S_3 are soluble in a solution of $(NH_4)_2S$, whereas HgS, PbS, Bi_2S_3, CuS, and CdS are insoluble. To speed up the rate of the solution process, the contents of the tube in Procedure 6 are kept hot. The molecular equations for the reactions that take place are

$$SnS_2 + (NH_4)_2S \rightleftharpoons (NH_4)_2SnS_3 \qquad \text{[ammonium trithiostannate (IV)]} \qquad (10.33)$$

$$As_2S_5 + 3(NH_4)_2S \rightleftharpoons 2(NH_4)_3AsS_4 \qquad \text{[ammonium tetrathioarsenate (V)]} \qquad (10.34)$$

The soluble compounds $(NH_4)_2SnS_3$, $(NH_4)_3AsS_3$, $(NH_4)_3AsS_4$, and $(NH_4)_3SbS_3$ are strong electrolytes and ionize to give NH_4^+ ions and SnS_3^{2-}, AsS_4^{3-}, and SbS_3^{3-} ions, respectively. The formation of the latter ions can be represented by the net equations

$$As_2S_3 + 3S^{2-} \rightleftharpoons 2AsS_3^{3-} \qquad \text{[trithioarsenate (III) ion]} \qquad (10.35)$$

$$Sb_2S_3 + 3S^{2-} \rightleftharpoons 2SbS_3^{3-} \qquad \text{[trithioantimonate (III) ion]} \qquad (10.36)$$

These ions are classed as complex ions (Chapter 5).

Mercury(II), lead, bismuth, copper, and cadmium do not form complex thio ions. Therefore, the separation of the sulfides of arsenic, antimony, and tin from the sulfides of mercury(II), lead, bismuth, copper, and cadmium depends on the fact that the sulfides of arsenic, antimony, and tin form complex ions with sulfide ion, whereas the sulfides of the other metals in the group do not form such complex ions.

A careful examination of the formulas for the thio salts of arsenic, antimony, and tin and the molecular equations for their formation suggests why these sulfides can be dissolved by $(NH_4)_2S$. In the sulfur system of compounds, $(NH_4)_2SnS_3$ is the ammonium salt of thiostannic acid, H_2SnS_3. It is the analog of the ternary salt $(NH_4)_2SnO_3$ in the oxygen system. Any ternary salt in the oxygen system may be considered to have been formed as a result of the reaction of a basic oxide with an acidic oxide. The oxides of metals are basic oxides; the oxides of nonmetals are acidic oxides.

Basic oxide	Acidic oxide		Salt
CaO	$+ CO_2$	$\rightleftharpoons$	$CaCO_3$
PbO	$+ N_2O_5$	$\rightleftharpoons$	$Pb(NO_3)_2$
$(NH_4)_2O$	$+ SnO_2$	$\rightleftharpoons$	$(NH_4)_2SnO_3$
$3(NH_4)_2O$	$+ As_2O_3$	$\rightleftharpoons$	$2(NH_4)_3AsO_3$

The sulfides of the elements bear the same relationship to H_2S that the oxides bear to H_2O.

$$\begin{array}{ll} CuO & H_2O \\ CuS & H_2S \\ As_2O_3 & H_2O \\ As_2S_3 & H_2S \end{array}$$

Therefore, CuS, PbS, HgS, CdS, SnS, Bi_2S_3, and $(NH_4)_2S$ are basic sulfides in the sulfur system of compounds, just as CuO, PbO, HgO, CdO, SnO, Bi_2O_3, and $(NH_4)_2O$ are basic oxides in the oxygen system. Likewise, SnS_2, As_2S_3, As_2S_5, and Sb_2S_3 are acidic sulfides in the sulfur system, just as SnO_2, As_2O_3, As_2O_5, and Sb_2O_3 are acidic oxides in the oxygen system.

Just as a basic oxide will react with an acidic oxide to form a salt in the oxygen system, so a basic sulfide will react with an acidic sulfide to form a salt in the sulfur system. The substances SnS_2, As_2S_3, As_2S_5, and Sb_2S_3 are acidic sulfides, while $(NH_4)_2S$ is a basic sulfide. Therefore, the former would be expected to react with the latter to form the salts $(NH_4)_2SnS_3$, $(NH_4)_3AsS_3$, $(NH_4)_3AsS_4$, and $(NH_4)_3SbS_3$ as, indeed, is observed.

In the oxygen system, one basic oxide does not ordinarily react with another basic oxide. By analogy, one basic sulfide will not react with another basic sulfide. Since $(NH_4)_2S$ is a basic sulfide, it would not be expected to react with the basic sulfides HgS, PbS, Bi_2S_3, CuS, CdS, and SnS. Therefore it seems justifiable to conclude that separation of the sulfides of arsenic, antimony, and tin(IV) from the sulfides of mercury, lead, bismuth, copper, and cadmium depends on the fact that the former, being *acidic sulfides, will dissolve in the basic sulfide* $(NH_4)_2S$, while the latter, being *basic sulfides, will not dissolve in the basic sulfide* $(NH_4)_2S$.

There remains the question of why SnS is basic, while SnS_2 is acidic. It is generally true that *as the oxidation number of a polyvalent element increases, it becomes more nonmetallic in character, and as an element becomes more nonmetallic in character, its compounds become more acidic.* Chromium and manganese illustrate this generalization. Chromium(III) forms the cation Cr^{3+}, which is definitely metallic in character and, therefore, basic; chromium(VI) forms the anion CrO_4^{2-}, which is an acid radical. Manganese(II) gives the metallic cation Mn^{2+}, while manganese(VII) forms the acidic anion MnO_4^-. Likewise, the Bi^{3+} ion is metallic, while the bismuthate ion BiO_3^- is definitely acidic. We can conclude, therefore, that in its divalent state tin is predominantly metallic; hence, SnS is a basic sulfide. Tin(IV), on the other hand, is nonmetallic; therefore, SnS_2 is acidic. Oxidation with H_2O_2 guarantees that the tin is present in its higher oxidation state. Antimony(V) does not ordinarily exist in solution, its oxidation state of $5+$ being very unstable.

PROCEDURE 6

Separation of the Copper Subgroup from the Arsenic Subgroup

To the test tube containing the precipitate from Procedure 5, add 10 drops of ammonium sulfide solution. Stir the contents of the tube well; then heat for 3 to 4 min in a boiling water bath, while stirring the contents. Avoid heating the tube to the point where excessive foaming of the contents occurs. Centrifuge, decant into a test tube, and save the decantate, which may contain AsS_4^{3-}, SbS_4^{3-}, SbS_3^{2-}, for Procedure 11. Repeat the treatment of the precipitate with a second 10-drop portion of ammonium sulfide solution, heating the tube in the boiling water bath for 2 min, while stirring. Centrifuge and decant, combining the second decantate with the first; save it in a stoppered test tube for Procedure 11.

Wash the precipitate twice with 20-drop portions of a hot solution prepared by mixing equal volumes of water and 1 M $NH_4C_2H_3O_2$; analyze this precipitate, which may consist of the sulfides of mercury(II), lead, bismuth, copper, and cadmium, according to Procedure 7.

SEPARATION OF MERCURY FROM LEAD, BISMUTH, COPPER, AND CADMIUM

The sulfides of copper, bismuth, cadmium, and lead are soluble in warm dilute HNO_3, while HgS is insoluble. The separation of mercury from lead, bismuth, copper, and cadmium is based on this fact.

The sulfides of lead, bismuth, copper, and cadmium dissolve in HNO_3 in accordance with the equation

$$3CuS + 2NO_3^- + 8H^+ \rightleftharpoons 3Cu^{2+} + 3S(s) + 2NO + 4H_2O \tag{10.37}$$

The detailed reactions involved are

$$CuS \rightleftharpoons Cu^{2+} + S^{2-} \tag{10.38}$$

$$3S^{2-} + 2NO_3^- + 8H^+ \rightleftharpoons 3S(s) + 2NO + 4H_2O \tag{10.39}$$

Reaction (10.39) removes the S^{2-} ions from the equilibrium described in Eq. (10.38) and causes it to shift to the right.

PROCEDURE 7

Separation and Detection of Mercury(II)

Add 15 drops of 3 M HNO_3 to the test tube containing the precipitate from Procedure 6, mix thoroughly, transfer to a casserole, and boil gently for about a minute (see Note 1). Replenish the HNO_3 when necessary. If a solid remains, transfer to a test tube, centrifuge, and decant into a test tube, saving this decantate for Procedure 8. Wash the precipitate [HgS(s) and S(s)] twice with 15-drop portions of water made acidic with a drop of 3 M HNO_3 (see Note 2). Add the first washing to the decantate in the test tube; discard the second washing.

Treat the precipitate with 6 drops of 12 M HCl and 2 drops of 16 M HNO_3, mix thoroughly, and heat for 1 min in the boiling water bath (see Note 3). Add 10 drops of hot water, transfer to a casserole, boil gently for about 30 s, and then transfer back to a test tube. Cool by holding the test tube under the water tap; then centrifuge and decant into another test tube. To the cool decantate in the test tube, add 2 to 5 drops of 0.2 M $SnCl_2$ solution. A black (Hg) or gray (Hg_2Cl + Hg) precipitate proves the presence of mercury(II) (see Note 4).

Notes

1. If the HNO_3 is too concentrated, it will dissolve some of the HgS. In addition, it may oxidize PbS to $PbSO_4$, which will remain as a residue mixed with the HgS.
2. HgS is black. The presence of a white or yellow residue after digestion with HNO_3 does not, however, eliminate the possibility of mercury being present. The residue, whatever its color, should be tested for mercury.

3. HgS is insoluble in both concentrated HNO_3 and concentrated HCl when each is used separately. However, a mixture of the two concentrated acids dissolves HgS readily and quickly. The detailed reaction is discussed in Chapter 8.

   The net equation for the overall reaction is

   $$3HgS + 2NO_3^- + 8H^+ + 12Cl^- \rightleftharpoons 3HgCl_4^{2-} + 2NO + 4H_2O + 3S(s)$$

   The pale yellow $HgCl_4^{2-}$ ion is largely dissociated into colorless Hg^{2+} and Cl^- ions when the solution is diluted with water. The Hg^{2+} ions can then be identified by means of the reaction represented in Note 4.

4. The extent to which Hg^{2+} is reduced by Sn^{2+} depends on the relative amounts of the two kinds of ions present. If Hg^{2+} is present in great excess, the reduction is mostly to Hg_2Cl_2, as follows:

   $$2Hg^{2+} + Sn^{2+} + 2Cl^- \rightleftharpoons Hg_2Cl_2(s) + Sn^{4+}$$
   $$\text{(white)}$$

   If more Sn^{2+} ions are added, the Hg_2Cl_2 that is first formed according to the reaction above is further reduced to Hg, as follows:

   $$Hg_2Cl_2(s) + Sn^{2+} \rightleftharpoons 2Hg(l) + Sn^{4+} + 2Cl^-$$
   $$\text{(black)}$$

   If Sn^{2+} ions are present in great excess, the following reaction will take place.

   $$Hg^{2+} + Sn^{2+} \rightleftharpoons Hg(l) + Sn^{4+}$$
   $$\text{(black)}$$

   Since $SnCl_2$ solution is added as a reagent, Sn^{2+} ions are generally present in excess. Therefore the precipitate that is formed is dark gray or black.

   These equations illustrate the fact that tin(II) compounds are strong reducing agents. The solution is boiled to remove chlorine formed by oxidation of Cl^- by NO_3^-. Chlorine would oxidize Sn^{2+} to Sn^{4+}, thus, destroying the reductant (Sn^{2+}) and preventing it from reducing the Hg^{2+}.

SEPARATION OF LEAD FROM BISMUTH, COPPER, AND CADMIUM

Lead sulfate is insoluble in water. The sulfates of bismuth, copper, and cadmium are soluble. This fact is the basis for the separation of lead ions from bismuth, copper, and cadmium ions.

PROCEDURE 8

Separation and Detection of Lead

Add 4 drops of 18 M H_2SO_4 to a casserole containing the decantate from Procedure 7 and evaporate **(in a hood)** carefully until the volume is about 1 drop and *dense white fumes of SO_3 are formed*. These fumes are so dense and heavy that they will fill the casserole and completely obscure the view of its inside walls (see Note 1). More pale fumes, which appear as the HNO_3 boils off during the evaporation, must not be mistaken for the SO_3 fumes; the latter are unmistakable when they do appear and begin to billow out of the casserole. Cool, add 15 drops of cold water, and stir the contents until all material in the casserole is dissolved or suspended; then transfer it quickly to a test tube before the suspended material has a chance to settle; swirl the casserole with 4 drops of cold water and transfer this washing to the same test tube. Cool under the water tap. A white precipitate ($PbSO_4$) in the form of a fine suspension shows the presence of lead (see Note 2). Centrifuge until the supernatant liquid is clear and decant into a test tube, saving this decantate for Procedure 9. Wash the precipitate twice with 10-drop portions of cold water. To the washed precipitate in the test tube, add 4 drops of 1 M $NH_4C_2H_3O_2$ and stir for 20 s; then add 2 drops of 0.2 M K_2CrO_4. A yellow precipitate ($PbCrO_4$) confirms the presence of lead (see Note 3).

Notes

1. $PbSO_4$ is appreciably soluble in concentrated HNO_3 due to formation of the hydrogen sulfate ion in the manner discussed in Note 2. For this reason, HNO_3 must be removed before $PbSO_4$ will precipitate. When a solution containing H_2SO_4, HNO_3, and water is boiled, the water and HNO_3 are first driven off because they boil at comparatively low temperatures (100–120°C). After they are removed, further heating results in boiling the H_2SO_4 that remains (boiling point of sulfuric acid is 338°C). At its boiling temperature the H_2SO_4 decomposes to a slight extent.

$$H_2SO_4 \rightleftharpoons H_2O + SO_3$$

SO_3 fumes strongly in moist air. Therefore, formation of dense white fumes of SO_3 at the end of the evaporation gives assurance that all HNO_3 has been removed.

2. When the solution that has been boiled down with concentrated H_2SO_4 is cooled, sulfates of bismuth, copper, and cadmium may crystallize out. However, they are soluble in dilute H_2SO_4 and will dissolve when water is added. $PbSO_4$, on the other hand, is quite soluble in concentrated H_2SO_4 due to formation of the HSO_4^- ion.

$$PbSO_4(s) \rightleftharpoons SO_4^{2-} + Pb^{2+}$$

$$SO_4^{2-} + H_2SO_4 \rightleftharpoons 2HSO_4^-$$

On dilution with water, the equilibria are shifted to the left, and $PbSO_4$ precipitates. The precipitate of $PbSO_4$ is very finely divided. The sulfates of bismuth, copper, and cadmium form relatively large crystals.

3. $PbSO_4$ dissolves in ammonium acetate because the complex ion $Pb(C_2H_3O_2)_4^{2-}$, which is formed in the following reaction, is very stable, thus giving a solution with a very low concentration of lead ions.

$$PbSO_4(s) + 4C_2H_3O_2^- \rightleftharpoons Pb(C_2H_3O_2)_4^{2-} + SO_4^{2-}$$

$PbCrO_4$ is less soluble than $PbSO_4$. Therefore, when K_2CrO_4 is added to the solution formed by adding $NH_4C_2H_3O_2$ to $PbSO_4$, a precipitate of $PbCrO_4$ is formed even though the concentration of lead ion in the solution is very low.

Since lead is largely removed in the silver group, only very small quantities will ordinarily appear in the copper-arsenic group. For this reason, the test for lead may not be as strong as the test for other cations in this group.

SEPARATION OF BISMUTH FROM COPPER AND CADMIUM

Addition of NH_4OH to a solution containing bismuth, copper, and cadmium ions first precipitates the hydroxides of all three metals. The hydroxides of copper and cadmium, however, dissolve in an excess of NH_4OH, whereas the hydroxide of bismuth does not; the separation of bismuth ions from copper and cadmium ions is based on this fact.

PROCEDURE 9

Separation and Detection of Bismuth

To the decantate from Procedure 8 add 15 M NH_3, dropwise and with constant mixing, until it becomes distinctly alkaline. Stir for 2 min (see Note 1). Centrifuge and decant, saving the decantate for Procedure 10. Wash the precipitate twice with 15-drop portions of hot water. To the washed precipitate add 3 drops of 8 M NaOH and 2 drops of 0.2 M $SnCl_2$ and stir (see Note 2). A jet-black precipitate (Bi) proves the presence of bismuth.

Notes

1. When NH_4OH is added to a solution containing copper ions, $Cu(OH)_2$ is first precipitated.

$$Cu^{2+} + 2NH_4OH \leftrightharpoons Cu(OH)_2(s) + 2NH_4^+$$
$$\text{(blue)}$$

An excess of NH_4OH, however, dissolves the $Cu(OH)_2$, to give a deep blue solution in which copper is present as the complex tetramminecopper(II) ion, $Cu(NH_4)_4^{2+}$.

$$Cu(OH)_2(s) + 4NH_4OH \leftrightharpoons Cu(NH_3)_4^{2+} + 2OH^- + 4H_2O$$

The detailed reaction in which $Cu(OH)_2$ is dissolved by excess NH_4OH proceeds in the manner already discussed in Chapter 7.

a. $Cu(OH)_2(s) \leftrightharpoons Cu^{2+} + 2OH^-$

b. $Cu^{2+} + 4NH_3 \leftrightharpoons Cu(NH_3)_4^{2+}$

Cadmium behaves in the same manner as copper; $Cd(NH_3)_4^{2+}$ is formed. $Cd(OH)_2$ (white) dissolves quite slowly in excess NH_4OH; therefore, the solution should be stirred for 2 min to ensure complete solution of $Cd(OH)_2$. Similar complexes will be met in the case of the ions of nickel, cobalt, and zinc. $Ag(NH_3)_2^+$ was formed in the analysis of the silver group (see Chapter 9).

2. The formation of the black precipitate of bismuth results from the action of stannate(II) ions $[Sn(OH)_4^{2-}]$ on $Bi(OH)_3$ (white):

$$2Bi(OH)_3(s) + 3Sn(OH)_4^{2-} \rightleftharpoons 2Bi(s) + 3Sn(OH)_6^{2-}$$
$$\text{(white)} \qquad\qquad\qquad \text{(black)}$$

The stannate(II) ions were formed when the $SnCl_2$, added as the final reagent, reacted with the excess of NaOH that had been added previously.

$$Sn^{2+} + 4OH^- \rightleftharpoons Sn(OH)_4^{2-}$$

The reaction of stannate(II) ions with $Bi(OH)_3$ to give stannate(IV) ions $[Sn(OH)_6^{2-}]$ and bismuth serves further to illustrate the reducing character of tin(II) compounds.

The formation of stannate(II) ions $[Sn(OH)_4^{2-}]$ by the action of NaOH on $SnCl_2$ is a two-stage reaction. If NaOH is added dropwise to a solution of $SnCl_2$, a white precipitate of $Sn(OH)_2$ is first formed.

$$Sn^{2+} + 2OH^- \rightleftharpoons Sn(OH)_2(s)$$
$$\text{(white)}$$

This white precipitate redissolves in more NaOH to form a clear, colorless solution containing tetrahydroxostannate(II) ions. The following reaction is believed to take place.

$$Sn(OH)_2(s) + 2OH^- \rightleftharpoons Sn(OH)_4^{2-}$$
$$\text{(white)}$$

If the white precipitate of $Sn(OH)_2$ that first forms when NaOH is added to a solution of $SnCl_2$ is treated with an acid, such as HCl, it dissolves to form a clear solution that can be shown to contain Sn^{2+} ions. The equation for the reaction that takes place is therefore

$$Sn(OH)_2(s) + 2H^+ \rightleftharpoons Sn^{2+} + 2H_2O$$
$$\text{(white)}$$

A metal hydroxide such as $Sn(OH)_2$, which will dissolve in either a strong acid (HCl) or a strong base (NaOH), is said to be *amphoteric* (see Chapter 5). In addition to $Sn(OH)_2$, the following hydroxides encountered in qualitative analysis are amphoteric: $Sn(OH)_4$, $Pb(OH)_2$, $Sb(OH)_3$, $Al(OH)_3$, $Cr(OH)_3$, $Zn(OH)_2$. The hydroxides of the other metals that we deal with in this scheme of analysis are not amphoteric.

It should be pointed out that there is not complete agreement among chemists about the composition of the stannate(II) ion. In addition to $Sn(OH)_4^{2-}$ and SnO_2^{2-}, the formulas $Sn(OH)_3^-$ and $HSnO_2^-$ have been suggested. Since the formula of the ion may change as the concentration of the solution changes, it may well be that all four of the ions above are present at one time or another. In fact, they may all be present, in equilibrium, at one time.

Stannate(II) ions will reduce the hydroxides of antimony, lead, copper, and cadmium to the corresponding metal. These hydroxides, however, are reduced slowly and the metallic deposit is not jet black, whereas $Bi(OH)_3$ is reduced instantly and forms a jet-black deposit of metallic bismuth.

On standing in contact with air, stannate(II) ions are rapidly oxidized to stannate(IV) ions [$Sn(OH)_6^{2-}$]. The equation for the reaction is (see Table 6.2)

$$2Sn(OH)_4^{2-} + O_2 + 2H_2O \leftrightharpoons 2Sn(OH)_6^{2-}$$

Furthermore, stannate(II) reacts with itself in solution, as follows:

$$2Sn(OH)_4^{2-} \leftrightharpoons Sn(s) + Sn(OH)_6^{2-} + 2OH^- \tag{10.40}$$

For these reasons, sodium stannate(II) is not kept on the shelf as a reagent but is formed by the action of NaOH on $SnCl_2$ at the time it is to be used.

A reaction of the type illustrated in Eq. (10.40), in which a substance acts as both an oxidizing agent and a reducing agent, is referred to as disproportionation. One mole of $Sn(OH)_4^{2-}$ is reduced to Sn, and the other is oxidized to $Sn(OH)_6^{2-}$ [see Eq. (9.9)].

DETECTION OF COPPER AND CADMIUM

PROCEDURE 10

Separation and Detection of Copper and Cadmium

(A) Detection of copper. If the decantate from Procedure 9 is colorless, copper may or may not be absent; solutions containing copper in amounts less than 1 part in 25,000 may appear colorless. If the decantate is deep blue, the $Ca(NH_3)_4^{2+}$ ion, copper is present (see Note 1). In either case, it is wise to verify the presence of copper. Place 5 drops of the decantate in a test tube, add 5 M $HC_2H_3O_2$ until the solution is decolorized and then add 2 drops of 0.2 M $K_4Fe(CN)_6$. A red precipitate [$Cu_2Fe(CN)_6$] confirms the presence of copper (see Note 2).

(B) Detection of cadmium. If copper is absent, treat the colorless decantate from Procedure 9 with 2 or 3 drops of ammonium sulfide solution, mix thoroughly, and allow to stand for about 1 min. The formation of a yellow precipitate (CdS) proves the presence of cadmium.

If copper is present, add 0.2 M KCN (**Caution! See Note 3**), dropwise, to a 10-drop portion of the blue decantate until the color disappears. Treat the solution with 2 or 3 drops of ammonium sulfide solution, mix thoroughly, and allow to stand for about 1 min (see Note 4). A yellow precipitate (CdS) proves the presence of cadmium. As soon as the test for cadmium has been completed (and, for the "known," approved), dispose of the contents of the tube to which the KCN was added in the manner described by your instructor; special provisions for the disposal of KCN wastes will have been made.

Notes

1. The blue color of $Cu(NH_3)_4^{2+}$ is visible when as little as 1 part of copper is present in 25000 parts of water. The red precipitate of $Cu_2Fe(CN)_6$ will detect 1 part of copper in 1 million parts of water.
2. $Cu_2Fe(CN)_6$ is soluble in strong acids, such as HCl and H_2SO_4, but precipitates readily in the presence of a weak acid, such as acetic acid. $Cd_2Fe(CN)_6$ precipitates under the same conditions as does $Cu_2Fe(CN)_6$, but it is white.

3. **KCN is poisonous;** for that reason, KCN solution is not kept on the regular reagent shelf but is dispensed by the stockroom clerk or by the laboratory instructor. A solution or precipitate containing cyanide must never be mixed with an acid or acid solution. An acid, even a very weak one, will react with it and liberate HCN gas. **HCN gas is poisonous.** The best way to dispose of the KCN mixture is to put it in a waste container that is provided for this purpose by your instructor.

4. When excess KCN is added to a solution containing $Cu(NH_3)_4^{2+}$ and $Cd(NH_3)_4^{2+}$, the complex ions $Cu(CN)_2^-$ and $Cd(CN)_4^{2-}$ are formed. The $Cu(CN)_2^-$ ion is very stable and is only very slightly dissociated into Cu^+ and CN^- ions; the resulting concentration of Cu^+ is so low that no precipitate of Cu_2S forms when sulfide ions are added. The $Cd(CN)_4^{2-}$ ion is less stable and is appreciably dissociated into Cd^{2+} and CN^- ions; the resulting concentration of Cd^{2+} is high enough for a precipitate of CdS to form when sulfide ions are added (see Table 5.4).

The reactions that take place when excess KCN is added are represented by the following equations:

$$2Cu(NH_3)_4^{2+} + 5CN^- + H_2O \rightleftharpoons 2Cu(CN)_2^- + CNO^- + 6NH_3 + 2NH_4^+$$

$$Cd(NH_3)_4^{2+} + 4CN^- \rightleftharpoons Cd(CN)_4^{2-} + 4NH_3$$

$$Cd(CN)_4^{2-} \rightleftharpoons Cd^{2+} + 4CN^-$$

PROCEDURE 11

Reprecipitation of the Sulfides of Arsenic, Antimony, and Tin

To the test tube that contains the decantate from Procedure 6 (see Note 1) as a solution of AsS_4^{3-}, SbS_3^{3-}, and SnS_3^{2-}, add 6 *M* HCl **in a hood**, with constant stirring, until the solution shows an acidic reaction when tested with litmus (see Note 2). As long as each drop of 6 *M* HCl keeps on bringing down more precipitate, the solution is still alkaline; when no more precipitate forms, the solution is probably acidic. A large excess of HCl must be avoided (see Note 3). Centrifuge and decant, discarding the decantate. Wash the precipitate three times, each time with 10 drops of a hot solution prepared by mixing equal volumes of water and 1 *M* $NH_4C_2H_3O_2$, and analyze according to Procedure 12 (see Note 4).

Notes

1. A dark-colored decantate from Procedure 6 means that some CuS, and perhaps HgS, has been put into a state of colloidal suspension by the $(NH_4)_2S$. If this decantate is allowed to stand for about 24 h, the CuS and HgS will ordinarily settle out. The yellow supernatant liquid can then be decanted, leaving the CuS and HgS behind. If the decantate is to be analyzed immediately, add 5 drops of 1 *M* $NH_4C_2H_3O_2$ to coagulate the CuS and HgS, mix thoroughly, centrifuge, and decant, discarding the precipitate.

2. The addition of dilute HCl to the soluble thio salts of the arsenic subgroup reprecipitates the sulfides of the three elements in accordance with the following molecular equation:

$$2(NH_4)_3AsS_4 + 6HCl \rightleftharpoons 6NH_4Cl + 2H_3AsS_4$$

The compound H_3AsS_4 is unstable and decomposes.

$$2H_3AsS_4 \rightleftharpoons 3H_2S + As_2S_5(s)$$

$(NH_4)_2SnS_3$ and $(NH_4)_3SbS_3$ react in a similar manner. The H_2SnS_3 and H_3SbS_3 that are formed decompose to give SnS_2 and Sb_2S_3, respectively.

$$2SnS_3^{2-} + 4H^+ \rightleftharpoons 2SnS_2 + 2H_2S \tag{10.41}$$

$$2SbS_3^{3-} + 6H^+ \rightleftharpoons Sb_2S_3 + 3H_2S \tag{10.42}$$

The net equations for the reactions are showing that H_2S plays the same role in the thio acid system that water plays in the oxygen acids and that As_2S_5, Sb_2S_3, and SnS_2 are acid anhydrides in the sulfur acid system.

3. In reprecipitating the arsenic group, a large excess of HCl must be avoided, since SnS_2 and Sb_2S_3 are appreciably soluble in high concentrations of this acid. Also, the liquid in contact with the precipitate of SnS_2, As_2S_5, and Sb_2S_3 should be decanted off at once since the SnS_2 and Sb_2S_3 will dissolve on long standing.

4. The precipitate that is formed in Procedure 11 may be only sulfur. Ammonium sulfide may contain some ammonium polysulfide, $(NH_4)_2S_2$. When HCl is added to a solution containing $(NH_4)_2S_2$, sulfur is liberated according to the reaction

$$(NH_4)_2S_2 + 2H^+ \rightleftharpoons 2NH_4^+ + H_2S + S(s)$$

Since $(NH_4)_2S_2$ is a moderately strong oxidizing agent, it will oxidize As(III) to As(V) in Procedure 6, according to the following molecular equation:

$$As_2S_3 + 3(NH_4)_2S_2 \rightleftharpoons 2(NH_4)_3AsS_4 + S(s)$$

SEPARATION OF ARSENIC FROM ANTIMONY AND TIN

The separation of arsenic from antimony and tin depends on the fact that arsenic sulfide is insoluble in concentrated HCl, whereas the sulfides of tin and antimony dissolve in the HCl to form soluble chlorides.

Concentrated HCl dissolves SnS_2 and Sb_2S_3 as follows:

$$SnS_2 + 4H^+ \rightleftharpoons Sn^{4+} + 2H_2S \tag{10.43}$$

$$Sb_2S_3 + 6H^+ \rightleftharpoons 2Sb^{3+} + 3H_2S \tag{10.44}$$

The Sn^{4+} and Sb^{3+} ions react with the Cl^- ions present to form the complex chloro ions $SnCl_6^{2-}$ and $SbCl_4^-$.

$$Sn^{4+} + 6Cl^- \rightleftharpoons SnCl_6^{2-} \tag{10.45}$$

$$Sb^{3+} + 4Cl^- \rightleftharpoons SbCl_4^- \tag{10.46}$$

PROCEDURE 12

Separation of Arsenic from Antimony and Tin

To the precipitate from Procedure 11, add 15 drops of 12 M HCl, mix thoroughly, and heat the test tube in the boiling water bath for 3 to 4 min, stirring frequently (see Note 1). Add 7 drops of hot water, mix well, and continue heating in the boiling water bath for another 20 to 30 s. Centrifuge, and decant into a test tube. Save this decantate, which may contain antimony and tin, for Procedure 14. Wash the precipitate once with 10 drops of 6 M HCl and then three times with 10-drop portions of a hot solution prepared by mixing equal volumes of water and 1 M $NH_4C_2H_3O_2$. Analyze the precipitate according to Procedure 13.

Note

1. The precipitate (As_2S_5) is first washed with 6 M HCl rather than with water to remove all traces of Sb^{3+}. Sb^{3+} ions will react with water to form insoluble SbOCl, which will remain with the As_2S_5. The final washing with hot water is to remove all traces of HCl. Chloride, if present, will interfere with the confirmatory test for arsenic (Procedure 13) by forming a white precipitate of AgCl.

PROCEDURE 13

Detection of Arsenic

Add 10 drops of 16 M HNO_3 to the test tube containing the precipitate from Procedure 12 and heat in the boiling water bath for about 1 min, or until the original precipitate is disintegrated and a deposit of sulfur is formed (see Note 1). Add 4 to 5 drops of water, centrifuge, and decant into a casserole, discarding any precipitate that remains in the test tube. Evaporate the solution in the casserole very carefully to complete dryness, but do not bake. Allow to cool. Add 4 drops of 0.2 M $AgNO_3$ and swish around in the casserole for about 10 s. A reddish brown precipitate (Ag_3AsO_4) proves the presence of arsenic (see Note 2). If no reddish brown precipitate appears, add 0.2 M $NaC_2H_3O_2$, a drop at a time and with thorough mixing, until precipitation is complete or a maximum of 30 drops has been added. A reddish brown or chocolate brown precipitate (Ag_3AsO_4) proves the presence of arsenic.

Notes

1. The As_2S_5 (and As_2S_3, if present) is dissolved by HNO_3 as follows:

$$3As_2S_5(s) + 10NO_3^- + 10H^+ + 4H_2O \leftrightharpoons 6H_3AsO_4 + 10NO + 15S(s) \quad (10.47)$$

$$3As_2S_3(s) + 10NO_3^- + 10H^+ + 4H_2O \leftrightharpoons 6H_3AsO_4 + 10NO + 9S(s) \quad (10.48)$$

If the H_3AsO_4 is heated to 160–200°C, As_2O_5 is formed.

$$2H_3AsO_4 \leftrightharpoons As_2O_5 + 3H_2O$$

Ag_3AsO_4 is re-formed when the aqueous solutions of $AgNO_3$ and $NaC_2H_3O_2$ are added.

$$H_3AsO_4 + 3Ag^+ \leftrightarrows Ag_3AsO_4 + 3H^+$$

2. Ag_3AsO_4 is formed in the confirmatory test for arsenic according to the equation

$$H_3AsO_4 + 3Ag^+ \leftrightarrows Ag_3AsO_4 + 3H^+$$

Ag_3AsO_4 is the salt of a weak acid (see Table 3.2) and is, therefore, soluble in the strong acid HNO_3; it is, however, insoluble in a weakly acidic or neutral solution. Evaporation to dryness should remove all the HNO_3. As a precaution against the possibility of some HNO_3 being left, $NaC_2H_3O_2$ is added to the reaction mixture. As the salt of a weak acid, $NaC_2H_3O_2$ will serve as a buffer for any HNO_3 that might have remained. The $C_2H_3O_2^-$ ions from the strong electrolyte $NaC_2H_3O_2$ combine with the H^+ ions from HNO_3 to form the weak acid $HC_2H_3O_2$. This reaction maintains the H^+ ion concentration at such a low value (makes the solution so weakly acidic) that Ag_3AsO_4 will precipitate.

The formation of a white precipitate (AgCl) means that the As_2S_5 was not washed free of chloride in Procedure 12. $AgC_2H_3O_2$ (white), being sparingly soluble, may also precipitate.

PROCEDURE 14

Detection of Antimony and Tin

(A) *Detection of antimony* (see Note 1). Transfer the decantate from Procedure 12 to a casserole, boil for 1 min to remove all H_2S, then add 4 to 5 drops of cold water and mix thoroughly. Place 4 drops of this solution in a test tube and add 5 M NH_4OH drop by drop with constant stirring until a litmus test shows that the solution is just barely alkaline (see Note 2; if either antimony or tin is present, a white precipitate or milky suspension, SbOOH and/or SO(OH)$_4$, will form when the solution is alkaline; if no precipitate or suspension forms, tin and antimony are absent). Then add 5 M $HC_2H_3O_2$ drop by drop with constant stirring until the solution is acidic to litmus (If a white precipitate or suspension is present in the alkaline solution, it will remain in the weakly acidic solution; it should be ignored, since it will not interfere with the test for antimony). Add a crystal of $Na_2S_2O_3 \cdot 5H_2O$ (see Note 3); then heat by placing the test tube in the boiling water bath for 2 to 3 min. An orange red precipitate (Sb_2OS_2) proves the presence of antimony.

(B) *Detection of tin* (see Note 4). To the remainder of the solution in the casserole, add a 1-in. piece of 26-gauge aluminum wire (see Note 5). Warm gently until wire has dissolved; then boil gently for about 2 min, or until the black precipitate either has all dissolved or appears not to be dissolving any more, replenishing the solution with 6 M HCl if necessary (if no black residue remains, antimony is absent; if a black residue, Sb, remains after boiling 2 min, antimony is present). Transfer the contents of the casserole immediately to a test tube, cool, centrifuge, and decant. Immediately add 2 to 3 drops of 0.1 M $HgCl_2$ solution to the decantate, mix thoroughly, and allow to stand 1 min; a white (Hg_2Cl_2) or gray (Hg_2Cl_2 + Hg) precipitate proves the presence of tin (see Note 6).

Notes

1. The solution used in making the confirmatory test for antimony must be free from H_2S; otherwise SnS_2, as well as SbS_3, will precipitate when it is made alkaline with NH_4OH.
2. Addition of NH_4OH precipitates $SbOOH$ and $Sn(OH)_4$ according to the following net equations:

$$SbCl_4^- + 3NH_4OH \rightleftharpoons SbOOH(s) + 3NH_4^+ + 4Cl^- + H_2O$$

$$SnCl_6^{2-} + 4NH_4OH \rightleftharpoons Sn(OH)_4(s) + 4NH_4^+ + 6Cl^-$$

The detailed reactions are between the Sb^{3+} and Sn^{4+} ions in equilibrium with $SbCl_4^-$ and $SnCl_6^{2-}$ and the OH^- ions in equilibrium with NH_4OH.
3. The thiosulfate ion $S_2O_3^{2-}$ disproportionates sparingly to yield H_2S.

$$S_2O_3^{2-} + H_2O \rightleftharpoons SO_4^{2-} + H_2S$$

Although the concentration of S^{2-} ions provided by this H_2S is not high enough to precipitate tin as SnS_2, it will precipitate antimony as Sb_2OS_2.

$$2SbOOH(s) + 2H_2S \rightleftharpoons Sb_2OS_2(s) + 3H_2O$$

4. A flame test for tin may be carried out as follows: Use cold water to fill a test tube that is clean on the outside as well as on the inside. Dip the bottom of the test tube into the solution to be tested; then hold the bottom of the test tube in a hot, nonluminous Bunsen flame. A blue coloration of the flame, which appears to cling to the wall of the test tube, proves the presence of tin. A trial flame test for tin should first be run on a sample of tin salt solution from the reagent shelf so that the characteristic blue coloration can be recognized.
5. The aluminum is added to reduce Sn^{4+} to Sn. The reaction is

$$4Al + 3Sn^{4+} \rightleftharpoons 4Al^{3+} + 3Sn(s)$$

Excess aluminum is dissolved by the HCl.

The tin then dissolves in HCl to form Sn^{2+} ions.

The tin does not dissolve until all the aluminum has dissolved or the excess is removed. Furthermore, since tin is not very active, it dissolves less readily than the aluminum. For that reason the mixture must be boiled.

Since the tin(II) formed by the reaction of tin with HCl may eventually be oxidized by the oxygen of the air to tin(IV), the test should be completed as rapidly as possible.
6. The chemistry of the final test for tin, in which Sn^{2+} reduces Hg^{2+} to $Hg + Hg_2Cl_2$, is the same as that involved in the confirmatory test for mercury (see Note 4, Procedure 7).

The metallic aluminum added to the solution containing Sn^{4+} and Sb^{3+} replaces antimony (black), as well as tin (gray). Antimony, being less active than hydrogen, does not dissolve in HCl, however.

PROBLEMS

10.1 Using the scheme of analysis as a guide, write net equations for all reactions that take place in the copper-arsenic group.

10.2 Upon what fact or facts is each of the following based?

    a. The separation of the cations of the copper-arsenic group from the cations of the groups that follow.

    b. The separation of the copper subgroup from the arsenic subgroup.

    c. The separation of mercury from lead, bismuth, copper, and cadmium.

    d. The separation of lead from bismuth, copper, and cadmium.

    e. The separation of bismuth from copper and cadmium.

    f. The identification of cadmium in the presence of copper.

    g. The confirmatory test for mercury.

    h. The confirmatory test for bismuth.

    i. The separation of arsenic from antimony and tin.

10.3 What difficulties, if any, would arise under the following conditions in the precipitation of the copper-arsenic group?

    a. If HNO_3 was used in place of HCl.

    b. If H_2SO_4 was used in place of HCl.

10.4 Give the reason for each of the following in the copper-arsenic group analysis.

    a. Addition of H_2O in the first procedure.

    b. Evaporation of the solution to dryness with 6 M HCl in the first procedure.

    c. Avoiding baking the precipitate in the first evaporation step.

    d. The addition of $NaC_2H_3O_2$ in the confirmatory test for arsenic.

10.5 Point out, in detail, what takes place in each of the following. Give the equations for the significant reactions.

    a. The confirmatory test for tin.

    b. The confirmatory test for arsenic.

    c. The confirmatory test for bismuth.

    d. The confirmatory test for cadmium in the presence of copper.

    e. The confirmatory test for lead.

10.6 The precipitate formed in the confirmatory test for mercury(II) is sometimes white, sometimes gray, and sometimes black. Explain.

10.7 Point out where the following take place in the copper-arsenic group analysis and give the equations for the reactions.

    a. Three reactions in which tin is oxidized from an oxidation state of 2 to an oxidation state of 4.

    b. One reaction in which mercury is reduced from 2 to 1.

    c. One reaction in which mercury is reduced from 1 to 0.

    d. One reaction in which bismuth is reduced from 3 to 0.

    e. One reaction in which tin is reduced from 4 to 0.

    f. One reaction in which sulfur is oxidized from -2 to 0.

10.8 A 1 M solution of HCl in water was saturated with H_2S. This solution was then diluted with four volumes of distilled water, and the diluted solution was saturated with H_2S. How does the concentrate of sulfide ions in the diluted solution compare with the concentration of sulfide ions in the original solution, both of which were saturated with H_2S? Explain.

10.9 Give the formula for a chemical substance that will cause each reaction.

  a. Form a precipitate with $SnCl_2$ solution, but not with $SnCl_4$.
  b. Form a precipitate with $Pb(NO_3)_2$ solution, but not with $Cu(NO_3)_2$.
  c. Form a precipitate with H_2S solution, but not with HCl.
  d. Form a precipitate with $(NH_4)_3AsS_4$ solution, but not with $CuCl_2$.
  e. Form a precipitate with $SnCl_2$ solution, but not with $HgCl_2$.
  f. Form a precipitate with $Bi_2(SO_4)_3$ solution, but not with $CuSO_4$.
  g. Form a precipitate with HNO_3 solution, but not with HCl.
  h. Form a precipitate with H_2SO_4 solution, but not with HNO_3.
  i. Form a precipitate with $Pb(NO_3)_2$ solution, but not with NaCl.
  j. Form a precipitate with $CuCl_2$ solution, but not with KCl.
  k. Dissolve As_2S_5, but not HgS.
  l. Dissolve CuS, but not HgS.
  m. Dissolve Sb_2S_3, but not As_2S_5.
  n. Dissolve SnS_2, but not PbS.
  o. Dissolve $Pb(NO_3)_2$, but not PbS.
  p. Dissolve $HgCl_2$, but not Hg_2Cl_2.
  q. Oxidize $Sn(OH)_4^{2-}$ to $Sn(OH)_6^{2-}$.
  r. Reduce Hg^{2+} to Hg.
  s. Reduce Bi(III) to Bi.
  t. Oxidize Sn^{2+} to Sn^{4+}.
  u. Dissolve $Cu(OH)_2$, but not $Bi(OH)_3$.
  v. Oxidize Cl^- to Cl_2.

10.10 How would you distinguish between the following compounds by means of one reagent? (Tell what happens to each substance.)

  (a) *Solids*

CuS and SnS_2	$Cu(OH)_2$ and $Bi(OH)_3$
SnS_2 and As_2S_5	FeS and CdS
CdS and HgS	SnS and SnS_2
$PbSO_4$ and $Bi_2(SO_4)_3$	$SbCl_3$ and $PbCl_2$
Hg_2Cl_2 and $HgCl_2$	ZnS and CuS

  (b) *Solutions*

$(NH_4)_2S$ and NH_4Cl	H_2S and HCl
$CdCl_2$ and $CuCl_2$	Na_3AsO_4 and NaCl
$Pb(NO_3)_2$ and $Cd(NO_3)_2$	$SbCl_3$ and $AlCl_3$
$SnCl_2$ and $HgCl_2$	$(NH_4)_2SnS_3$ and $SnCl_2$
HNO_3 and HCl	$SnCl_2$ and $SbCl_3$
H_2SO_4 and HNO_3	H_2SO_4 and HCl

10.11 How do you account for the following facts?

a. SnS_2 is soluble in $(NH_4)_2S$, while SnS is insoluble.

b. $PbSO_4$ is soluble in a solution of $NH_4C_2H_3O_2$.

c. $PbSO_4$ is soluble in concentrated H_2SO_4.

d. $Sn(OH)_2$ is soluble in excess NaOH.

e. $Cd(OH)_2$ is soluble in excess NH_4OH.

f. Ag_3AsO_4 will not precipitate from a solution containing a small amount of HNO_3, but will precipitate from a solution containing a small amount of HNO_3 and also some dissolved $NaC_2H_3O_2$.

g. HgS is insoluble in HNO_3, but CuS is soluble.

h. CdS is insoluble in HCl, but soluble in HNO_3.

i. HgS is insoluble in HNO_3, but soluble in aqua regia.

j. Sb_2S_3 and As_2S_3 are acidic in character, while PbS and CuS are basic.

k. Iron(III) compounds are oxidizing agents.

l. Copper sulfide will precipitate when H_2S is added to a solution containing $Cu(NH_3)_4^{2+}$ ions, but will not precipitate when H_2S is added to a solution containing $Cu(CN)_2^-$ ions.

10.12 Give the detailed equations to account for each of the following.

a. When NaOH is added, drop by drop, to a solution of $SnCl_2$, a white precipitate is first formed; it disappears on further addition of NaOH to form a water-clear solution. When HCl is added, dropwise, to this water-clear solution, a precipitate is formed; it disappears on further addition of HCl.

b. When NH_4OH is added, dropwise, to a solution of $CdCl_2$, a white precipitate is first formed; it disappears on further addition of NH_4OH to form a water-clear solution. When HCl is added, dropwise, to the water-clear solution, a white precipitate is first formed; it disappears on further addition of HCl to give a clear solution.

10.13 For each of the following pairs of solids, give the formula of a single chemical reagent that will dissolve one solid but not the other. Give the formula of the predominant soluble species that is formed when the solid dissolves.

(a) $PbSO_4$ and $PbCrO_4$ (b) SnS and SnS_2

(c) Cu_2S and CdS   (d) Ag_3AsO_4 and AgCl

(e) Sn and Sb   (f) $Cd(OH)_2$ and $Bi(OH)_3$

(g) HgS and As_2S_5 (h) $Pb(OH)_2$ and $Bi(OH)_3$

10.14 Select the most stable, complex ion from the following.

a. $Cd(CN)_4^{2-}$ with instability constant of 14×10^{-19}.

b. $Cd(NH_3)_4^{2+}$ with instability constant of 7.5×10^{-8}.

c. $Cu(NH_3)_4^{2+}$ with instability constant of 4.7×10^{-15}.

d. HgI_4^{2-} with instability constant of 5×10^{-31}.

10.15 Select the compound with the smallest molar solubility from the following (assume an activity coefficient of 1 for each ion).

a. Ag_3AsO_4 with K_{sp} of 1×10^{-23}.

b. ZnS with K_{sp} of 3×10^{-23}.

c. Hg_2Br_2 with K_{sp} of 1×10^{-23}.

d. $Cu_3(PO_4)_2$ with K_{sp} of 1×10^{-23}.

10.16 In the following four statements, choose the one statement that has no connection with the other three statements.

    a. SnS is insoluble in 4 M $(NH_4)_2S$, but SnS_2 is soluble in 4 M $(NH_4)_2S$.

    b. In its higher oxidation states an element is more nonmetallic, while in its lower oxidation states it is more metallic.

    c. When SnS and SnS_2, respectively, are treated with concentrated HNO_3, white SnO_2 is formed in each instance.

    d. SnS is predominantly basic in character, while SnS_2 is predominantly acidic in character.

10.17 Using the scheme of analysis for the copper-arsenic group as your only source of information, select, in each part, the correct substance.

    a. Compound with smallest solubility product: CdS, HgS, ZnS.

    b. Most stable ion: $Cu(NH_3)_4^{2+}$, $Cu(CN)_2^-$, $Cd(CN)_4^{2-}$.

    c. Most electropositive: Sb, Sn, Al.

    d. Most acidic sulfide: CuS, SnS, Sb_2S_3.

    e. Solution that, when saturated with H_2S gas, will contain the highest $[S^{2-}]$: 1 M HCl, 2 M HCl, 3 M HCl.

    f. Reaction with the highest oxidation potential: $Sb \leftrightharpoons Sb^{3+} + 3e^-$, $Sn \leftrightharpoons Sn^{2+} + 2e^-$, $Al \leftrightharpoons Al^{3+} + 3e^-$.

    g. Strongest reducing agent: Sn^{2+}, Hg_2^{2+}, Hg.

10.18 The mixture of sulfides obtained when a copper-arsenic group unknown was completely precipitated with H_2S was completely insoluble in $(NH_4)_2S$ and completely soluble in 3 M HNO_3, sulfur being formed in the latter case. The solution obtained with HNO_3 gave a colorless solution with no precipitate or milky suspension when treated with excess NH_4OH. What conclusions can be drawn?

10.19 A mixture of solid chlorides, known to contain only cations of the copper-arsenic group, dissolved readily and completely in cold water to give a clear, pale blue solution. The regular H_2S precipitation gave a brown precipitate. This precipitate was partially soluble in $(NH_4)_2S$; the residue, which did not dissolve in $(NH_4)_2S$, was completely soluble in warm dilute HNO_3. What conclusions can be drawn?

10.20 A solution known to contain only cations of the copper-arsenic group was divided into four parts, which were treated as described below.

    a. One part, when diluted with water, gave a white precipitate; this precipitate dissolved on addition of HCl.

    b. The second part gave no precipitate when warmed and treated with $SnCl_2$ solution.

    c. The third part, on evaporation with H_2SO_4 to white SO_3 fumes and dilution with water, gave a finely divided white precipitate.

    d. The fourth part was treated with H_2S in the usual manner. A brownish black precipitate formed. This precipitate was partially soluble in $(NH_4)_2S$. The decantate from the $(NH_4)_2S$ treatment, on being acidified with dilute HCl, gave a precipitate that dissolved completely in 12 M HCl. What conclusions can be drawn?

10.21 A solution, A, which was known to contain only those cations of the silver and copper-arsenic groups listed below, gave a white precipitate, B, and a colorless supernatant liquid, C, when treated with excess NH_4OH. The decanted supernatant liquid, C, upon addition of 6 M HCl, gave no precipitate. The white precipitate, B, upon treatment with excess NaOH, dissolved completely to give a colorless solution. When this colorless solution was acidified with dilute H_2SO_4, a white precipitate, D, was formed. When $(NH_4)_2S$ was added to the decantate from D, a brown precipitate, E, was formed. The addition of concentrated HCl to the decantate from E resulted in the evolution of a gas; when this HCl solution was evaporated to dryness, the addition of water to the residue gave a white precipitate. Indicate whether each of the following cations is present, absent, or undetermined: Ag^+, Hg_2^{2+}, Pb^{2+}, Hg^{2+}, Bi^{3+}, Cu^{2+}, Cd^{2+}, Sb^{3+}, Sn^{2+}, Sn^{4+}.

10.22 A solid unknown is known to be a mixture of two or more of the following compounds: $AgNO_3$, Na_3AsO_4, $Pb(NO_3)_2$, $Cu(NO_3)_2$, $BiCl_3$, $Cd(NO_3)_2$, $Hg(NO_3)_2$, $SnSO_4$.

   a. One sample of the solid dissolved completely in cold water to give a clear blue solution; there was no precipitate of any kind in the beaker. When $0.1\ M$ $SnSO_4$ was added to this clear solution, a precipitate formed.

   b. After a second sample of the solid contained in a beaker was treated with hot dilute HCl, the beaker was found to contain a blue solution and a white precipitate.

   c. A third sample of the solid was found to dissolve completely in dilute H_2SO_4 to give a blue solution.

   d. After a fourth sample of the solid was treated with excess $2\ M$ NH_4OH, the beaker was found to contain a deep blue solution and a white precipitate.

   e. After a fifth sample of the solid was treated with excess $2\ M$ NaOH, the beaker was found to contain a colorless solution and a dark precipitate. Indicate each compound that is present, each compound that is absent, and each compound that is undetermined.

10.23 A copper-arsenic unknown solution, when treated with an excess of $(NH_4)_2S$, yielded a dark brown precipitate, A, and a pale yellow solution, B. Precipitate A, when treated with a mixture of HNO_3 and HCl, dissolved to give a clear, colorless solution and a residue of sulfur. The colorless solution was first boiled with $6\ M$ HCl until all HNO_3 was destroyed; it was then treated with an excess of $(NH_4)_2S$. A yellow precipitate and a pale yellow solution formed; this yellow solution gave a yellow precipitate when treated with $2\ M$ HCl.

The pale yellow solution, B, when treated dropwise with $12\ M$ HCl, formed a precipitate; this precipitate dissolved completely in excess $12\ M$ HCl to form a solution, C. When a piece of tin was placed in solution C, the only sign of reaction was the evolution of hydrogen gas bubbles on the surface of the tin.

Indicate whether each of the copper-arsenic group ions is present, absent, or indeterminate.

The Aluminum-Nickel Group

Addition of NH_4Cl, NH_4OH, and $(NH_4)_2S$ to a solution containing all the cations not precipitated in the preceding groups results in the precipitation of aluminum, chromium, and iron(III) as hydroxides and manganese, nickel, cobalt, iron(II), and zinc as sulfides (Figure 11.1). Under these conditions the hydroxides and sulfides of calcium, barium, magnesium, potassium, and sodium are soluble. This solubility permits a separation of the cations of the aluminum-nickel group from those of the barium-magnesium group.

THE CHEMISTRY OF THE ALUMINUM-NICKEL GROUP IONS

Iron, Fe^{3+} and Fe^{2+}

Iron in its compounds is ordinarily found in the $2+$ (ferrous) or the $3+$ (ferric) state. The latter is more common, since most ferrous compounds oxidize in the air, particularly if water is present. Iron(II) compounds are usually found as hydrates and are light green. Iron(III) salts are also ordinarily obtained as hydrates and are often yellow or orange. Both Fe^{2+} and Fe^{3+} form many complexes; perhaps the most stable are those with cyanide, $Fe(CN)_6^{4-}$ and $Fe(CN)_6^{3-}$. Ferrithiocyanate complexes such as $Fe(SCN)_6^{3-}$ have very characteristic deep red colors. Metallic iron is a good reducing agent, dissolving readily in 6 M HCl with evolution of hydrogen. Conversion of Fe(II) to Fe(III) or the reverse is easily accomplished by common oxidizing agents (air or H_2O_2 in acid) or by reducing agents such as H_2S, Sn^{2+}, and I^-.

Cobalt, Co^{2+}

Like iron, cobalt can exist in two common oxidation states in aqueous solutions, Co^{2+} and Co^{3+}. The $3+$ state is stable only in the presence of strong coordinating liquids such as NH_3; thus, the general chemistry of cobalt in schemes of qualitative analysis is usually that of Co^{2+}. Cobalt(II) salts in water solution are characteristically pink, the color of the hydrated cobalt ion, $Co(H_2O)_6^{2+}$. The pink color is usually too delicate to be used to characterize the ion. Cobalt(II) forms several complex ions. When heated to boiling, the color of cobalt chloride solutions turns

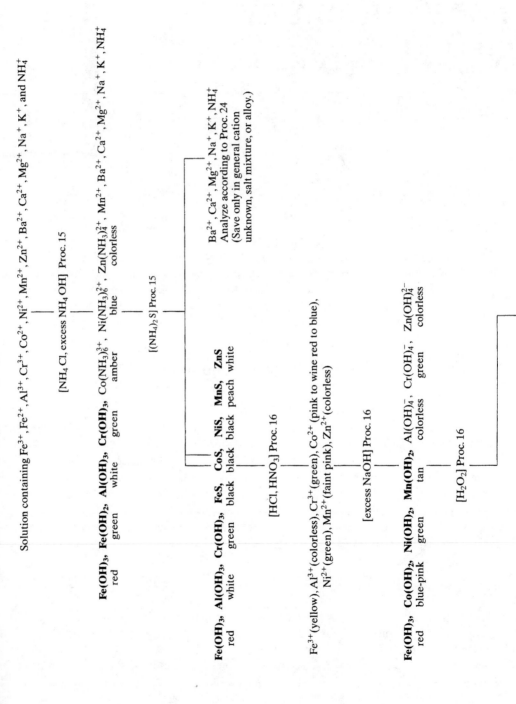

FIGURE 11.1 Analysis scheme for the aluminum-nickel group.

Solution containing Fe^{3+}, Fe^{2+}, Al^{3+}, Cr^{3+}, Co^{2+}, Ni^{2+}, Mn^{2+}, Zn^{2+}, Ba^{2+}, Ca^{2+}, Mg^{2+}, Na^+, K^+, and NH_4^+

[NH_4Cl, excess NH_4OH] Proc. 15

Fe(OH)$_3$, **Fe(OH)$_2$**, **Al(OH)$_3$**, **Cr(OH)$_3$**, $Co(NH_3)_6^{3+}$, $Ni(NH_3)_6^{2+}$, $Zn(NH_3)_4^{2+}$, Mn^{2+}, Ba^{2+}, Ca^{2+}, Mg^{2+}, Na^+, K^+, NH_4^+
red   green   white   green   amber   blue   colorless

[(NH$_4$)$_2$S] Proc. 15

Ba^{2+}, Ca^{2+}, Mg^{2+}, Na^+, K^+, NH_4^+
Analyze according to Proc. 24
(Save only in general cation unknown, salt mixture, or alloy.)

Fe(OH)$_3$, **Al(OH)$_3$**, **Cr(OH)$_3$**, **FeS**, **CoS**, **NiS**, **MnS**, **ZnS**
red   white   green   black   black   black   peach   white

[HCl, HNO$_3$] Proc. 16

Fe^{3+} (yellow), Al^{3+} (colorless), Cr^{3+} (green), Co^{2+} (pink to wine red to blue),
Ni^{2+} (green), Mn^{2+} (faint pink), Zn^{2+} (colorless)

[excess NaOH] Proc. 16

Fe(OH)$_3$, **Co(OH)$_2$**, **Ni(OH)$_2$**, **Mn(OH)$_2$**, $Al(OH)_4^-$, $Cr(OH)_4^-$, $Zn(OH)_4^{2-}$
red   blue-pink   green   tan   colorless   green   colorless

[H$_2$O$_2$] Proc. 16

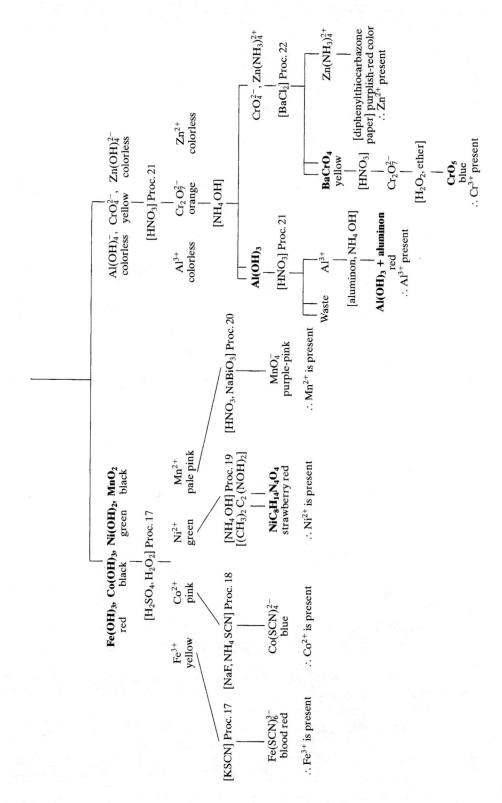

from pink to blue. The color change occurs because of a change in coordination number; the pink form arises from octahedrally coordinated Co(II), whereas the blue form is tetrahedrally coordinated Co(II). Cobalt sulfide does not dissolve readily in 6 M HCl even when heated. Co(II) reacts with thiocyanate anion, SCN^-, to form a blue complex, $Co(SCN)_4^{2-}$, which is much more stable in ethanol than in water. Addition of KNO_2 to solutions of Co^{2+} produces a characteristic yellow precipitate of $K_3Co(NO_2)_6$ in which the metal has been oxidized [Eq. (11.1)].

$$Co^{2+} + 7NO_2^- + 3K^+ + 2H^+ \leftrightarrows K_3Co(NO_2)_6(s) + NO(g) + H_2O \qquad (11.1)$$

Under strongly oxidizing conditions, Co(II) can be converted to Co(III), which has a stability that is enhanced in a complex species like $Co(NH_3)_6^{3+}$ or an insoluble substance such as the yellow cobaltinitrite produced in the reaction described by Eq. (11.1).

Nickel, Ni^{2+}

Nickel salts, like those of most of the other members of analysis Group III, are typically colored; the hydrates are green, which is the color of $Ni(H_2O)_6^{2+}$ complex ion. Nickel forms several complex ions, many of which have characteristic colors; the blue $Ni(NH_3)_6^{2+}$ ion is perhaps the most common of these. Nickel sulfide, NiS, when first precipitated tends to be colloidal and difficult to settle by centrifuging; it is not very soluble in 6 M HCl, even though it cannot be precipitated from acidic solutions, a behavior that is reminiscent of the chemistry of cobalt. Nickel forms a very characteristic rose red precipitate with dimethylglyoxime (**1**), an organic chelating agent—a bidentate ligand. This product is useful also in the quantitative analysis of nickel.

$$
\begin{array}{c}
H \\
| \\
H-C-H \\
| \\
C=NOH \\
| \\
C=NOH \\
| \\
H-C-H \\
| \\
H
\end{array}
$$

1

Chromium, Cr^{3+}

Chromium is ordinarily encountered in either the 3+ or the 6+ oxidation state. The common cation Cr^{3+} forms several complexes, all of which are colored; thus, $Cr(H_2O)_6^{3+}$ is reddish violet in solution.

The chromium-containing precipitate obtained in the Group III precipitation scheme is the hydroxide $Cr(OH)_3$ rather than sulfide. Chromium(III) hydroxide is amphoteric; it dissolves in excess base to form the green chromite $Cr(OH)_4^-$ ion and in acid to form the hydrated chromium(III) ion. Chromium(III) can be oxidized to Cr(VI) with several oxidizing agents, such as ClO_3^- in 16 M HNO_3, H_2O_2 in 6 M NaOH, and ClO^- in 6 M NaOH. Chromium is usually identified in the form of a Cr(VI) species in neutral or basic solutions where Cr(VI) exists as the bright yellow chromate ion, CrO_4^{2-}. If acid is added to this species, orange dichromate ion, $Cr_2O_7^{2-}$, is produced [Eq. (11.2)].

$$2CrO_4^{2-} + 2H^+ \rightleftharpoons Cr_2O_7^{2-} + H_2O \tag{11.2}$$

If H_2O_2 is added to a solution containing the $Cr_2O_7^{2-}$ ion, a transitory blue color due to CrO_5 is observed. This reaction forms the basis for the confirmatory test for chromium.

Structure **(2)** has been proposed for CrO_5. Note that chromium has a covalence of 6, the same as in CrO_4^{2-} or $Cr_2O_7^{2-}$.

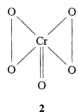

2

The oxidation number of chromium in CrO_5 is 10, the value used in balancing redox equations. It should be pointed out, however, that an oxidation number of 10 does not mean, in this instance, that 10 electrons of the chromium atom are involved in the bonding.

Aluminum, Al^{3+}

The only oxidation state of aluminum that is important in aqueous chemistry is the 3+ state. Since aluminum is not a member of a transition series of elements, all its simple salts are colorless; most aluminum compounds, except the oxide and hydroxide, are soluble in water. Aqueous solutions containing Al^{3+} contain the hydrated ion $Al(H_2O)_6^{3+}$, which is colorless. Aqueous solutions of aluminum salts are extensively hydrolyzed [Eq. (11.3)] and accordingly are acidic.

$$Al(H_2O)_6^{3+} + H_2O \rightleftharpoons Al(H_2O)_5(OH)^{2+} + H_3O^+ \tag{11.3}$$

In extreme cases, hydrolysis can be sufficiently extensive to yield cloudy solutions containing hydrated aluminum hydroxides. Hydrolysis can be suppressed by making these solutions acidic, thus reversing the process shown in Eq. (11.3).

Manganese, Mn^{2+}

The aqueous chemistry of manganese incorporates several common oxidation states; Mn(II), which appears in many simple binary salts; Mn(IV), as in the insoluble MnO_2; and Mn(VII), in the form of the deeply colored permanganate ion MnO_4^-.

Aqueous solutions of Mn^{2+} are colored pink, an observation that is not useful in the qualitative analysis scheme except in very rare cases. Manganous sulfide, MnS, which is precipitated from basic solution, is also pink.

Manganese(II) is easily oxidized to Mn(IV) in basic solution [Eq. (11.4)], but with stronger oxidizing agents in acid solution, oxidation to the Mn(VII) state can occur [Eq. (11.5)].

$$Mn(OH)_2(s) + H_2O_2 \leftrightharpoons MnO_2(s) + 2H_2O \tag{11.4}$$

$$2Mn^{2+} + 5BiO_3^- + 14H^+ \leftrightharpoons 2MnO_4^- + 5Bi^{3+} + 7H_2O \tag{11.5}$$

The formation of the characteristic color of the permanganate ion makes the reaction in Eq. (11.5) a sensitive test for manganese in mixtures containing many other cations.

Zinc, Zn^{2+}

Compounds containing the Zn^{2+} ion are usually colorless and many are soluble in water; aqueous solutions contain the colorless complex ion $Zn(H_2O)_4^{2+}$, which can be extensively hydrolyzed [Eq. (11.6)].

$$Zn(H_2O)_4^{2+} + H_2O \leftrightharpoons Zn(H_2O)_3OH^+ + H_3O^+ \tag{11.6}$$

 Zinc(II) forms complexes with many ligands, such as NH_3, H_2O, and OH^-. Zinc hydroxide is amphoteric, dissolving in both acid and base.

PRECIPITATION AND ANALYSIS OF THE ALUMINUM-NICKEL GROUP

The conditions for the precipitation of the aluminum-nickel group of ions must be carefully maintained to keep ions from the following groups (mainly Mg^{2+}) from precipitating prematurely. This is accomplished by manipulating the equilibrium associated with the weak base, aqueous ammonia. In the precipitation of the aluminum-nickel group, ammonium chloride, NH_4Cl, is added to prevent the precipitation of magnesium as $Mg(OH)_2$. If $Mg(OH)_2$ were to precipitate, the subsequent analysis of the group would be more complicated; hence it is desirable to carry magnesium over to the next group.

If a solution contains enough Mg^{2+} ions and enough OH^- ions so that the two, when they combine, will give more than enough $Mg(OH)_2$ to form a saturated solution, the extra $Mg(OH)_2$—over and above what will ordinarily dissolve in the solution—will precipitate. A solution contains enough Mg^{2+} and OH^- to form a saturated solution, and hence a precipitate, when the product of the concentration of the Mg^{2+} ions and the OH^- ions exceeds the solubility product for $Mg(OH)_2$ ($K_{sp} = [Mg^{2+}][OH^-]^2 = 8.9 \times 10^{-12}$).

Aqueous NH_3 is a weak base; its ionization may be represented according to Eq. (11.7).

$$NH_3 + H_2O \leftrightharpoons NH_4^+ + OH^- \tag{11.7}$$

Even though aqueous NH_3 is a weak base, it ionizes to give enough OH^- ions to form a precipitate with the concentration of Mg^{2+} present in the usual known or unknown containing Mg^{2+}; $Mg(OH)_2$ is precipitated.

Ammonium chloride is a salt and, hence, a strong electrolyte. Consequently, NH_4Cl in solution provides an abundance of NH_4^+ ions. If NH_4Cl is present in a solution also containing ammonium hydroxide, the NH_4^+ ions from NH_4Cl will shift the equilibrium in Eq. (11.7) to the left. This shift will greatly reduce the concentration of OH^- ions, and there are not enough OH^- ions present from Eq. (11. 7) to form a precipitate of $Mg(OH)_2$ with the Mg^{2+} ions present in the average solution. In other words, NH_4Cl prevents the precipitation of $Mg(OH)_2$ by NH_4OH; it reduces the concentration of the OH^- ion to such a low value that the product of the concentration of the Mg^{2+} ions and the OH^- ions squared is less than the solubility product for $Mg(OH)_2$. The decrease of the ionization of aqueous NH_3 by ammonium ions from aqueous NH_3 is an example of the common ion effect.

Ammonium chloride also prevents the precipitation of $Mn(OH)_2$ by NH_4OH, and in exactly the same way. However, MnS is quite insoluble, so that manganese precipitates as MnS when $(NH_4)_2S$ is added; MgS is soluble and does not precipitate.

The hydroxides of aluminum, chromium, and iron(III) are so insoluble that they precipitate even in the presence of the low concentration of OH^- ions represented in Eq. (11.7). The hydroxides of cobalt, nickel, and zinc are likewise so insoluble that they also will precipitate in the presence of the OH^- in Eq. (11.7); they, however, redissolve in excess NH_4OH because those cations form stable complexes with NH_3.

The behavior of iron(II) in the presence of NH_4^+ and OH^- is complicated by the fact that Fe(II) is readily oxidized by air to Fe(III). The net result is that a green precipitate, probably $Fe(OH)_2 \cdot Fe(OH)_3$, is formed when NH_4OH is added to Fe^{2+}. This green precipitate gradually darkens as more $Fe(OH)_2$ is oxidized to $Fe(OH)_3$.

$$4Fe(OH)_2(s) + O_2(air) + 2H_2O \leftrightarrows 4Fe(OH)_3(s) \tag{11.8}$$

The colors of the precipitates in this group are as follows:

$Fe(OH)_3$	Brownish red	FeS	Black
$Cr(OH)_3$	Grayish green	CoS	Black
$Al(OH)_3$	White	MnS	Peach
$Fe(OH)_2$	Green	NiS	Black
		ZnS	White

Careful observation of the colors of the precipitate formed, first when NH_4OH is added and later when $(NH_4)_2S$ is added, may give useful information about the presence or absence of certain ions.

If NH_4Cl and aqueous NH_3 give no precipitate, aluminum, chromium, iron(II), and iron(III) ions are definitely absent. If a white precipitate forms, the presence of aluminum is indicated; a green precipitate indicates iron or chromium, and a brownish red precipitate indicates iron. In interpreting colors we must recognize that a dark color will obscure, or cover up, a lighter color. Thus, brownish red $Fe(OH)_3$ will cover up green $Cr(OH)_3$ and white $Al(OH)_3$, and green $Cr(OH)_3$ will cover up white $Al(OH)_3$.

If the addition of $(NH_4)_2S$ gives no further precipitation, cobalt, nickel, manganese, and zinc ions are absent. If a white precipitate is formed, zinc is probably present; manganese, which forms a peach-colored sulfide, is probably absent; and cobalt and nickel, which form black sulfides, are definitely absent.

The formation of a black precipitate on addition of $(NH_4)_2S$ may be due to the conversion of red $Fe(OH)_3$ to black FeS according to the equation

$$2Fe(OH)_3(s) + 6NH_4^+ + 3S^{2-} \leftrightarrows 2FeS(s) + 6NH_3 + 6H_2O + S \tag{11.9}$$

If aqueous NH_3 and $(NH_4)_2S$ were added together instead of being added separately, the hydroxides and sulfides would immediately precipitate together; it would then obviously be impossible to make the observations of color noted above.

The behavior of zinc, nickel, and cobalt ions when treated with aqueous NH_3 is the same as that already noted for copper and cadmium ions (see Procedure 9 in Chapter 10). The hydroxides of the three metals are first formed. Excess of NH_4OH dissolves these hydroxides with the formation of $Zn(NH_3)_4^{2+}$, $Ni(NH_3)_6^{2+}$, and $Co(NH_3)_6^{3+}$; $Zn(NH_3)_4^{2+}$ is colorless, $Ni(NH_3)_6^{2+}$ is pale violet blue, and $Co(NH_3)_6^{3+}$ has a characteristic deep amber color.

Note that cobalt(II) has been oxidized to cobalt(III). This results from the fact that, in alkaline solution, cobalt(II) is a fairly strong reducing agent and is slowly oxidized to cobalt(III) by atmospheric oxygen. The reaction can be represented in either of two ways—Eqs. (11.10) to (11.12) or Eqs. (11.13) and (11.14).

$$Co^{2+} + 2NH_3 + 2H_2O \leftrightarrows Co(OH)_2(s) + 2NH_4^+ \tag{11.10}$$

$$4Co(OH)_2(s) + O_2 + 2H_2O \leftrightarrows 4Co(OH)_3(s) \tag{11.11}$$

$$Co(OH)_3(s) + 6NH_3 \leftrightarrows Co(NH_3)_6^{3+} + 3OH^- \tag{11.12}$$

$$Co(OH)_2(s) + 6NH_3 \leftrightarrows Co(NH_3)_6^{2+} + 2OH^- \tag{11.13}$$

$$4Co(NH_3)_6^{2+} + O_2 + 2H_2O \leftrightarrows 4Co(NH_3)_6^{3+} + 4OH^- \tag{11.14}$$

Ammonium sulfide reacts with $Zn(NH_3)_4^{2+}$, $Ni(NH_3)_6^{2+}$, and $Co(NH_3)_6^{3+}$ to form ZnS, NiS, and CoS, as follows:

$$S^{2-} + Ni(NH_3)_6^{2+} \leftrightarrows NiS(s) + 6NH_3 \tag{11.15}$$

$$3S^{2-} + 2Co(NH_3)_6^{3+} \leftrightarrows 2CoS(s) + S + 12NH_3 \tag{11.16}$$

Also, MnS and FeS are formed by direct combination of the two ions involved.

$$Mn^{2+} + S^{2-} \leftrightarrows Mns \tag{11.17}$$

Ferrous sulfide may also be formed by reduction of Fe(III), as described in Eq. (11.9).

If $(NH_4)_2S$ is added to a cold neutral solution containing aluminum, chromium, and iron(III) ions, $Al(OH)_3$, $Cr(OH)_3$, and Fe_2S_3 will be precipitated.

$$2Fe^{3+} + 3S^{2-} \leftrightarrows Fe_2S_3(s) \tag{11.18}$$

$$2Al^{3+} + 3S^{2-} + 6H_2O \leftrightarrows 2Al(OH)_3(s) + 3H_2S \tag{11.19}$$

$$2Cr^{3+} + 3S^{2-} + 6H_2O \leftrightarrows 2Cr(OH)_3(s) + 3H_2S \tag{11.20}$$

At higher temperatures, Fe_2S_3 is hydrolyzed; brownish red $Fe(OH)_3$ is precipitated.

$$Fe_2S_3 + 6H_2O \leftrightarrows 2Fe(OH)_3(s) + 3H_2S \tag{11.21}$$

As described above, FeS will be formed.

PROCEDURE 15

Precipitation of the Aluminum-Nickel Group in the Absence of Phosphates and Borates

If the solution to be analyzed is an aluminum-nickel group, known or unknown, follow (**A**) (see Note 1). If the decantate from the copper-arsenic group precipitation, Procedure 5 in Chapter 10, is to be analyzed, follow (**B**).

(**A**) Place 3 drops of the aluminum-nickel group known or unknown solution in a test tube (see Note 2). Add 4 drops of 2 M NH_4Cl (see Note 3), mix thoroughly, and then add 15 M aqueous NH_3, drop by drop and with constant stirring, until the solution is just alkaline; for the known solution, 3 drops of aqueous NH_3 is generally sufficient (see Note 4). Then add 1 drop extra of 15 M aqueous NH_3 and 20 drops of hot water and mix thoroughly. Centrifuge, but do not decant; however, note carefully the color of the precipitate and of the supernatant liquid. Then add 8 or 9 drops of ammonium sulfide solution and mix thoroughly.

Heat the tube carefully for 1 or 2 min in the boiling water bath; avoid allowing the contents of the tube to overflow because of foaming. Centrifuge, and test for complete precipitation with a drop of ammonium sulfide solution. When precipitation with $(NH_4)_2S$ (see Note 5) is complete, wash down the sides of the tube with a few drops of hot water, centrifuge, and note carefully the color of the precipitate and of the supernatant liquid; then decant, saving the decantate for Procedure 24 in Chapter 12 (see Note 6). Wash the precipitate three times with 20-drop portions of a hot solution prepared by mixing equal volumes of water and 1 M $NH_4C_2H_3O_2$ (see Note 3), and analyze according to Procedure 16.

(**B**) Place the entire decantate from Procedure 5 in Chapter 10 in a casserole and evaporate down carefully to a volume of 8 to 10 drops (see Note 7). Transfer to a test tube, centrifuge, and decant into another test tube, discarding the precipitate. Treat the decantate in the test tube according to the directions beginning with the second sentence of Method (**A**).

Notes

1. If phosphates, borates, fluorides, oxalates, silicates, or tartrates are present in an unknown to be analyzed and if this unknown also contains calcium, barium, or magnesium ions as well as cations of the aluminum-nickel group, a special series of procedures — different from those presented in this chapter — must be used. Since the combination of ions just mentioned is rare and will not ordinarily be encountered in solutions being analyzed, the average analysis is carried out most successfully by using a more general form of the procedure such as that presented in this chapter.

    Fluorides, oxalates, silicates, and tartrates are not among the anions discussed in this book; therefore they will not be present in solutions ordinarily submitted for analysis in this course.

    If an unknown solution to be analyzed does, in fact, contain the combination of ions listed at the beginning of this note, the student should refer to the basic acetate method in a reference textbook assigned by your instructor.

2. If the aluminum-nickel group unknown is issued in the form of a finely divided solid, proceed as follows: In a test tube place as much of the solid as can be carried on $\frac{1}{4}$ in. of the tip of a spatula. Add 3 to 4 drops of 6 M HCl, warm gently, and then add 5 to 6 drops of hot

water. Mix thoroughly and heat carefully until a clear solution is obtained, replenishing the water and acid if necessary. Proceed with the analysis of 3 drops of this solution as directed in the second sentence of Procedure 15(A).

If the solid unknown is not finely divided when received, the entire sample should first be pulverized in a clean mortar.

3. NH_4Cl, a strong electrolyte, helps coagulate any hydroxides or sulfides, thereby preventing them from becoming colloidal. Similarly, the addition of $NH_4C_2H_3O_2$ to the wash water at various points in the analysis prevents peptization of the precipitate being washed.

4. A large excess of NH_4OH tends to cause dispersion of the precipitated sulfides and hydroxides, making them difficult to settle out. If a very large excess of aqueous NH_3 has been added, the mixture should be boiled for 1 min before centrifuging.

5. Ammonium sulfide is completely ionized and provides a high concentration of S^{2-} ions. This high concentration of sulfide ions ensures complete precipitation of zinc, manganese, cobalt, nickel, and iron(II) as sulfides.

6. If the substance being analyzed is an aluminum-nickel group known or unknown, this decantate can be discarded.

7. The decantate from the copper-arsenic group is boiled down before the aluminum-nickel group is precipitated in order to drive off all H_2S and to precipitate any sulfides of the copper-arsenic group that may have gone into the decantate. If the H_2S were not removed, the sulfides and hydroxides of the aluminum-nickel group would all precipitate together when aqueous NH_3 was added. As a result, the desirable valuable observation of colors of precipitates to which we referred earlier could not be made.

SEPARATION OF THE ALUMINUM SUBGROUP FROM THE NICKEL SUBGROUP

The hydroxides of aluminum, chromium, and zinc are amphoteric and are, therefore, soluble in NaOH. In contrast, the hydroxides of iron, manganese, cobalt, and nickel are not amphoteric; they are not soluble in NaOH. The separation of the cations of the aluminum subgroup (Al^{3+}, Cr^{3+}, and Zn^{2+}) from the cations of the nickel subgroup (Fe^{3+}, Co^{2+}, Ni^{2+}, and Mn^{2+}) is based upon this fact.

Of the species precipitated as the aluminum-nickel group of ions, $Al(OH)_3$, $Fe(OH)_3$, $Cr(OH)_3$, MnS, FeS, and ZnS are readily soluble in HCl; NiS and CoS are not readily soluble in HCl, but are soluble in aqua regia (HCl + HNO_3).

$$NiS(s) \leftrightharpoons Ni^{2+} + S^{2-} \tag{11.22}$$

$$3S^{2-} + 2NO_3^- + 8H^+ \leftrightharpoons 3S + 2NO + 4H_2O \tag{11.23}$$

$$Ni^{2+} + 4Cl^- \leftrightharpoons NiCl_4^{2-} \tag{11.24}$$

Aqua regia oxidizes Fe(II) to Fe(III).

$$3Fe^{2+} + NO_3^- + 4H^+ \leftrightharpoons 3Fe^{3+} + NO + 2H_2O \tag{11.25}$$

$$Fe^{3+} + 4Cl^- \leftrightharpoons FeCl_4^- \tag{11.26}$$

When NaOH is added to a solution containing the members of the aluminum-nickel group, the hydroxides of all seven metals are first precipitated. The hydroxides of aluminum, chromium, and zinc are amphoteric (see Note 2 of Procedure 9 in Chapter 10, and Chapter 5) and dissolve in an excess of NaOH to form the complex ions $Al(OH)_4^-$, $Cr(OH)_4^-$, and $Zn(OH)_4^{2-}$, respectively. The hydroxides of iron, manganese, cobalt, and nickel are not amphoteric and do not dissolve in excess NaOH. The overall procedure is shown below.

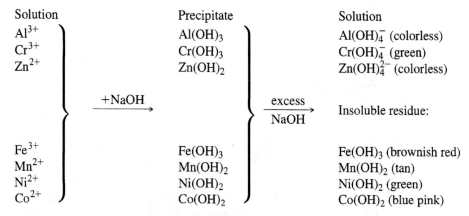

Solution	Precipitate	Solution
Al^{3+}	$Al(OH)_3$	$Al(OH)_4^-$ (colorless)
Cr^{3+}	$Cr(OH)_3$	$Cr(OH)_4^-$ (green)
Zn^{2+}	$Zn(OH)_2$	$Zn(OH)_4^{2-}$ (colorless)

$+$NaOH → , excess NaOH → Insoluble residue:

Fe^{3+}	$Fe(OH)_3$	$Fe(OH)_3$ (brownish red)
Mn^{2+}	$Mn(OH)_2$	$Mn(OH)_2$ (tan)
Ni^{2+}	$Ni(OH)_2$	$Ni(OH)_2$ (green)
Co^{2+}	$Co(OH)_2$	$Co(OH)_2$ (blue pink)

A careful observation of colors of solutions and precipitates formed in Procedure 16 may give valuable information regarding the cations present. The colors not already given above are as follows.

Al^{3+}	Colorless
Zn^{2+}	Colorless
Mn^{2+}	Colorless to faint pink
Fe^{3+}	Reddish brown to yellow
Cr^{3+}	Green
Ni^{2+}	Green
Co^{2+}	Reddish pink
$Al(OH)_3(s)$	White
$Cr(OH)_3(s)$	Grayish green
$Zn(OH)_2(s)$	White

Pure, freshly precipitated $Mn(OH)_2$ is tan, but it rapidly turns brown in contact with air because of oxidation to MnO_2 and Mn_2O_3.

The precipitate formed when NaOH is added to Co^{2+} may vary in color from blue to pink to tan to light brown. $Co(OH)_2$ exists in two forms, one blue and the other pink. The blue form is believed to be more finely dispersed. It changes to the coarser pink form on standing; heat speeds up the change. Air oxidizes some $Co(OH)_2$ to black $Co(OH)_3$.

A pink $CoCl_2$ solution turns blue when heated to boiling but regains its pink color when cooled. A pink solution of $CoCl_2$ turns blue when treated with concentrated HCl; the pink color is restored by dilution with water. These color changes are due to changes in the composition and structure of the complex ions present in the solution. The pink color is due to the $Co(H_2O)_6^{2+}$ ion; the blue color is due to the $Co(H_2O)Cl_3^-$ and $CoCl_4^{2-}$ ions. It should be noted that the ions responsible for the colors of the solutions given above are probably all complex in a manner similar to that already noted for the cobalt ion. They will, however, be treated as simple ions.

The fact that nickel and cobalt do not precipitate in the copper-arsenic group indicates that NiS and CoS are soluble in dilute HCl. In the analysis of the aluminum-nickel group, however, CoS and NiS do not dissolve in concentrated HCl. This contradictory behavior suggests that freshly precipitated NiS and CoS exist in a form readily dissolved by dilute HCl but, on standing, this soluble modification changes into a second form that is insoluble in both dilute and concentrated HCl.

PROCEDURE 16

Separation of the Aluminum Subgroup from the Nickel Subgroup

Treat the precipitate from Procedure 15 with 10 drops of 12 M HCl, mix thoroughly, transfer to a casserole, and boil gently for 30 s. If the precipitate is not completely dissolved, add 3 drops of 16 M HNO$_3$, mix thoroughly, and boil until a clear solution is obtained. Add 10 drops of cold water, transfer to a test tube, centrifuge to remove any precipitate of sulfur, and decant into a casserole. Note the color of the solution. Make the solution strongly alkaline with 8 M NaOH and mix thoroughly. If the quantity of precipitate is so large that the product is mushy or nonfluid, add 10 to 20 drops of water. Note the color of the precipitate and the solution. Then add 2 drops of 3 percent H$_2$O$_2$ (see Note 1), stir for 1 min, and then boil for 2 min (see Note 2), replenishing the water lost. Transfer to a test tube before the precipitate has had a chance to settle, and centrifuge. Decant, saving the decantate for Procedure 21. Note the color of the precipitate and of the decantate. Wash the precipitate three times with hot water and analyze according to Procedure 17.

Notes

1. Hydrogen peroxide is added to oxidize chromate(III) [Cr(OH)$_4^-$] to chromate(VI) [CrO$_4^{2-}$]. The reason for this is discussed in Note 1, Procedure 21. The equation for the reaction is

$$2Cr(OH)_4^- + 3H_2O_2 + 2OH^- \rightleftharpoons 2CrO_4^{2-} + 8H_2O \qquad (11.27)$$

$$\text{green} \hspace{5.5cm} \text{yellow}$$

Cr(OH)$_4^-$ is green; CrO$_4^{2-}$ is yellow. Therefore, if the color of the solution changes from green to yellow on treatment with H$_2$O$_2$, chromium is present.

Hydrogen peroxide in alkaline solution readily oxidizes Mn(OH)$_2$ to MnO$_2$ and Co(OH)$_2$ to Co(OH)$_3$, as follows (see Table 6.2).

$$H_2O_2 + Mn(OH)_2(s) \rightleftharpoons MnO_2(s) + 2H_2O \qquad (11.28)$$

$$\text{tan} \hspace{2.8cm} \text{black}$$

$$2Co(OH)_2(s) + H_2O_2 \rightleftharpoons 2Co(OH)_3(s) \qquad (11.29)$$

$$\text{blue/pink} \hspace{3.2cm} \text{black}$$

MnO$_2$ and Co(OH)$_3$ are black; hence, a blackening of the color of the precipitated hydroxides on addition of H$_2$O indicates the presence of manganese or cobalt.

2. Any excess H$_2$O$_2$ present in the solution must be decomposed by boiling; otherwise it will interfere with the subsequent analysis as discussed in Note 2, Procedure 21.

DETECTION OF IRON, COBALT, NICKEL, AND MANGANESE

Up to this point in the analysis, the general practice has been to separate a cation from all other cations before identifying it. The exceptions were Cu^{2+} and Cd^{2+}, which were identified in each other's presence, and Sb^{3+}, which was identified in the presence of Sn^{4+}. In the identification of Fe^{3+}, Co^{2+}, Ni^{2+}, and Mn^{2+}, separate samples of the same solution will be used in making each confirmatory test; the four ions will not be separated from each other prior to the test. This can be done because, in this particular quartet, none of any three interferes seriously with the confirmatory test for the fourth.

Iron(III) and nickel hydroxides, $Fe(OH)_3$ and $Ni(OH)_2$, dissolve readily in 2 M H_2SO_4 to give Fe^{3+} and Ni^{2+}, but MnO_2 and $Co(OH)_3$ are very slowly dissolved by 2 M H_2SO_4. Hydrogen peroxide speeds up the solution process by reducing MnO_2 and $Co(OH)_3$ to MnO and $Co(OH)_2$; these latter compounds, in which Mn and Co are in their lower oxidation states, are more basic in character and, as a result, are readily dissolved by H_2SO_4. The net equations for the reduction by H_2O_2 and the subsequent solution by H_2SO_4 are

$$MnO_2(s) + H_2O_2 \leftrightharpoons MnO(s) + H_2O + O_2 \tag{11.30}$$

$$2Co(OH)_3(s) + H_2O_2 \leftrightharpoons 2Co(OH)_2(s) + 2H_2O + O_2 \tag{11.31}$$

$$MnO(s) + 2H^+ \leftrightharpoons Mn^{2+} + H_2O \tag{11.32}$$

$$Co(OH)_2(s) + 2H^+ \leftrightharpoons Co^{2+} + 2H_2O \tag{11.33}$$

Attention has already been called to the fact that H_2O_2, in basic solution, *oxidizes* $Mn(OH)_2$ and $Co(OH)_2$ to MnO_2 and $Co(OH)_3$, respectively. In Procedure 17 we find that, in acid solution, H_2O_2 functions as a reducing agent and *reduces* MnO_2 and $Co(OH)_3$ to Mn^{2+} and Co^{2+}, respectively. This dual behavior of H_2O_2 can be understood from the data in Table 6.1 and in Table 6.2. H_2O_2, in both acid and basic solution, can function as either a reducing agent [Eqs. (37) in Table 6.1 and (10) in Table 6.2] or an oxidizing agent [Eqs. (59) in Table 6.1 and (22) in Table 6.2]. The role it plays depends on the relative reducing and oxidizing power of the substance with which it reacts and the relative rates of the reactions involved. In basic solution, the potential of the half-reaction $Mn(OH)_2(s) + 2OH^- \leftrightharpoons MnO_2(s) + 2H_2O + 2e^-$ (0.05 v) is only slightly lower than that of the half-reaction $H_2O_2 + 2OH^- \leftrightharpoons O_2 + 2H_2O + 2e^-$ (0.076 v), but very much higher than that of the half-reaction $2OH^- \leftrightharpoons H_2O_2 + 2e^-$ (−0.88 v).

Accordingly, the tendency for H_2O_2 to oxidize $Mn(OH)_2(s)$. to $MnO_2(s)$ will be very much greater than its tendency to reduce $MnO_2(s)$ to $Mn(OH)_2(s)$. This is equivalent to saying that, in basic solution, the reducing power of Mn(II) toward H_2O_2 is much greater than the oxidizing power of Mn(IV) toward H_2O_2. In acid solution, the situation is essentially reversed; although the potential of the half-reaction $Mn^{2+} + 2H_2O \leftrightharpoons MnO_2(s) + 4H^+ + 2e^-$ (−1.23 v) differs from that of the half-reaction $H_2O_2 \leftrightharpoons O_2 + 2H^+ + 2e^-$ (−0.682 v) by about the same amount as it does from that of the half-reaction $2H_2O \leftrightharpoons H_2O_2 + 2H^+ + 2e^-$ (−1.77 v), the rate of the reaction of H_2O_2 involving the former is considerably greater than that of the latter. As a result, the tendency for H_2O_2 to reduce MnO_2 to Mn^{2+} is greater than its tendency to oxidize Mn^{2+} to MnO_2. The same rationale applies to the behavior of Co(II) and Co(III). The potential of the half-reaction $Co^{2+} \leftrightharpoons Co^{3+} + e^-$ in acid solution [Eq. (60) in Table 6.1] is −1.82 v; which means that, in acid solution, H_2O_2 will not oxidize Co^{2+} to Co^{3+}. The tendency for Co^{3+} to be reduced to Co^{2+} will be very great. In basic solution the potential of the half-reaction $Co(OH)_2(s) + OH^- \rightleftharpoons Co(OH)_3(s) + e^-$ [Eq. (14)] is -0.17 v; accordingly, the tendency for H_2O_2 to oxidize Co(II) to Co(III) will be very much greater than its tendency to reduce Co(III) to Co(II).

It should be noted in the equations above that when H_2O_2 functions as a reducing agent, O_2 gas is given off. In effect, the H_2O_2 molecule gives off one atom of oxygen. This atom then picks up one atom of oxygen from the compound that functions as the oxidizing agent [$MnO_2(s)$ and $Co(OH)_2(s)$ in the equations above] to form a molecule of O_2 gas.

PROCEDURE 17

Preparation of the Solution; Detection of Iron

To the precipitate from Procedure 16, add 20 drops of 2 M H_2SO_4, mix thoroughly, and transfer to a casserole. Boil gently for 1 min, add a drop of 3 percent H_2O_2, and continue boiling for 1 min after the precipitate is completely dissolved. Add 10 drops of water, allow to cool, note the color of the solution, and divide into four approximately equal portions to use to test for iron, cobalt, nickel, and manganese.

Test for iron (see Note 1). To one portion, add 1 or 2 drops of 0.2 M KSCN. A blood red solution, due to $Fe(SCN)_6^{3-}$ (see Note 2), proves the presence of iron.

Notes

1. To test for iron in a solid, proceed as follows. Dissolve a small sample of the solid in 10 to 15 drops of dilute HCl. Divide the solution into three parts. Test one part for iron(III) ions according to Procedure 17. To the second part, add a few drops of potassium hexacyanoferrate(II), $K_4Fe(CN)_6$. A dark blue precipitate (Prussian blue) proves the presence of Fe^{3+} ions.

$$4Fe^{3+} + 3Fe(CN)_6^{4-} \rightleftharpoons Fe_4[Fe(CN)_6]_3(s) \quad \text{(dark blue)}$$

To the third part, add a few drops of potassium hexacyanoferrate(III), $K_3Fe(CN)_6$. A dark blue precipitate (Turnbull's blue) proves the presence of Fe^{2+} ions.

$$3Fe^{2+} + 2Fe(CN)_6^{3-} \rightleftharpoons Fe_3[Fe(CN)_6]_2(s) \quad \text{(dark blue)}$$

Note particularly that, whereas potassium hexacyanoferrate(II) (potassium ferrocyanide), $K_4Fe(CN)_6$, gives a dark blue precipitate with Fe^{3+}, potassium hexacyanoferrate(III) (potassium ferricyanide), $K_3Fe(CN)_6$, gives a dark blue precipitate with Fe^{2+}.

2. Although the $Fe(SCN)_6^{3-}$ ion is widely accepted as the predominant species responsible for the red color in the test for Fe^{3+}, there is little doubt that complexes with a smaller number of SCN^- ions, such as $FeSCN^{2+}$, $Fe(SCN)_2^+$, $Fe(SCN)_4^-$, and $Fe(SCN)_5^{2-}$, are also present. When the concentration of SCN^- is high, as is the case when 0.2 M KSCN is added in making the test, the predominant species is likely to be $Fe(SCN)_6^{3-}$.

Traces of iron are often introduced as impurities in the course of the manufacture of the compounds used in making up the unknown and compounds used as reagents in the analysis. Hence, the student should learn to distinguish between a weak test that indicates a trace of iron, probably as an impurity, and a strong positive test. In case of doubt, tests for iron(II) and iron(III) should be made on the original sample (see Note 1).

PROCEDURE 18

Detection of Cobalt (see Note 1)

To the second portion of the solution prepared in Procedure 17, add enough solid NaF to form a saturated solution; mix well by stirring. Then add 10 to 20 drops of a saturated solution of NH_4SCN in ethyl alcohol (see Note 2). The formation of a blue solution due to $Co(SCN)_4^{2-}$ proves the presence of Co^{2+}.

Notes

1. The identification of Fe^{3+} and Co^{2+} in Procedures 17 and 18 is unique in that each is identified by the use of the same reagent and each forms the same type of complex ion, $Fe(SCN)_6^{3-}$ and $Co(SCN)_4^{2-}$. The test for iron is carried out in dilute aqueous solution, in which the red complex ion $Fe(SCN)_6^{3-}$ is very stable; the blue complex ion $Co(SCN)_4^{2-}$ is unstable in aqueous solution. As a result, cobalt ions do not interfere with the test for iron. The test for cobalt is carried out in alcoholic solution and in the presence of fluoride ions. The blue complex ion $Co(SCN)_4^{2-}$ is stable in alcoholic solution; the Fe^{3+} ions are tied up in the form of the colorless stable complex ion FeF_6^{3-} (or, possibly, FeF^{2+}, FeF_2^+, FeF_4^-, or FeF_5^{2-}) and cannot, therefore, interfere with the test for cobalt.
2. The reaction of Co^{2+} with SCN^- is incomplete, due to the instability of the complex ion.

$$Co^{2+} + 4SCN^- \rightleftharpoons Co(SCN)_4^{2-}$$

An excess of SCN^- is necessary to give a good test.

PROCEDURE 19

Detection of Nickel

Make the third portion of the solution prepared in Procedure 17 basic with 5 M aqueous NH_3. If a precipitate of $Fe(OH)_3$ or $Mn(OH)_2$ forms, centrifuge and decant; to the clear decantate, add 2 to 4 drops of dimethylglyoxime, mix thoroughly, and allow to stand for 1 min. A strawberry red precipitate [$NiC_8H_{14}N_4O_4$; see Structure (1)] proves the presence of nickel (see Note 1).

Notes

1. The following reaction takes place in the confirmatory test for nickel:

$$2(CH_3)_2C_2(NOH)_2 + Ni(NH_3)_6^{2+} \rightleftharpoons 2NH_4^+ + NiC_8H_{14}N_4O_4(s) + 4NH_3$$
$$\text{red}$$

Since the red compound is readily soluble in acids, the reaction must be carried out in alkaline solution. Iron(II) ions react with $(CH_3)_2C_2(NOH)_2$ to form a soluble red compound; however, no precipitate is formed.

PROCEDURE 20

Detection of Manganese

Dilute the fourth portion of the solution prepared in Procedure 17 with an equal volume of water, add 2 drops of 3 M HNO_3, mix thoroughly, then add a few grains of solid sodium bismuthate (see Note 1), mix thoroughly, and allow to stand for 1 min. Centrifuge. A pink to reddish purple solution, due to the permanganate ion (MnO_4^-), proves the presence of manganese (see Note 2).

Notes

1. Bismuth is predominantly metallic in character; its common oxidation number is 3 and its compounds ionize to give Bi^{3+} ions. In the compound $NaBiO_3$, bismuth has an oxidation number of 5 and has the property of a nonmetal in that it is present in the negative ion BiO_3^-. This behavior of bismuth illustrates the rule that as the oxidation number of an element increases, it becomes more nonmetallic in character.

   The behavior of $NaBiO_3$ illustrates the rule that those compounds of polyvalent elements in which the element exists in one of its higher oxidation states generally can act as oxidizing agents. Other compounds (oxidizing agents) that illustrate this rule are $KMnO_4$ and $K_2Cr_2O_7$.

2. The oxidation of Mn^{2+} to MnO_4^- by sodium bismuthate ($NaBiO_3$) takes place as follows (see Table 6.1):

$$2Mn^{2+} + 5HBiO_3(s) + 9H^+ \leftrightharpoons 2MnO_4^- + 5Bi^{3+} + 7H_2O$$

   Reducing agents of any kind, such as chlorides and sulfides, will interfere with the confirmatory test for manganese because they reduce the violet MnO_4^- to the practically colorless Mn^{2+}.

$$2MnO_4^- + 10Cl^- + 16 H^+ \leftrightharpoons 2Mn^{2+} + 5Cl_2 + 8H_2O$$

$$2MnO_4^- + 5H_2S + 6H^+ \leftrightharpoons 2Mn^{2+} + 5S + 8H_2O$$

PROCEDURE 21

Separation and Detection of Aluminum

Treat the decantate from Procedure 16 (see Note 1) with 16 M HNO_3 until slightly acid (see Note 2); then add 15 M aqueous NH_3 until distinctly alkaline (see Note 3). Continue stirring the ammoniacal solution for 1 min. Centrifuge and decant, saving the decantate for Procedure 22. Because the $Al(OH)_3$ that precipitates is gelatinous, highly translucent, very finely divided, and the color of opaque, bluish white glass, its presence, suspended in the solution, is not easy to detect. On centrifuging, however, it will appear in the bottom of the test tube as a whitish, jellylike, opaque precipitate. Wash the precipitate three times with hot water; then dissolve it in 4 to 5 drops of 3 M HNO_3 (see Note 4). [If any precipitate fails to dissolve, remove it by centrifuging and decantation (see Note 5).] Add 2 drops of aluminon solution (see Note 6), mix thoroughly, make *just barely alkaline* with 5 M aqueous NH_3, again mix thoroughly, and then centrifuge. A cherry red precipitate [a so-called lake of $Al(OH)_3$ and adsorbed aluminon (see Note 7)] proves the presence of aluminum.

Notes

1. It is imperative that all chromate(III) ions $[Cr(OH)_4^-]$ be completely oxidized to chromate(VI) ions (CrO_4^{2-}) in Procedure 16. Any unoxidized $Cr(OH)_4^-$ will react as follows in Procedure 21:

$$Cr(OH)_4^- + 4H^+ \text{ (from HNO}_3) \rightleftharpoons Cr^{3+} + 4H_2O$$

$$Cr^{3+} + 3NH_3 + 3H_2O \rightleftharpoons Cr(OH)_3(s) + 3NH_4^+$$

This reaction of chromate(III) ions is the same as that which the aluminate ion $[Al(OH)_4^-]$ will undergo in the course of the separation.

$$Al(OH)_4^- + 4H^+ \text{ (from HNO}_3) \rightleftharpoons Al^{3+} + 4H_2O$$

$$Al^{3+} + 3NH_3 + 3H_2O \rightleftharpoons Al(OH)_3(s) + 3NH_4^+$$

It is, thus, evident that if $Cr(OH)_4^-$ is not oxidized to CrO_4^{2-}, $Cr(OH)_3$ will precipitate when the test for aluminum is made. It is obvious, therefore, that oxidation of $Cr(OH)_4^-$ to CrO_4^{2-} by H_2O_2 is necessary if a separation of chromium from aluminum is to be accomplished. Although $Cr(OH)_3$ is green, in small amounts the color is not marked, and the $Cr(OH)_3$ precipitate may be mistaken for $Al(OH)_3$.

    $Cr(OH)_3$ does not form a red lake with aluminon. Excessive amounts of $Cr(OH)_3$ (green) will, however, mask the red aluminon lake. Therefore, if much $Cr(OH)_3$ precipitates with the $Al(OH)_3$, it should be reoxidized to CrO_4^{2-} with H_2O_2.

2. When HNO_3 is added in the first step in Procedure 21, a precipitate sometimes forms when the solution is just about neutral and then redissolves when more HNO_3 is added. This precipitate may be either $Al(OH)_3$ or $Zn(OH)_2$, both of which are white; in case chromium is not completely oxidized to CrO_4^{2-}, $Cr(OH)_3$ (green) may precipitate. Two reactions account for the formation and disappearance of this precipitate.
   a. When HNO_3 is added to neutrality,

$$Al(OH)_4^- + H^+ \text{ (from HNO}_3) \rightleftharpoons Al(OH)_3(s) + H_2O$$

   b. When more HNO_3 is added,

$$Al(OH)_3(s) + 3H^+ \rightleftharpoons Al^{3+} + 3H_2O$$

    Excess H_2O_2, if present in the decantate from Procedure 16, will interfere with the separation and identification of aluminum and chromium. When HNO_3 is added in the first step in Procedure 21, the following reactions will take place.

$$2CrO_4^{2-} + 2H^+ \rightleftharpoons Cr_2O_7^{2-} + H_2O$$

$$Cr_2O_7^{2-} + 3H_2O_2 + 8H^+ \rightleftharpoons 2Cr^{3+} + 3O_2 + 7H_2O$$

The chromate is, thus, reduced to Cr^{3+}, which will then interfere with the test for aluminum, as already discussed in Note 1.

In alkaline solution, the H_2O_2 oxidizes trivalent chromium to chromate, whereas in acid solution it reduces chromate to trivalent chromium. The reason for this difference in behavior was discussed earlier.

3. When aqueous ammonia is added in Procedure 21, a white precipitate sometimes forms; it redissolves when more aqueous NH_3 is added. This precipitate is probably $Zn(OH)_2$. The reactions that account for its formation and redissolution are

$$Zn^{2+} + 2NH_3 + 2H_2O \rightleftharpoons Zn(OH)_2(s) + 2NH_4^+$$

$$Zn(OH)_2(s) + 4NH_3 \rightleftharpoons Zn(NH_3)_4^{2+} + 2OH^-$$

4. Since HCl may reduce chromate to Cr^{3+}, the solution is acidified with HNO_3 rather than with HCl.

5. If lead, tin, and antimony are not completely precipitated in the copper-arsenic group, they appear as insoluble hydroxides or basic salts in the final test for aluminum. The hydroxides of lead, tin, and antimony, like $Al(OH)_3$, are amphoteric. Also, their hydroxides, like $Al(OH)_3$, do not dissolve in excess NH_4OH. They do not, however, form the characteristic red lake with aluminon.

6. Aluminon is an organic dye. It is the ammonium salt of aurintricarboxylic acid, and its chemical formula is

$$(C_6H_3OHCOONH_4)_2C{:}C_6H_3OCOONH_4$$

Its structural formula is shown in (3).

3

7. A *lake* is formed when a colored compound (a dye) combines with or is adsorbed from solution by an insoluble gelatinous precipitate. The dye aluminon is preferentially adsorbed by $Al(OH)_3$; this dye is not adsorbed by $Cr(OH)_3$ or $Zn(OH)_2$ or by the hydroxides of lead, tin, and antimony.

PROCEDURE 22

Separation and Detection of Chromium and Zinc

Follow **(A)** if the decantate from Procedure 21 is colorless; follow **(B)** if it is yellow.

(A) *Decantate is colorless; therefore chromium is absent.* Place a drop of the decantate on a piece of diphenylthiocarbazone paper and allow to stand for 1 min. A purplish red coloration proves the presence of zinc ions (see Note 4; in the absence of the Zn^{2+} ion, NH_4OH will give a yellowish brown color). A confirmatory test for Zn^{2+} is given in Note 5.

(B) *Decantate is yellow; chromium is probably present* (see Note 1). Add 6 to 7 drops of 0.2 M $BaCl_2$ to the yellow decantate, mix thoroughly, centrifuge until the supernatant liquid is clear, and decant, saving the decantate for Part **(C)**. Wash the precipitate, $BaCrO_4$ mixed with some $BaSO_4$ (see Note 2), twice with hot water, add 3 drops of 3 M HNO_3, heat gently, but do not boil vigorously, and stir for about 1 min. Add 10 drops of cold water, mix thoroughly, cool under the cold water tap, and then add 10 drops of ether and 1 drop of 3 percent H_2O_2. Mix well by vigorously stirring and allow to settle. A blue coloration of the ether layer due to chromium peroxide (CrO_5) proves the presence of chromium.

(C) Place a drop of the decantate from **(B)** on a piece of diphenylthiocarbazone paper (see Note 3) and allow to stand for about 1 min. A purplish red coloration proves the presence of zinc ions (see Note 4; in the absence of the Zn^{2+} ion, NH_4OH will give a yellowish brown color). A confirmatory test for Zn^{2+} is given in Note 5.

Notes

1. The final test for chromium depends on the fact that in dilute acid solution $Cr_2O_7^{2-}$ interacts with H_2O_2 to form a deep indigo blue compound, chromium peroxide (CrO_5). Yellow chromate is first changed to orange dichromate $(Cr_2O_7^{2-})$.

$$BaCrO_4(s) \leftrightharpoons Ba^{2+} + CrO_4^{2-}$$

$$2CrO_4^{2-} + 2H^+ \leftrightharpoons 2Cr_2O_7^{2-} + H_2O$$

This change of chromate to dichromate, which always takes place when the solution is acidified, has already been mentioned in Note 2, Procedure 21.

The dichromate is then converted to CrO_5 by H_2O_2.

$$Cr_2O_7^{2-} + 4H_2O_2 + 2H^+ \leftrightharpoons 2CrO_5(s) + 5H_2O$$

The compound CrO_5, structure (**2**), is unstable and decomposes on standing to form Cr^{3+}. If the concentration of HNO_3 is low, the CrO_5 decomposes very slowly; if, however, the concentration of HNO_3 is high, CrO_5 decomposes so rapidly that the blue color may not be noticed. Furthermore, CrO_5 is very rapidly reduced to Cr^{3+} by excess H_2O_2. Hence, high concentrations of HNO_3 and excess H_2O_2 must be avoided.

$$4CrO_5(s) + 12H^+ \rightleftharpoons 4Cr^{3+} + 6H_2O + 7O_2$$

$$2CrO_5(s) + 6H^+ + 7H_2O_2 \rightleftharpoons 2Cr^{3+} + 10H_2O + 7O_2$$

CrO_5 is very soluble in ether; HNO_3 is not. Treating with ether partially separates the CrO_5 from the HNO_3 and concentrates it in the ether layer. Since separation from HNO_3 will increase the stability of CrO_5, extraction with ether preserves as well as concentrates CrO_5.

2. The precipitate obtained on addition of $BaCl_2$ may contain varying amounts of $BaSO_4$ because of oxidation of H_2S and S to H_2SO_4 in earlier procedures. $BaSO_4$ is white, whereas $BaCrO_4$ is light yellow. Very finely divided $BaCrO_4$ is such a pale yellow, however, that the precipitate may appear white even though it consists largely of $BaCrO_4$. The confirmatory test for chromium should therefore be completed as directed, even if the precipitate appears white.

   $BaCrO_4$, being the salt of the relatively weak acid H_2CrO_4, is soluble in HNO_3, whereas $BaSO_4$, being the salt of a strong acid, is insoluble.

3. Diphenylthiocarbazone paper is also called *dithizone paper*.

4. The purplish red color is believed to be due to the complex ion formed by interaction of Zn^{2+} ions with diphenylthiocarbazone molecules, which have the structure shown in (4).

$$C_6H_5-N\!=\!\!N-\overset{\overset{\displaystyle S}{\|}}{C}-\overset{\overset{\displaystyle H}{|}}{N}-\overset{\overset{\displaystyle H}{|}}{N}-C_6H_5$$

4

5. Make the decantate from (**B**) acid with 6 M HCl, then add 3 to 4 drops of 0.2 M $K_4Fe(CN)_6$ and mix thoroughly. The resulting mixture should be acidic. A grayish white to bluish green precipitate, $Zn_3K_2[Fe(CN)_6]_2$, proves the presence of zinc.

PROBLEMS

11.1 Using the scheme of analysis as a guide, write net equations for all reactions that take place in the precipitation and analysis of the aluminum-nickel group.

11.2 Upon what fact or facts is each of the following based?

   a. The separation of the cations of the aluminum-nickel group from the cations of the barium-magnesium group.

   b. The separation of the aluminum subgroup from the nickel subgroup.

11.3 Give the reason or reasons for each of the following.

   a. Addition of NH_4Cl before precipitating the aluminum-nickel group.

   b. Separate addition, first of NH_4OH, then of $(NH_4)_2S$, in the precipitation of the aluminum-nickel group.

    c. Separate addition of HCl and HNO_3, followed in each case by boiling, in dissolving the aluminum-nickel group precipitate (Procedure 16).

    d. Addition of H_2O_2 (Procedure 16).

    e. Boiling of the solution before centrifuging in the separation of the aluminum subgroup from the nickel subgroup (Procedure 16).

    f. Use of dilute HNO_3 rather than concentrated HNO_3 in the confirmatory test for chromium.

    g. Use of ether in the confirmatory test for chromium.

11.4 In the precipitation and analysis of the aluminum-nickel group, what difficulties, if any, would arise under the following conditions?

    a. If $Cr(OH)_4^-$ was not completely oxidized to CrO_4^{2-} (Procedure 16).

    b. If excess H_2O_2 was not decomposed (Procedure 16).

    c. If 1 cc of H_2O_2 was used instead of 1 drop in the confirmatory test for chromium.

11.5 Write an equation or equations for a reaction occurring in the precipitation and analysis of the aluminum-nickel group in which each reaction occurs.

    a. Fe(II) is oxidized to Fe(III).

    b. Mn(II) is oxidized to Mn(IV).

    c. Mn(III) is oxidized to Mn(VII).

    d. Mn(IV) is reduced to Mn(II).

    e. Mn(VII) is reduced to Mn(II).

    f. Cr(III) is oxidized to Cr(VI).

    g. Bi(V) is reduced to Bi(III).

    h. Co(II) is oxidized to Co(III).

    i. Fe(III) is reduced to Fe(II).

    j. CrO_5 is converted into Cr^{3+}.

    k. $Cr_2O_7^{2-}$ is converted into Cr^{3+}.

    l. CrO_4^{2-} is converted into $Cr_2O_7^{2-}$.

    m. $Al(OH)_3$ exhibits amphoterism.

    n. $Zn(OH)_4^{2-}$ is converted into Zn^{2+}.

    o. Ni^{2+} is converted into $Ni(NH_3)_4^{2+}$.

11.6 Give the formula for a chemical substance that will yield a precipitate (if excess of substance is required, so state).

    a. With $CrCl_3$ solution, but not with Na_2CrO_4.

    b. With $MnCl_2$ solution, but not with $KMnO_4$.

    c. With $FeCl_3$ solution, but not with $AlCl_3$.

    d. With Na_2CrO_4 solution, but not with $Zn(NO_3)_2$.

    e. With $NiCl_2$ solution, but not with $CrCl_3$.

    f. With $MnCl_2$ solution, but not with $AlCl_3$.

    g. With $FeCl_3$ solution, but not with $CoCl_2$.

    h. With $CoCl_2$ solution, but not with $ZnCl_2$.

    i. With $AlCl_3$ solution, but not with $ZnCl_2$.

    j. With Na_2CrO_4 solution, but not with $FeCl_3$.

    k. With $AlCl_3$ solution, but not with $CaCl_2$.

    l. With $NiCl_2$ solution, but not with $BaCl_2$.

11.7 Give the formula for a reagent that will react as described.

   a. Dissolve MnS, but not NiS.

   b. Dissolve ZnS, but not CuS.

   c. Dissolve $Al(OH)_3$, but not $Fe(OH)_3$.

   d. Dissolve $Ni(OH)_2$, but not $Fe(OH)^3$.

   e. Dissolve $Cr(OH)_3$, but not Bi_2S_3.

   f. Dissolve $Zn(OH)_2$, but not $Ni(OH)_2$.

   g. Dissolve Na_2CrO_4, but not $Al(OH)_3$.

11.8 Give the formula for a solid substance in the aluminum-nickel group that will react as described.

   a. Dissolve readily in 15 M NH_4OH and also in 8 M NaOH.

   b. Dissolve in 8 M NaOH, but not in 15 M NH_4OH.

   c. Dissolve in 15 M NH_4OH, but not in 8 M NaOH.

   d. Dissolve in 6 M HCl, but not in water.

   e. Dissolve in 6 M HCl and also in 8 M NaOH.

   f. Dissolve in HCl + HNO_3, but not in 6 M HCl.

11.9 How would you distinguish, by means of one reagent, between the following? (Tell what happens to each substance.)

   (a) *Solutions:*

Cr^{3+} and Al^{3+}	Fe^{3+} and Fe^{2+}
Cr^{3+} and Ni^{2+}	Fe^{3+} and CrO_4^{2-}
Mn^{2+} and Zn^{2+}	$Zn(NH_3)_4^{2+}$ and Zn^{2+}
Al^{3+} and Zn^{2+}	Zn^{2+} and Ba^{2+}
$Al(OH)_4^-$ and $Cr(OH)_4^-$	Mn^{2+} and Mg^{2+}

   (b) *Solids:*

ZnS and CoS	$Al(OH)_3$ and $Mn(OH)_2$
$Ni(OH)_2$ and $Cr(OH)_3$	MnO_2 and $Fe(OH)_3$
MnS and NiS	$Zn(OH)_2$ and $Al(OH)_3$
Na_2S and ZnS	CuS and FeS

11.10 Using as few operations as possible, how do you test for the following?

   a. Zn^{2+} in a solution known to contain Ni^{2+}.

   b. Mn^{2+} in a solution known to contain Co^{2+} and Cl^-.

   c. Al^{3+} in a solution known to contain Cr^{3+}.

   d. Fe^{2+} in a solution known to contain Fe^{3+}.

   e. Ni^{2+} in a solution known to contain Fe^{2+}.

11.11 Account for each of the following facts.

   a. MnO is readily soluble in H_2SO_4; MnO_2 is not readily soluble in H_2SO_4.

   b. In the presence of fluoride ions, Fe^{3+} ions do not give the characteristic red coloration with thiocyanate ions.

   c. When a deep blue alcoholic solution containing $Co(SCN)_4^{2-}$ ions is diluted with water, the blue color disappears.

d. When either NaOH or NH_4OH is added drop by drop to a solution of $ZnCl_2$, a white precipitate first appears, but redissolves—on addition of more of the reagent—to give a colorless, water-clear solution. When HCl is added, dropwise, to the water-clear solution, a white precipitate first appears, but redissolves—when more HCl is added—to give a water-clear solution.

e. When $(NH_4)_2S$ is added to a solution formed by dissolving solid $FeCl_3$ in water, a black precipitate forms.

f. When 5 M NH_4OH is added to a solution obtained by dissolving $MgCl_2$ in water, a white precipitate of $Mg(OH)_2$ forms; when 5 M NH_4OH is added to a solution obtained by dissolving a mixture of $MgCl_2$ and NH_4Cl in water, no precipitate forms.

g. When a solution prepared by mixing 1 M NH_4OH and 1 M NH_4Cl is added to a solution containing Mn^{2+} ions, there is at first no visible sign of reaction. After about 20 s, a tan coloration begins to appear; eventually a light brown precipitate forms.

h. When 2 M H_2SO_4 is added to a solid mixture of MnO_2 and $Co(OH)_3$, there is no sign of reaction. When 2 M H_2SO_4 and some 3 percent H_2O_2 are added, the solid mixture is quickly dissolved.

i. $Mg(OH)_2$, which is insoluble in water, is soluble in an aqueous solution of NH_4Cl.

11.12 Using the aluminum-nickel group scheme of analysis as your only source of information, select, in each line, the correct substance.

a. Weakest reducing agent in alkaline solution: Fe(II), Co(II), Ni(II).

b. Strongest reducing agent in acid solution: Mn^{2+}, Co^{2+}, Ni^{2+}.

c. Most stable complex ion: $Mn(OH)_4^{2-}$, $Zn(OH)_4^{2-}$, $Ni(OH)_4^{2-}$.

d. Sulfide with smallest solubility product: ZnS, CoS, MgS.

e. Amphoteric hydroxide: $Co(OH)_2$, $Zn(OH)_2$, $Ni(OH)_2$.

f. Substance with smallest instability constant: $Fe(OH)_6^{3-}$, FeF_6^{3-}, $Fe(SCN)_6^{3-}$.

11.13 A colorless solution known to contain only cations of the aluminum-nickel group was divided into three parts, which were treated as follows. What conclusions can be drawn?

a. To one part, NaOH was added, slowly and with constant stirring. A light-colored precipitate formed, part of which redissolved in an excess of NaOH. The solution was centrifuged and decanted. The precipitate was observed to darken on standing exposed to the air.

b. The second part gave no precipitate on being treated with NH_4Cl and excess NH_4OH.

c. The third part gave a light-colored precipitate when treated with NH_4Cl, NH_4OH, and $(NH_4)_2S$.

11.14 An aluminum-nickel group unknown solution was divided into three parts.

a. One part was treated with excess NaOH and H_2O_2. A light green precipitate formed.

b. One part was treated with NH_4Cl and excess NH_4OH. A white precipitate formed.

c. One part was treated with a solution of $(NH_4)_2S$ in 2 M NH_4OH. A black precipitate formed.

List the cations that are shown to be absent, those shown to be present, and those that are undetermined.

11.15 An unknown solid is made up of one or more of the following compounds: CuS, ZnS, $AlCl_3$, $NiCl_2$, MnO. Treatment of the solid with water leaves a black residue and a colorless decantate. The residue is partially soluble in 6 M HCl; the decantate from the HCl treatment yields a tan precipitate and colorless decantate when treated with an excess of NH_4OH. List the compounds shown to be absent, those shown to be present, and those that are undetermined.

11.16 On the addition of NH_4Cl and excess NH_4OH to a solution known to contain only cations of the aluminum-nickel group, a white, flocculent precipitate forms. On addition of $(NH_4)_2S$, a black precipitate forms. The combined precipitate does not dissolve completely in HCl, but dissolves completely in HCl + HNO$_3$. When this solution is treated with excess NaOH and then with H$_2$O$_2$ and centrifuged, the decantate is colorless and the precipitate is light green. What conclusions can be drawn?

11.17 On the addition of NH$_4$Cl and excess NH$_4$OH to a solution known to contain only cations of the aluminum-nickel group, a dark red precipitate forms. On addition of (NH$_4$)$_2$S, a black precipitate forms. The combined precipitate dissolves readily and completely in HCl. The resulting green solution, having first been boiled with a few drops of concentrated HNO$_3$, gives a reddish brown precipitate and a green solution when treated with excess NaOH. On being treated with H$_2$O$_2$, the solution turns yellow, but the precipitate undergoes no change in color. What conclusions can be drawn?

11.18 A solid material that was known to contain only cations of the aluminum-nickel group dissolved in HCl to give a green solution. The addition of NH$_4$Cl and excess NH$_4$OH gave a green precipitate, and the addition of (NH$_4$)$_2$S gave a black precipitate. The total precipitate was dissolved in HCl + HNO$_3$. When the solution thus formed was treated with excess NaOH followed by H$_2$O$_2$, a colorless solution and a reddish brown precipitate were formed. The precipitate was found to be soluble in HCl, H$_2$SO$_4$, or HNO$_3$. What conclusions can be drawn?

11.19 A solid unknown was known to be a mixture of two or more of the following solids: Ni(NO$_3$)$_2$, NH$_4$NO$_3$, Sn(NO$_3$)$_2$, Pb(NO$_3$)$_2$, Cr(NO$_3$)$_3$, Hg(NO$_3$)$_2$, CuSO$_4$, Zn(NO$_3$)$_2$, Fe(NO$_3$)$_2$, Mn(NO$_3$)$_2$, CoCl$_2$, SbCl$_3$. The solid dissolved readily and completely in cold water to give a clear, colored solution with no precipitate. Separate samples of this colored solution showed the following behavior.

a. One sample, when treated with sulfuric acid, gave a white precipitate.

b. A second sample, when treated with excess NH$_4$OH, gave a precipitate and a colored solution.

c. A third sample, when treated with SnCl$_2$, yielded a dark gray precipitate.

d. A fourth sample, when treated with excess NaOH, gave a precipitate and a colorless solution. When this colorless solution was treated, dropwise, with nitric acid, a precipitate first formed; it redissolved when more acid was added.

e. A fifth sample, when treated with excess NaOH and excess H$_2$O$_2$, gave a light green precipitate and a colorless solution.

Indicate for each compound whether it is present, absent, or indeterminate.

11.20 A student prepared the following six solutions and placed each in a separate unlabeled beaker.

1. 0.1 M CdCl$_2$.

2. 0.1 M SnCl$_4$.

3. 0.1 M AgNO$_3$.

4. A solution 0.1 M in FeCl$_3$ and 0.6 M in HCl.

5. A solution 0.1 M in NaCl and 0.2 M in Na$_2$CrO$_4$.

6. A solution 0.1 M in Cu(NH$_3$)$_4^{2+}$ and 0.1 M in SO$_4^{2-}$ and also containing excess NH$_3$.

While the student was gone from the laboratory, the six unlabeled beakers were moved and placed in a disordered group at the end of the desk. In order to identify the solution in each beaker, the student proceeded as follows: She labeled the six beakers A, B, C, D, E, and F. She then tested the solution in each beaker and found the following:

Solution A, on being treated drop by drop with 5 M NH$_4$OH, yielded a dark precipitate that dissolved in excess NH$_4$OH to give a clear solution. This clear solution yielded a black precipitate when treated with (NH$_4$)$_2$S.

Solution B showed no sign of any reaction when treated drop by drop with 5 M NH$_4$OH but yielded a black precipitate when treated with (NH$_4$)$_2$S.

Solution C, when treated drop by drop with (NH$_4$)$_2$S, yielded a yellow precipitate that dissolved in excess (NH$_4$)$_2$S to give a clear, colorless solution.

Solution D, when treated drop by drop with 5 M NH$_4$OH, yielded a white precipitate that dissolved in excess NH$_4$OH to give a clear, colorless solution.

Solution E, when treated drop by drop with 0.2 M AgNO$_3$, gave a precipitate that was white at first but turned red when more AgNO$_3$ was added.

Solution F, when treated with (NH$_4$)$_2$S, yielded a finely divided yellowish white precipitate; when NH$_4$OH was then added, a black precipitate formed.

Identify each of the solutions A, B, C, D, E, and F.

The Barium-Magnesium Group

After removal of the preceding groups, the remaining ions—barium, calcium, magnesium, sodium, potassium, and ammonium—constitute the ions of the barium-magnesium group.

THE CHEMISTRY OF THE BARIUM-MAGNESIUM GROUP IONS

Barium Ion, Ba^{2+}

Barium exists in its compounds only in the 2+ state and is not reducible to the metal in aqueous systems. Most of the common barium salts are soluble in water or dilute strong acids and are colorless in solution. The main exception is the sulfate $BaSO_4$, a finely divided white powder that is essentially insoluble in all common reagents. Barium chromate, $BaCrO_4$, is insoluble in acetic acid, but will dissolve in solutions of strong acids. The yellow color of $BaCrO_4$ is caused by the chromate ion. The oxalate and phosphate are soluble in acidic systems at a pH of 3 or less. Barium hydroxide is a strong base and is moderately soluble in water ($\sim$0.2 M) (Table 12.1). Barium salts impart a green yellow color to a flame, a property that is useful in qualitative analysis.

Calcium Ion, Ca^{2+}

Calcium in its compounds occurs in the 2+ state, and the ion is very difficult to reduce to the metal; typical insoluble calcium compounds are $CaCO_3$, $Ca(OH)_2$, CaC_2O_4, and $Ca(PO_4)_2$.

TABLE 12.1 THE SOLUBILITY PRODUCTS OF THE BARIUM-MAGNESIUM GROUP HYDROXIDES

Substance	K_{sp}
$Ba(OH)_2$	5×10^{-3}
$Ca(OH)_2$	6×10^{-6}
$Mg(OH)_2$	8.9×10^{-12}

Calcium salts, which are all colorless unless the anion is colored, are typically soluble in water or dilute acids. Like barium, calcium does not form many common complex ions. Although the hydroxide is less soluble ($\sim 0.02\ M$) than $Ba(OH)_2$, it will not precipitate in ammonia, but it can be precipitated by 6 M NaOH.

Calcium oxalate, CaC_2O_4, is white and not appreciably soluble in acetic acid, but it will dissolve in dilute strong acid solutions.

Calcium salts impart an orange red color to a Bunsen burner flame.

Magnesium Ion, Mg^{2+}

Magnesium exists in its compounds only in the 2+ oxidation state. Magnesium compounds are white (except those containing colored anions); the sulfate, sulfide, and sulfite are soluble in water. Magnesium(II) does not form any complex compounds that are important in qualitative analysis.

Magnesium hydroxide is the least soluble of the barium-magnesium group hydroxides, Table 12.1 ($K_{sp} = 8.9 \times 10^{-12}$), and can be precipitated by either NaOH or aqueous ammonia. Other typical insoluble magnesium salts are $MgCO_3$ and $Mg_3(BO_3)_2$. Probably the least soluble salt of magnesium is $MgNH_4PO_4$.

Sodium Ion, Na^+

Virtually all sodium salts are soluble in water, and they are nearly all colorless. Sodium ion forms no complex ions. The virtual inertness of Na^+ makes this a difficult species to detect qualitatively using conventional strategy. The best known insoluble salt of sodium is zinc uranyl acetate, a pale yellow substance, $NaZn(UO_2)_3(C_2H_3O_2)_9$. Sodium salts impart a characteristic yellow color to a flame; this flame test is very sensitive. The ubiquitous nature of sodium salts and the sensitivity of the flame test make the test useless unless great care is taken to keep equipment clean and carefully rinsed with distilled water.

Potassium Ion, K^+

The properties of potassium compounds are much the same as those of sodium. If anything, potassium salts tend to be more soluble than their sodium counterparts. Two of the more common, less soluble potassium salts are the yellow potassium sodium hexanitrocobaltate(III), $K_2Na[Co(NO_2)_6]$, and the yellow potassium hexachloroplatinate(IV), K_2PtCl_6. These find limited use for identification purposes. Potassium salts impart a characteristic violet color to a burner flame, but the test is much less sensitive than that for sodium and is easily masked by trace amounts of sodium.

Ammonium Ion, NH_4^+

The ammonium ion, NH_4^+, is not a metallic cation; however, it does form a large number of salts with properties very similar to those of the corresponding potassium salt. Solutions of ammonium salts hydrolyze and are acidic [Eq. (12.1)].

$$NH_4^+ + H_2O \leftrightharpoons NH_3 + H_3O^+ \qquad (12.1)$$

Unlike the alkali metal ions, NH_4^+ does not give a flame test.

DETECTION OF AMMONIUM SALTS

Since ammonium salts have been added throughout the analysis, the identification of ammonium ion must be carried out on a sample of the original unknown; this is done in Procedure 23. Accordingly, the ammonium ion is not included in the systematic barium-magnesium group procedure (Figure 12.1).

When an ammonium salt is heated with a strong base, ammonia gas evolves. This fact is the basis for the detection of ammonium salts.

PROCEDURE 23

Detection of Ammonium

Place 5 drops of 8 M NaOH in a casserole. Add 5 drops of the original solution to be tested, or a small amount of the original solid. Cover the casserole with a watch glass, on the bottom side of which has been attached a piece of moist red litmus paper; the moist litmus paper should stick easily to the watch glass. Warm the casserole gently, but *do not boil*. The evolution of NH_3 gas, detected by its action on the piece of red litmus or by its odor, confirms the presence of ammonium ions. The test can also be carried out in a test tube (see Note 1), with the piece of moist red litmus paper held near the mouth of the test tube while its contents are heated in a boiling water bath or open flame.

Note

1. In testing for escaping NH_3 gas, do not touch the casserole or the end of the test tube with the litmus paper; it may come in contact with NaOH.

THE SEPARATION AND DETECTION OF BARIUM

Barium sulfate is very insoluble; its solubility product is 1.5×10^{-9}. Calcium sulfate is sparingly soluble; its solubility product is 2.4×10^{-5}. The sulfates of magnesium, sodium, potassium, and ammonium are very soluble.

The separation of barium from the other members of the group is based on these facts.

PROCEDURE 24

Separation and Detection of Barium

If the solution to be analyzed is a barium-magnesium group known or unknown, follow (**A**). If the decantate from Procedure 15 in Chapter 11 is being analyzed, follow (**B**).

(**A**) First make a flame test on 2 drops of the solution to be analyzed in the manner described in Note 1. Record the results of this test. Place 6 drops of the solution to be analyzed in a test tube, add 1 drop of 0.2 M $(NH_4)_2SO_4$, and mix thoroughly (see Note 2). A white precipitate ($BaSO_4$) shows the presence of Ba^{2+} ions (if no precipitate forms, barium is absent; in

FIGURE 12.1 Analysis of the barium-magnesium group.

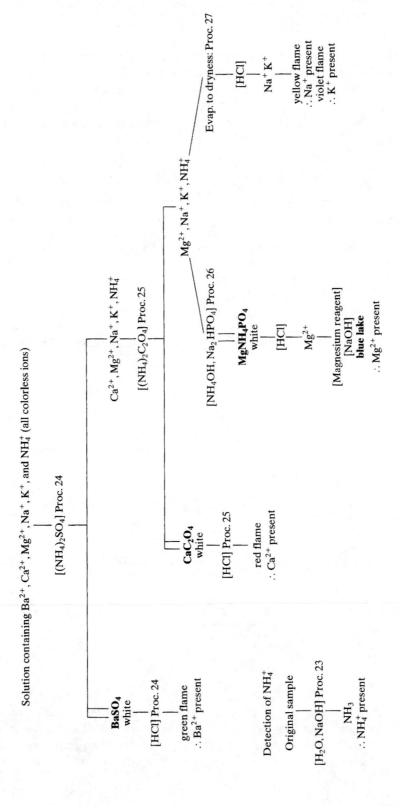

that case, go directly to the separation and detection of calcium, Procedure 25). Continue adding 0.2 M $(NH_4)_2SO_4$ dropwise, with thorough mixing, until precipitation of $BaSO_4$ is complete. Centrifuge and decant, saving the decantate for the separation and detection of calcium, Procedure 25. Wash the precipitate ($BaSO_4$) twice with 10-drop portions of hot water, then add 3 drops of 12 M HCl, mix well, and make a flame test on the supernatant liquid as directed in Note 1. A green flame proves the presence of barium (see Note 3).

(B) Place the entire decantate from Procedure 15 in Chapter 11 in a casserole, evaporate it down to a volume of 7 to 8 drops, transfer to a test tube, and centrifuge (see Note 4). Decant into a casserole, discarding any precipitate. Add 2 drops of 6 M HCl, evaporate to dryness, and bake until absence of sublimation indicates that excess ammonium salts have been driven off. Cool. Then add 7 to 8 drops of hot water, stir for about 1 min or until most of the solid has dissolved, and transfer the solution to a test tube (see Note 5). If the solution is clear, proceed as directed in **(A)**. If the solution is not clear, then centrifuge, decant the clear supernatant liquid into a test tube, and proceed as directed in **(A)**.

Notes

1. To make a flame test, seal a 1-cm piece of platinum wire into one end of a 3-cm piece of glass tubing; a small loop in the end of the wire will retain more of the unknown and give a stronger flame test. This platinum wire will serve for all subsequent flame tests (a 1.5-cm piece of *chromel*, or nichrome, wire with one end inserted in a cork stopper for a handle may be used instead of a platinum wire). Clean the platinum wire by alternately heating it to redness and thrusting it into concentrated HCl contained in a test tube until it gives no coloration to the nonluminous flame of the Bunsen burner.

    Dip the clean wire into the solution to be tested; then hold it in the nonluminous flame and note any colorations.

    The following ions give characteristic flame colorations: barium, green; borate, pale green; calcium, orange red; copper, green; potassium, violet; sodium, yellow.

    Since the chlorides of metals are in general more readily volatized than other salts, chlorides are used in making the flame test.

    A flame test performed on an original sample of the known solution for the barium-magnesium group will not be informative since the color of the flame will be a mixture of the colors produced by barium, calcium, sodium, and potassium. Therefore, when the barium-magnesium known solution is being analyzed, this flame test should be omitted. When a barium-magnesium group unknown or a salt or salt mixture is being analyzed, however, this preliminary flame test may be very helpful. Thus, if the flame test shows no coloration whatever—thereby indicating the absence of Ca^{2+}, Ba^{2+}, K^+, and Na^+—the separation and identification of these ions can be omitted. Likewise, the test may show the presence of K^+ and the absence of Ca^{2+}, Ba^{2+}, and Na^+, and so on. The fact that one flame color may cover up or modify another color must not be overlooked.

    By performing the flame test on a sample of the original material, the possibility of an erroneous test due to contamination is greatly reduced. Contamination with Na^+ during the analytical procedures is inevitable, so a flame test for sodium should always be made on the original substance.

2. If the barium-magnesium group unknown is issued in the form of a finely divided solid, proceed as follows. In a test tube place as much of the solid as can be carried on $\frac{1}{4}$ in. of the tip of a spatula. Add 3 to 4 drops of 6 M HCl, warm gently, and then add 10 to 12 drops

of water. Mix thoroughly and heat carefully until a clear solution is obtained. Proceed with the analysis of 6 drops of this solution as directed in Procedure 24(**A**).

     If the solid unknown is not finely divided when received, the entire sample should first be pulverized in a clean mortar.

3. Only a trace of $BaSO_4$ will dissolve in 12 M HCl; however, this trace will give a green flame test for Ba^{2+}.

4. Concentrating the decantate from the aluminum-nickel group will remove any cations of preceding groups as sulfides or hydroxides.

5. If no residue remains in the casserole after the ammonium salts are sublimed, then barium, calcium, magnesium, sodium, and potassium are absent.

THE SEPARATION AND DETECTION OF CALCIUM

Calcium oxalate is very insoluble; its solubility product is 1.3×10^{-9}. The oxalates of magnesium, sodium, potassium, and ammonium are very soluble. The separation and detection of calcium is based on these facts.

PROCEDURE 25

Separation and Detection of Calcium

Test the reaction of the decantate from Procedure 24 with litmus; if it is acidic, add 5 M NH_4OH, carefully, until it is slightly basic (see Note 1). Then add 1 drop of 0.2 M $(NH_4)_2C_2O_4$ (ammonium oxalate). Formation of a white precipitate (CaC_2O_4) shows the presence of Ca^{2+} ions (if no precipitate forms, calcium is absent; in that case, go directly to the analysis for Mg^{2+}, Na^+, and K^+ according to Procedure 26). Continue adding 0.2 M $(NH_4)_2C_2O_4$, with thorough stirring, until precipitation of CaC_2O_4 is complete. Centrifuge and decant, saving the decantate for analysis for Mg^{2+}, Na^+, and K^+ according to Procedure 26. Wash the precipitate (CaC_2O_4) twice with

10-drop portions of hot water, add 2 drops of 6 M HCl, mix well, and run a flame test on the solution as directed in Note 1 of Procedure 24. An orange red flame further confirms the presence of calcium (the calcium flame test is more elusive than that for barium; it shows up as a brick-red fluffiness as soon as the wire is placed in the flame and again as a deeper red at the end of the heating).

Note

1. Since CaC_2O_4 is the salt of a moderately weak acid, it is soluble in strong acids. Therefore, the solution should be basic when CaC_2O_4 is being precipitated.

PROCEDURE 26

Separation and Detection of Magnesium

Save half of the decantate from Procedure 25 for the detection of sodium and potassium, Procedure 27; treat the other half with 1 drop of 5 M aqueous NH_3 and 4 drops of 0.2 M Na_2HPO_4, mix well, warm gently, and allow to stand for 1 min. A white precipitate ($MgNH_4PO_4$) indicates the presence of magnesium (see Note 1). Centrifuge and decant, discarding the decantate. Wash the precipitate three times with hot water, dissolve in 2 to 3 drops of 2 M HCl, and add 3 or 4 drops of magnesium reagent. Then add 8 M NaOH, with constant mixing, until the solution is distinctly alkaline and centrifuge, but do not decant. A blue lake (flocculent precipitate) proves the presence of magnesium (see Note 2).

Notes

1. The precipitate of $MgNH_4PO_4$ sometimes forms very slowly. If a precipitate does not form within 1 min, heat to boiling, cool, and allow it to stand for 1 min. If no precipitate forms, magnesium is absent.

    The white precipitate of magnesium ammonium phosphate is formed as follows:

$$Mg^{2+} + HPO_4^{2-} + NH_3 \rightleftharpoons MgNH_4PO_4(s) \tag{12.2}$$

2. Since the phosphates of other metals besides magnesium may precipitate at this point, it is necessary to show that the precipitate is a magnesium compound. The special magnesium reagent is a solution of the complex organic dye, p-nitrobenzeneazoresorcinol, $C_6H_3(OH)_2—N{=}N—C_6H_4NO_2$. In this particular test, the OH^- ions from the 8 M NaOH react with the Mg^{2+} ions to form insoluble $Mg(OH)_2$. This $Mg(OH)_2$ then combines with the dye, or possibly absorbs it, to give a blue lake.

 Nickel and cobalt hydroxides, if present, form similar blue lakes. Excess ammonium ions reduce the sensitivity of the test by interfering with the formation of the lake.

DETECTION OF SODIUM AND POTASSIUM

All cations that have been discussed up to this point form a number of highly insoluble compounds; several of them form complex ions of distinctive color. For that reason it has been comparatively easy to separate and identify them. The cases of potassium and sodium are quite different. Each of these ions forms only one or two compounds that are sufficiently insoluble to make them worthy of consideration for separation and identification. Furthermore, these insoluble compounds are not highly insoluble; therefore, the tests that they provide are not nearly so sensitive as are the tests for the cations already discussed. The insoluble potassium compound formed in such a procedure is $K_2NaCo(NO_2)_6$; the insoluble sodium compound is $NaZn(UO_2)_3(C_2H_3O_2)_9$. The insoluble potassium salt is formed with a fairly low concentration of potassium ions; the insoluble sodium salt forms only if the concentration of sodium ions is high; both salts may form slowly. In each case, the reagent used to form the insoluble compound must be made up according to exact specifications if it is to be satisfactory; this is particularly true of the sodium reagent.

In summary, the identification of sodium and potassium by precipitation tests is not very successful and, for that reason, will not be used in this book. The precipitation tests are, however, given as notes in Procedure 27.

Fortunately, sodium and potassium have very sensitive and distinctive flame tests. These flame tests are used for their identification.

The sodium flame test is so sensitive that even a trace of sodium ion gives a characteristic fluffy yellow coloration. Traces of sodium ion get into solutions as a result of contact with glassware; sodium chloride that is continually being deposited on the surface of the skin as a result of evaporation of perspiration is transferred to anything touched by the hands. The net result is that every solution will give a positive test for sodium ion. The question to be decided is not whether sodium is absent or present, but whether it is present in a small amount or a large amount. You are urged not to get discouraged by the lack of definiteness of results in the test for sodium ions in an unknown. If a more definite test were available, it would be used.

The fluffy yellow sodium flame will cover up the color due to any other ion. In particular, it will completely cover the lavender color of the potassium flame. A cobalt blue glass will screen out the yellow sodium light, but will not affect the potassium coloration. Potassium can, therefore, be detected in the presence of sodium by viewing the flame through cobalt blue glass.

PROCEDURE 27

Detection of Sodium and Potassium

Evaporate the remaining half of the decantate designated in Procedure 26 to dryness and bake in the casserole at the maximum temperature of the flame until the absence of white smoke or vapor indicates that all NH_4Cl has sublimed off (see Note 1). If there is no solid residue in the casserole, sodium and potassium are absent. Cool, then treat the solid residue with 2 drops of 6 M HCl, and make a flame test as directed in Note 1 of Procedure 24. A fluffy yellow flame proves the presence of sodium; a lavender flame proves the presence of potassium and the absence of sodium. In reporting the results of this test, note whether the yellow flame is fairly diffuse and of short duration, indicating the presence of sodium as impurity, or if it is dense, fluffy, and of long duration, indicating the presence of sodium ions as a *bona fide* component of the sample being tested. To aid you in making a decision, run a flame test on a sample solution of the original unknown before it has been submitted to any of the procedures. To further aid you, compare the test that you get with that given by a 0.2 M solution of NaCl and by a sample of distilled water.

If sodium is present, observe the yellow flame through the blue cobalt glass; it may be desirable to have a fellow student thrust the platinum wire into the flame so that you can give all your attention to observing the color. A violet coloration of short duration, when viewed through the blue glass, proves the presence of potassium (see Note 2). In reporting the results of this test, note whether the violet coloration is a small flash, indicating a trace present as impurity, or is a strong and extensive coloration of reasonable duration, indicating a *bona fide* presence. To aid you in making a decision, perform the test on a sample of the original unknown solution, on a sample of 0.2 M KCl from the reagent shelf, and on a sample of distilled water.

Precipitation tests for sodium and potassium may, if desired, be made on the solution remaining in the casserole in the manner described in Note 3.

Notes

1. Ammonium salts sublime when heated and, consequently, make it difficult to detect the color of the sodium and potassium flames; hence, their removal by sublimation.
2. A violet coloration is observed when a green flame is viewed through the blue glass. Since barium, copper, and borate ions give a green coloration to the flame, they must be absent when the flame test for potassium is made.

The glowing of the red-hot wire must not be mistaken for a potassium test. Potassium, if present, will give a violet coloration to the flame the moment the wire is introduced.

3. *Precipitation tests for sodium and potassium.* Evaporate the solution on which the flame tests for sodium and potassium have been made to dryness, allow to cool, add 3 or 4 drops of water, and swish around for 2 min. Transfer to a dry test tube and, if necessary, centrifuge to give a clear supernatant liquid. Add 1 drop of this supernatant liquid and 2 drops of zinc uranyl acetate solution (sodium reagent) to a dry test tube, mix thoroughly, and allow to stand for 5 min. A pale yellow crystalline precipitate $[NaZn(UO_2)_3(C_2H_3O_2)_9]$ proves the presence of sodium. Failure of this precipitate to form does not positively prove the absence of sodium.

Transfer the remainder of the clear supernatant liquid to a dry test tube, add 1 drop of 5 M $HC_2H_3O_2$, and mix thoroughly. Then add a volume of sodium hexanitrocobaltate(III) solution, $Na_3Co(NO_2)_6$, equal to the volume of the solution in the test tube, mix thoroughly, and allow to stand for 5 min. A yellow precipitate $[K_2Na[Co(NO_2)_6]]$ proves the presence of potassium. Absence of a yellow precipitate does not positively prove the absence of potassium.

It should be pointed out that ammonium ions will give a yellow precipitate $[(NH_4)_2NaCo(NO_2)_6]$. Therefore, ammonium salts must be completely removed by sublimation; otherwise this test is no good.

PROBLEMS

12.1 Using the scheme of analysis as a guide, write net equations for all reactions that take place in the precipitation and analysis of the barium-magnesium group.

12.2 How do you account for the fact that there are no oxidation-reduction reactions among the barium-magnesium group equations?

12.3 Upon what fact or facts are the following based?

    a. The separation of barium ions from calcium and magnesium ions.

    b. The confirmatory test for ammonium ions.

12.4 Consider the precipitation and analysis of the barium-magnesium group.

    a. Why run a flame test on the original solid or solution?

    b. Why test for ammonium on a sample of the original material?

12.5 How would you distinguish between the following solids?

    a. NH_4Cl and $NaCl$

    b. $(NH_4)_2C_2O_4$ and $(NH_4)_2SO_4$

    c. $BaCl_2$ and $CaCl_2$

    d. $(NH_4)_2C_2O_4$ and NH_4Cl

12.6 How would you test for each?

    a. Mg^{2+} in a solution known to contain Ca^{2+}.

    b. K^+ in a solution known to contain Ba^{2+}.

    c. Ca^{2+} in a solution known to contain Ba^{2+}.

    d. Ba^{2+} in a solution known to contain Mg^{2+}.

12.7 Using the scheme of analysis of the barium-magnesium group as a guide, select, in each of the following pairs, the substance with the smaller solubility product.

    a. $CaSO_4$ and $BaSO_4$

    b. CaC_2O_4 and $CaSO_4$

    c. MgC_2O_4 and CaC_2O_4

12.8 An unknown is made up of equivalent amounts of two or more of the six solid compounds listed below. On the basis of the following information, indicate whether the substance is absent, present, or impossible to determine.

$$Ba(NO_3)_2, \ CaCl_2, \ Na_2CrO_4, \ NH_4Cl, \ MgSO_4, \ K_2Cr_2O_7$$

    a. A white residue remained when the solid was mixed with an adequate amount of water.

    b. The mixture of residue and solution was separated.

    c. The residue was completely insoluble in dilute HCl.

    d. A sample of the solution yielded no precipitate when treated with aqueous NH_3.

    e. A sample of the solution gave an orange red flame.

12.9 A solid unknown contains equivalent amounts of two or more of the six compounds listed below. On the basis of the following information, indicate whether a compound is present, absent, or impossible to determine.

$$Ca(NO_3)_2, \ NaCl, \ (NH_4)_2SO_4, \ BaSO_4, \ MgC_2O_4, \ BaCl_2$$

    a. A white precipitate remained when the solid was dissolved in water. The precipitate and solution were separated.

    b. The precipitate from part (a) was completely soluble in dilute HCl.

    c. The solution from part (a) yielded no precipitate when treated with aqueous NH_3.

12.10 A solid is known to consist of equimolar quantities of two or more of the water-soluble compounds listed below. On the basis of the information given below, indicate which of the following compounds are present, which are absent, and which are impossible to determine.

$$BaCl_2, \ Ca(NO_3)_2, \ MgCl_2, \ K_2CrO_4, \ NaCl, \ (NH_4)_2SO_4, \ (NH_4)_2C_2O_4$$

    a. When the solid, contained in a beaker, is treated with enough water to yield a 0.1 M solution of each substance that dissolves, a white solid and a colorless solution remain in the beaker. The white solid (S_1) is separated from the solution (S_2) by filtration.

    b. Solid S_1 dissolves readily and completely in dilute HCl.

    c. When a sample of solution S_2 is treated with 0.1 M $Ba(NO_3)_2$, a white precipitate insoluble in dilute HCl is formed.

    d. A second sample of solution S_2 gives no precipitate when treated with aqueous NH_3.

12.11 A mixture of dry salts is known to consist of equivalent amounts of two or more of the following salts: $BaCl_2, \ AgNO_3, \ Na_2C_2O_4, \ ZnSO_4, \ Mg(NO_3)_2, \ NH_4Cl, \ K_2CrO_4$.

    When sufficient water was added to form 0.20 M solutions of all soluble species, a white precipitate (A) and a colorless solution (B) were present in the beaker.

    The white precipitate (A) was divided into three parts. One part formed a clear solution when treated with 3 M HNO_3. A second part formed a clear and colorless solution when treated with 3 M HCl. A third part left a white precipitate when treated with 3 M H_2SO_4.

The colorless solution B formed a colorless solution with no precipitate when treated with excess aqueous NH_3.

Indicate whether each substance is absent, present, or indeterminate.

12.12 A mixture of dry salts is known to consist of equivalent amounts of two or more of the following substances: $Na_2C_2O_4$, $(NH_4)_2CrO_4$, $BaSO_4$, $Ca(NO_3)_2$, $MgCl_2$, NH_4Cl.

When sufficient water was added to form 0.20 M solutions of all soluble species, a white solid (A) and a colorless solution (B) were present in the beaker. The white solid was found to be readily and completely soluble in 3 M HCl. When the colorless solution (B) was treated with NH_4OH, a white precipitate formed.

Indicate whether each substance is absent, present, or indeterminate.

12.13 A solid unknown consists of equimolar amounts of two or more of the following compounds: $Ca(NO_3)_2$, $BaCl_2$, $Mg(C_2H_3O_2)_2$, Na_2SO_4, $(NH_4)_2C_2O_4$, K_2CrO_4.

When the mixture is treated with sufficient water to give a 0.10 M solution of each compound that dissolves, a white precipitate (A) and a colored solution (B) are present in the beaker.

The precipitate (A) is found to be readily and completely soluble in 2 M HCl. When exactly 50 mL of solution B are mixed with exactly 50 mL of 0.20 M Na_2HPO_4, a white precipitate (C) is formed.

On the basis of the information given above, state whether each compound is absent, present, or impossible to determine.

12.14 Suppose you wish to include an additional cation among those to be separated and identified in this course. What properties of this cation and its compounds must you know in order to decide where it will fit in the total scheme of analysis?

The Analysis of Alloys

When two or more metals are melted together and the resulting liquid is cooled until completely solidified, the solid product is called an *alloy*. Nonmetals are frequently found in alloys, sometimes as impurities and sometimes as intentional additions, designed to give certain desirable properties. The procedure outlined in this chapter, however, includes only the separation and identification of the metallic constituents of an alloy.

Obviously, the dissolution of the sample is of prime importance if the ideas of qualitative analysis developed in this book are to be applied successfully to the analysis of alloys. The choice of reagent to dissolve an alloy is dependent upon the nature of the alloy, but then if we knew the composition, there would be probably no reason to perform a qualitative analysis. From a practical point of view, aqua regia is used as a general solvent for alloys. This mixture is a more powerful solvent than other concentrated acids because of the combined action of the NO_3^- and Cl^- ions present. The NO_3^- ion acts as an oxidizing agent and the Cl^- ion forms complex ions with many metal ions. Aqua regia is usually prepared, as needed, by mixing 2 or 3 volumes of concentrated HCl with one volume of concentrated HNO_3; the proportions are usually not critical.

The following net equations represent the reactions that are believed to take place when the metals that constitute an alloy are dissolved.

1. *Silver forms insoluble AgCl.*

$$3Ag(s) + NO_3^- + 3Cl^- + 4H^+ \rightleftharpoons 3AgCl(s) + NO + 2H_2O \tag{13.1}$$

2. *Metals of periodic Groups Ia and IIa form simple cations.*

$$3Ca(s) + 2NO_3^- + 8H^+ \rightleftharpoons 3Ca^{2+} + 2NO + 4H_2O \tag{13.2}$$

3. *Arsenic forms arsenic acid.*

$$3As(s) + 5NO_3^- + 5H^+ + 2H_2O \rightleftharpoons 3H_3AsO_4 + 5NO \tag{13.3}$$

4. *All other metals are likely to form complex chloro ions. The reaction of lead is an example.*

$$3Pb(s) + 2NO_3^- + 12Cl^- + 8H^+ \rightleftharpoons 3PbCl_4^{2-} + 2NO + 4H_2O \qquad (13.4)$$

The composition of the complex chloro ion formed by a specific cation generally will vary with the concentration of the Cl^- ions. For example, with increasing concentration of Cl^- ions, the composition of the iron(III) complex will change from $FeCl_4^-$ to $FeCl_5^{2-}$ to $FeCl_6^{3-}$; it is probable that all three of these ions exist in equilibrium with each other, and it is not a simple matter to decide which one predominates.

Complex chloro ions that are present exist in equilibrium with the cation and chloride ions and are largely dissociated into these simple ions when the solution is diluted with water as exemplified by the copper complex [Eq. (13.5)].

$$CuCl_4^{2-} + 4H_2O \rightleftharpoons Cu(OH_2)_4^{2+} + 4Cl^- \qquad (13.5)$$

Therefore, the solution formed by dissolving an alloy in aqua regia can, for all practical purposes, be looked upon as a solution of the simple cations of all the metals; arsenic is present as arsenic acid.

Hot concentrated HNO_3 alone will react with all metals that are considered in this course. Antimony and tin are converted to white, insoluble Sb_2O_5 and SnO_2, respectively; arsenic forms soluble arsenic acid, H_3AsO_4. The other metals are all converted to soluble nitrates.

$$2Sb + 10NO_3^- + 10H^+ \rightleftharpoons Sb_2O_5 + 10NO_2 + 5H_2O \qquad (13.6)$$

$$As + 5NO_3^- + 5H^+ \rightleftharpoons H_3AsO_4 + 5NO_2 + H_2O \qquad (13.7)$$

$$Cu + 2NO_3^- + 4H^+ \rightleftharpoons Cu^{2+} + 2NO_2 + 2H_2O \qquad (13.8)$$

Less concentrated HNO_3 will yield NO rather than NO_2 as the reduction product for NO_3^-. SnO_2 and Sb_2O_5 will dissolve in HCl to form $SnCl_6^{2-}$ and $SbCl_4^-$, respectively.

$$Sb_2O_5 + 10H^+ + 12Cl^- \rightleftharpoons 2SbCl_4^- + 2Cl_2 + 5H_2O \qquad (13.9)$$

The $SnCl_6^{2-}$ and $SbCl_4^-$ ions are in equilibrium with Sn^{4+} and Sb^{3+} ions.

$$SnCl_6^{2-} + 6H_2O \rightleftharpoons Sn(OH_2)_6^{4+} + 6Cl^- \qquad (13.10)$$

Sodium, potassium, calcium, and barium are seldom found in commercial alloys. Alloys containing high concentrations of these metals react with water to evolve hydrogen.

$$2Na + 2H_2O \rightleftharpoons 2Na^+ + 2OH^- + H_2 \qquad (13.11)$$

Since ammonium, NH_4^+, exists only as an ion, there are no stable ammonium alloys.

From these facts, it should be apparent that an alloy can be analyzed by a plan in which concentrated HNO_3 alone serves as the initial solvent. The plan is represented, schematically, in Figure 13.1.

In Eqs. (13.1) to (13.11), observe that the metals in an alloy react exactly as if they were separate pure metals. These reactions might lead to the conclusion that an alloy is simply a homogeneous mixture of metals. Some alloys are, in fact, solid solutions of two or more individual metals. In a great many alloys, however, the component metals exist as compounds. In general, it may be stated that most alloys are very complex in structure. Since alloy components (other than trace components) are present in relatively large proportions, the solutions we obtain after dissolution of an alloy sample usually contain the metal ions in relatively higher concentrations than are present in the usual artificial unknown solutions. Accordingly, it is important to test for completeness of precipitation in each step to make certain that the ion in question is removed before the next step is taken.

PROCEDURE 28

Dissolving an Alloy

The general strategy for dissolving an alloy is shown by the flow diagram in Figure 13.1.

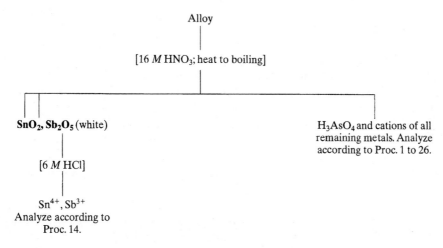

FIGURE 13.1 Analysis scheme for an alloy.

Place a quantity of the alloy about one-tenth the volume of a drop of water in a casserole (see Note 1). Add 10 drops of 16 M HNO$_3$ and 10 drops of 12 M HCl, and warm gently under a hood. If gentle heating does not cause a reaction to take place, the liquid should be boiled. Replenish the acids in the ratio of 1 drop of 16 M HNO$_3$ to 1 drop of 12 M HCl as fast as the liquid boils away, continuing the boiling until the alloy is completely dissolved (see Note 2) or disintegrated (see Note 3). If the alloy dissolves but a finely divided solid, either crystalline or curdy and white, remains in the bottom of the casserole, add water drop by drop with continued careful heating until the solid either is completely dissolved or becomes curdy and white (see Note 4). If the solid is completely dissolved, the absence of silver is proved; in that case, follow the directions given in (**A**). If a white, curdy precipitate remains in the casserole, the presence of silver is indicated; in that case, follow the directions given in (**B**).

(**A**) Add 12 M HCl, drop by drop, with continued boiling, until no more brown fumes of NO$_2$ are given off; then evaporate the solution to a volume of about 10 drops; if a crystalline solid begins to form before the volume has been reduced to 10 drops, discontinue the evaporation and add water, drop by drop and with heating, until the solid is all dissolved; analyze 4 drops of this solution as directed in Procedure 5(A) in Chapter 10.

(B) If a white curdy precipitate remains after the alloy has dissolved, transfer the contents of the casserole to a test tube, centrifuge at once, and decant immediately into a casserole. Wash the precipitate twice with 10-drop portions of 2 *M* HCl and analyze it as directed in Procedure 4 in Chapter 9. Treat the decantate in the casserole as described in **(A)** above.

Notes

1. Commercial alloys are usually found in the form of castings, machined articles, instrument parts, rods, plates, bolts, wire, and so on. When such an alloy is to be analyzed, a small amount is filed off or a hole is drilled in the specimen, or a sample is machined on a lathe or milling machine and the filings, drillings, borings, turnings, or millings thus obtained are subjected to analysis. Some of the alloys analyzed in this course are authentic commercial specimens; some are prepared solely for the purpose of giving the student experience in alloy analysis.

2. The color of the solution obtained when an alloy is dissolved in aqua regia may give valuable clues as to the metals present. The following metals form colored chloride solutions: $CuCl_2$, green; $FeCl_3$, yellowish brown; $CrCl_3$, green; $NiCl_2$, green; $CoCl_2$, reddish pink or blue; $MnCl_2$, faint pink. The yellow brown color of $FeCl_3$ masks the other colors if it is reasonably concentrated.

3. Dissolving an alloy is often a slow process, requiring long boiling and numerous acid replenishings. If the alloy appears to be partly or wholly insoluble, the instructor should be consulted.

4. Silver chloride is insoluble and will, accordingly, be present in the casserole as a white, curdy residue. Lead chloride is only sparingly soluble and, unless present in very small amounts, will be found in the casserole as a white crystalline precipitate.

    Aqua regia, a powerful oxidizing agent, will oxidize mercury to its high oxidation state. Consequently, no Hg_2Cl_2 will precipitate, even though mercury is present in the alloy. Tin, iron, and arsenic, like mercury, will be converted to their high oxidation states.

Analysis of the Solubilized Alloy

Having dissolved the alloy using aqua regia and dealt with any insoluble residue remaining according to the details described in Procedure 28, we now turn our attention to the ions of the metals that remain in the spent aqua regia solution. We can treat this solution like a general soluble unknown from which silver (as AgCl) has been removed. The dissolution process has also yielded important clues concerning the identity of the other possible metallic constituents from the colors observed during the dissolution process; see Note 2 of Procedure 28. Accordingly, it is logical to try to establish the identities of the soluble metallic species by picking up the analysis at Procedure 5(A) in Chapter 10 as if you were investigating a general soluble unknown. Now go to Procedure 5(A) in Chapter 10 and continue.

PROBLEMS

13.1 Show, by means of outlines, the complete analysis of alloys containing the following metals.

   a. Sn, Ag, Cu, Fe, Zn, Mg

   b. Pb, Hg, Bi, Mn, Al

   c. As, Sb, Cd, Ni, Cr

13.2 Show, by appropriate equations, what happens when each of the following is dissolved in aqua regia: mercury, iron, copper, tin, arsenic, antimony, chromium, silver.

13.3 Show, by appropriate equations, what happens when each of the following is treated with hot concentrated HNO_3: arsenic, antimony, tin, mercury, lead, silver, iron, copper.

13.4 Using the simplest chemical method possible, how would you distinguish between the following metals?

    a. Tin and zinc.

    b. Nickel and silver.

    c. Iron and antimony.

    d. Arsenic and antimony.

    e. Lead and bismuth.

    f. Magnesium and zinc.

    g. Chromium and nickel.

    h. Calcium and magnesium.

13.5 Using the simplest method possible, how would you determine whether an alloy is brass (Cu–Zn) or bronze (Cu–Sn)?

13.6 An alloy showed the following behavior.

    a. It was attacked by HNO_3 to give a colorless supernatant liquid and a finely divided white residue. This residue was soluble in HCl.

    b. It was attacked by aqua regia to give a colorless solution and a white, curdy residue.

    c. It was not acted upon by water. What conclusions can be drawn regarding its composition?

13.7 An alloy was found to be completely dissolved (a) by cold, moderately concentrated HNO_3, (b) by cold dilute HCl, and (c) by cold dilute H_2SO_4. In each case it gave a colorless solution with no precipitate. What conclusions can be drawn regarding its composition?

13.8 a. An alloy that is known to contain two or more of the eight elements listed below, but not others, is completely disintegrated (dissolved) by HNO_3, but yields a white precipitate. The precipitate and solutions are separated.

Ag, Mg, As, Sn, Pb, Cu, Sb, Ni

    b. The precipitate from part (a) dissolves completely in HCl, but a white precipitate forms when this solution is diluted with water.

    c. The solution from (a) is green in color, but changes to blue upon addition of H_2O.

    d. A precipitate is formed when HCl is added to the solution from (a); none of this precipitate dissolves when heated.

    e. The solution from part (d) yields a precipitate when H_2S is added. This precipitate is partially soluble in $(NH_4)_2S$. Indicate the elements that are present, absent, and impossible to determine.

Analysis of Salts and Salt Mixtures

Up to this point we have been concerned mainly with the chemistry of cations as it pertains to the detection of these species. The general analysis of salts, on the other hand, involves a consideration of the chemistry of *anions*. The properties of salts obviously depend upon the nature of the anion that is combined with the cations present. For example, AgCl is distinguished from $AgNO_3$ by the relative solubility of the two substances; AgCl is insoluble in water, whereas $AgNO_3$ is soluble. Indeed, the insolubility of AgCl was used to separate and tentatively identify Ag^+ in the qualitative analysis scheme developed earlier. Thus, we can imagine the possibility of identifying anions by the properties of the compounds they form with certain cations in much the same way the cation scheme was developed.

Detection of the cations was accomplished by a very logical sequence of procedures involving, first, *separation* and second, *identification*. Detection of the anions does not proceed in quite the same way; that is, it does not consist of a methodical sequence of separations and identifications. First, an effort is made to eliminate or verify the presence of certain anions on the basis of the color and solubility of the sample and the results of the cation analysis. Then, the material being analyzed is submitted to a series of "preliminary tests."

As a result of these observations and tests, certain of the anions may be definitely shown to be present or absent, thereby paring down the number of possibilities from the maximum possible in one case, 13, to some lesser number. Next, specific tests are carried out for those anions not definitely eliminated in the preliminary tests and observations. Although this sequence of preliminary tests and specific tests is not quite so well ordered as the procedures for cation analysis, it is very logical and makes possible a rapid and systematic identification of the anions.

Only the following 13 anions are considered in this scheme: acetate, $C_2H_3O_2^-$; nitrate, NO_3^-; chloride, Cl^-; bromide, Br^-; iodide, I^-; sulfide, S^{2-}; sulfate, SO_4^{2-}; sulfite, SO_3^{2-}; carbonate, CO_3^{2-}; chromate, CrO_4^{2-}; phosphate, PO_4^{3-}; arsenate, AsO_4^{3-}; and borate, BO_2^- (or BO_3^{3-}). This list represents less than one-third of the anions that we might encounter in a course in general chemistry; it does, however, include the more common ones, and the methods and principles used in the identification of these 13 will serve to illustrate the methods and principles used for the identification of all others.

As in the case of cation identification, the physical (usually color and solubility) and chemical (reactions to form precipitates) properties of the compounds formed by the anions provide the basis for their identification. The following general procedures give helpful information.

EXAMINATION OF THE SOLID

Often, the color of a substance offers a clue to its constituent ions. Certain metals, such as copper, nickel, manganese, chromium, cobalt, and iron, form colored salts. Others form white salts. Certain anions, such as chromates, dichromates, and permanganates, have characteristic colors. It is important, therefore, to note the color of the solid, since it may give indications of the presence of one or more of the ions listed above.

The characteristic colors of the more common ions follow:

CrO_4^{2-}	Yellow	Fe^{2+}	Grayish-green
$Cr_2O_7^{2-}$	Orange-red	Cr^{3+}	Green to bluish-gray to black
MnO_4^-	Violet-purple	Cu^{2+}	Blue to green to brown
Ni^{2+}	Green	Co^{2+}	Wine-red to blue
Fe^{3+}	Reddish-brown to yellow	Mn^{2+}	Pink to tan

SOLUBILITY OF THE SALT

The solubility of a substance in water and/or other aqueous solutions, together with a knowledge of the cations present in the substance, often permits us to narrow down the choice of anions present. For example, a white substance containing Ag^+ that is insoluble in acid solution (H_2SO_4) strongly suggests that the substance also contains Cl^- from our list of anions. The ambiguity in this conclusion lies in the realization that Ag_2SO_4 has all the necessary properties except that it is slightly soluble. Thus, a review of the general solubility rules (Chapter 4) and the systematic chemistry of the cations at this point will be beneficial in following the arguments based on the observations of attempts to dissolve a substance in solvents of various kinds. Obviously, to determine the cations present in a salt requires that a sample be put into solution before the necessary systematic scheme can be applied.

Careful observation of the solid and the solution that remains when the sample is dissolved in water often gives significant and important clues to the identity of the species present in the sample. The color of the solid may indicate the identity of the cations and/or anions present; the color of the solvent can also be a useful indicator of the species present. Indeed, even if all of the sample does not dissolve, the color of the remaining solid compared with the color of the original solid can provide revealing clues to the identity of the species present in the original solid sample. For example, if a dark-colored solid produces a yellow solution and an insoluble red-brown solid it suggests that the yellow solution is a soluble chromate (CrO_4^{2-} is yellow) and the red-brown insoluble substance remaining might be an Fe^{3+} compound; since the latter is insoluble, it is also suggestive concerning the identity of the associated anion—perhaps OH^- or O_2^-.

REACTIONS WITH SPECIFIC REAGENTS

If the unknown substance is soluble in water, it is possible to obtain clues to the anions present by treating this aqueous solution with reagents that should give characteristic reactions for the anions present. Two reagents, $AgNO_3$ and $BaCl_2$, give information on the presence or absence of certain groups of anions. Unlike the cation scheme, the analysis of anions depends more heavily on specific tests.

Treatment of a Solution of Anions with AgNO₃

The silver salts of all 13 anions that form the potential unknown group, except nitrate, acetate, and sulfate, are insoluble in water; $AgNO_3$ is very soluble, Ag_2SO_4 is moderately soluble, and $AgC_2H_3O_2$ is sparingly soluble. The insoluble silver salts have the following characteristic colors:

Salt	Color	Solubility in dilute HNO₃
$AgCl$	White	Insoluble
$AgBr$	Cream	Insoluble
AgI	Pale yellow	Insoluble
Ag_2S	Black	Insoluble
Ag_3AsO_4	Chocolate brown	Soluble, colorless solution
Ag_3PO_4	Yellow	Soluble, colorless solution
Ag_2CrO_4	Brown-red	Soluble, yellow solution
Ag_2CO_3	White	Soluble, bubbles form
Ag_2SO_3	White	Soluble, colorless solution
$AgBO_2$	White	Soluble, colorless solution

Four (4) of the silver salts are insoluble in dilute HNO_3, but the remainder are soluble. Accordingly, if silver ions in the form of a solution of silver nitrate are added to water solutions containing the possible 13 anions, the 10 insoluble silver salts listed above will precipitate; $AgC_2H_3O_2$ may precipitate if the concentration of acetate ions is fairly high. If each precipitate is, in turn, treated with dilute HNO_3, all but $AgCl$, $AgBr$, AgI, and Ag_2S will dissolve readily. If the solution containing the anions is acidified with dilute HNO_3 before addition of $AgNO_3$, only $AgCl$, $AgBr$, AgI, and Ag_2S will precipitate. A summary of the observations of a systematic precipitation of insoluble silver salts and a test of the solubility of the precipitated species is given in Figure 14.1.

The anions whose silver salts are insoluble in nitric acid (Cl^-, Br^-, I^-, S^{2-}) are commonly referred to as the *Hydrochloric Acid Group* of anions.

Treatment of a Solution of Anions with BaCl₂

The barium salts, $BaCl_2$, $BaBr_2$, BaI_2, BaS, $Ba(C_2H_3O_2)_2$, and $Ba(NO_3)_2$, are soluble in water and in alkaline solution; $BaSO_4$, $BaSO_3$, $BaCO_3$, $BaCrO_4$, $Ba_3(AsO_4)_2$, $Ba_3(PO_4)_2$, and $Ba(BO_2)_2$ are insoluble. Barium sulfate, $BaSO_4$, is insoluble in strong acids, such as dilute HNO_3 or dilute HCl; the other water-insoluble barium salts are salts of weak acids and are, therefore, soluble in dilute HNO_3 or dilute HCl. These facts form the basis for the use of barium chloride as a reagent for the elimination of anions.

Barium chromate, $BaCrO_4$, is yellow; the other barium salts are white.

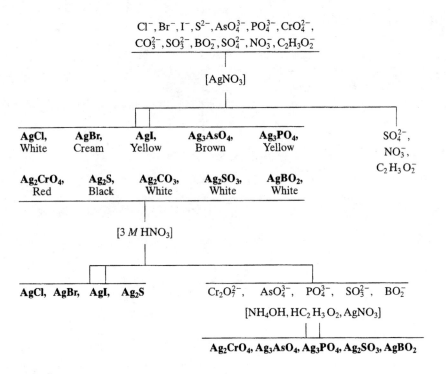

FIGURE 14.1 Anion analysis scheme based on $AgNO_3$.

The anions that have barium salts insoluble in water and alkaline solution (SO_4^{2-}, SO_3^{2-}, CO_3^{2-}, CrO_4^{2-}, BO_2^-, AsO_4^{3-}, and PO_4^{3-}) are commonly referred to as the *Sulfuric Acid Group* of anions.

The anions that do not fall in either the Hydrochloric Acid Group or Sulfuric Acid Group (NO_3^-, $C_2H_3O_2^-$) are commonly referred to as the *Nitric Acid Group* of anions. A scheme of analysis based on the behavior of barium salts appears in Figure 14.2.

Treatment of Solid Unknown with H_2SO_4

Several of the anions in our scheme of analysis form volatile weak acids with, or they are oxidized by, concentrated sulfuric acid. Careful observations of the action of concentrated sulfuric acid on a sample of the solid can give an insight into the possible anions that might be present. A summary of the pertinent observations is given in Table 14.1. The remaining anions, $C_2H_3O_2^-$, SO_4^{2-}, PO_4^{3-}, BO_2^-, AsO_4^{3-}, and NO_3^-, provide no useful evidence of their identity when they react with sulfuric acid. On the other hand, the use of hot, concentrated H_2SO_4 gives some additional information.

Nitrates give brown fumes of NO_2 when heated with H_2SO_4:

$$2NaNO_3 + H_2SO_4 \leftrightharpoons Na_2SO_4 + 2HNO_3 \tag{14.1}$$

When HNO_3 is heated, a part of it decomposes as follows:

$$4HNO_3 \leftrightharpoons 2H_2O + O_2 + 4NO_2 \text{ (brown gas)} \tag{14.2}$$

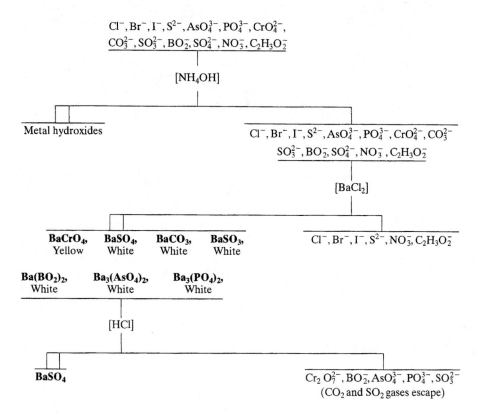

FIGURE 14.2 Anion analysis scheme based on barium salts.

TABLE 14.1 REACTIONS OF SALTS WITH COLD, CONCENTRATED H_2SO_4

Anion	Observation	Equation
Cl^-	Effervescence. The gas evolved is colorless, has a sharp odor (HCl), fumes in moist air, turns blue litmus red.	$NaCl + H_2SO_4 \rightleftharpoons NaHSO_4 + HCl$
Br^-	Effervescence. The gas evolved is brown (Br_2), has a characteristic sharp odor, fumes in moist air, turns blue litmus red.	$NaBr + H_2SO_4 \rightleftharpoons NaHSO_4 + HBr$ $H_2SO_4 + 2HBr \rightleftharpoons 2H_2O + SO_2 + Br_2$
I^-	Effervescence. Solid turns dark brown instantly, slight evolution of gas that fumes in moist air, odor of H_2S, violet fumes of iodine (I_2).	$NaI + H_2SO_4 \rightleftharpoons NaHSO_4 + HI$ $H_2SO_4 + 8HI \rightleftharpoons H_2S + 4H_2O + 4I_2$ $H_2SO_4 + 2HI \rightleftharpoons 2H_2O + SO_2 + I_2$
S^{2-}	Effervescence. Odor of H_2S gas, free sulfur deposited.	$ZnS + H_2SO_4 \rightleftharpoons ZnSO_4 + H_2S$ $H_2SO_4 + H_2S \rightleftharpoons 2H_2O + SO_2 + S(s)$
CO_3^{2-}	Effervescence. Colorless, odorless gas (CO_2).	$Na_2CO_3 + H_2SO_4 \rightleftharpoons Na_2SO_4 + H_2O + CO_2$
SO_3^{2-}	Effervescence. Colorless gas with a sharp, choking odor (SO_2).	$Na_2SO_3 + H_2SO_4 \rightleftharpoons Na_2SO_4 + H_2O + SO_2$
CrO_4^{2-}	Color changes from yellow (CrO_4^{2-}) to orange-red ($Cr_2O_7^{2-}$).	$2K_2CrO_4 + H_2SO_4 \rightleftharpoons K_2Cr_2O_7 + H_2O + K_2SO_4$

Acetates yield acetic acid when warmed with H_2SO_4. Acetic acid has the characteristic odor of vinegar.

$$2NaC_2H_3O_2 + H_2SO_4 \rightleftharpoons Na_2SO_4 + 2HC_2H_3O_2 \qquad (14.3)$$

Chlorides, bromides, iodides, sulfides, carbonates, sulfites, and chromates show no additional reactions on being heated, and sulfates, phosphates, borates, and arsenates show no reaction on being heated with H_2SO_4.

Specific Tests for Anions

Information gained in the course of the cation analysis, together with information gained from observation of the solubility of the solid in water, may show the presence or absence of certain anions. The three preliminary tests may also show the absence of certain anions and the presence of others. For each salt mixture, however, there usually are several anions whose presence or absence is not established by any of the procedures noted above. Specific tests must then be carried out for each. These specific tests are discussed in Procedures 36 to 48.

The specific tests are independent of the cation analysis and, in general, can be made whether or not the tests for cations are carried out. They make it possible to test for nitrate in a fertilizer or for phosphate in a baking powder without going through the systematic cation analysis.

PRELIMINARY EXPERIMENTS

Up to this point you have learned how to separate and identify the cations (metallic ions) and you have applied this information to the analysis of four group unknowns and, possibly, a cation mixture or an alloy. The next unknown that you will analyze will be a single salt or a mixture of salts. To analyze a salt you must identify the anions (acid radicals) as well as the cations. Consequently, before you can proceed with the analysis of your first salt or salt mixture, you must learn how the anions are identified. To help familiarize you with anion identification, you will perform Experiments 1, 2, 3, and 4, as described below.

EXPERIMENT 1

Treatment of a Solid Salt of Each Acid with H_2SO_4

Solid salts of each of the 13 acids for which tests are to be made can be found on the reagent shelf. Treat a very small amount of each solid salt with concentrated sulfuric acid as directed in Procedure 33. Note exactly what happens, record your observations in the table below, and compare these observations with those given in Table 14.1. Write the equations for the reactions that take place. Have the results of this experiment approved by your instructor before proceeding with the next experiment.

Salt	Observations	Equations
$NaNO_3$	_____	_____
	_____	_____
$NaCl$	_____	_____
	_____	_____

NaBr

_____   _____

NaI

_____   _____

ZnS

_____   _____

NaCO$_3$

_____   _____

Na$_2$SO$_3$

_____   _____

Na$_2$CrO$_4$

_____   _____

Na$_2$SO$_4$

_____   _____

NaBO$_2$

_____   _____

Na$_3$PO$_4$

_____   _____

Na$_3$AsO$_4$

_____   _____

NaC$_2$H$_3$O$_2$

_____   _____

EXPERIMENT 2

Treatment of a Solution of the Anions with AgNO$_3$

(A) Place 3 to 4 drops of a 0.2 M solution of the sodium salt of each of the anions in 13 separate test tubes. Add 1 to 2 drops of 0.2 M AgNO$_3$ to each. Note the colors of the precipitates formed. Centrifuge and decant, saving the precipitate for (**B**). In the following spaces, write the net equations for the formation of the precipitates obtained with AgNO$_3$. Indicate the color of each precipitate (see Procedure 34).

$C_2H_3O_2^-$ _____

NO_3^- _____

Cl^- _____

Br^- _____

I^- _____

S^{2-} _____

CO_3^{2-} _____

SO_3^{2-} _____

CrO_4^{2-} _____

BO_2^- _____

PO_4^{3-} _____

AsO_4^{3-} _____

SO_4^{2-} _____

(B) Attempt to dissolve each of the precipitates obtained in part **(A)** in 5 to 6 drops of 3 M HNO_3. In the spaces below, write net equations for those cases in which the silver salts are *dissolved* by 3 M HNO_3. Indicate gases evolved by an upward arrow.

_____

_____

_____

_____

_____

_____

_____

_____

_____

List the silver salts *not dissolved* by 3 M HNO_3.

_____

Taking into account the results obtained in Experiment 2**(A)** and 2**(B)**, list the anions that will react with $AgNO_3$ to form precipitates in 3 M nitric acid solution.

_____

Have the net results of this experiment approved by your instructor before going on with the next experiment.

EXPERIMENT 3

Treatment of a Solution of the Anions with $BaCl_2$

(A) Place 3 to 4 drops of a 0.2 M solution of the sodium salt of each of the anions in 13 individual test tubes. Make each solution alkaline with a drop of 5 M aqueous NH_3 and then add 1 to 2 drops of 0.2 M $BaCl_2$ to each. Note the colors of all precipitates formed. Centrifuge, decant, and wash twice with cold water. Save each precipitate for (**B**).

In the following spaces write the net equations for the formation of the precipitates obtained with $BaCl_2$ in alkaline (aqueous NH_3) solution. Indicate the color of each precipitate (see Procedure 35).

$C_2H_3O_2^-$ _____

NO_3^- _____

Cl^- _____

Br^- _____

I^- _____

S^{2-} _____

CO_3^{2-} _____

SO_3^{2-} _____

CrO_4^{2-} _____

BO_2^- _____

PO_4^{3-} _____

AsO_4^{3-} _____

SO_4^{2-} _____

(**B**) Attempt to dissolve each of the precipitates from (**A**) in 4 to 5 drops of 2 M HCl. In the spaces below, write net equations for those cases in which the barium salts are *dissolved* by 2 M HCl. Note in which cases gases are evolved.

_____

_____

_____

_____

_____

_____

List the barium salts that are *not dissolved* by 2 M HCl.

_____

List the anions that will react with Ba^{2+} to form precipitates in alkaline solution.

_____

_____

Have the results of this experiment approved by your instructor before proceeding with the next experiment.

EXPERIMENT 4

The Specific Tests for the Anions

Use solutions of the sodium salts of each of the 13 acids or a small amount of a solid salt of each of these acids. Carry out a specific test for each anion, as directed in Procedures 36 to 48.

Note exactly what happens in each test and compare your observations with those noted in each procedure. In the spaces below, write the net equations for the reactions that take place. Have these equations approved by your instructor before proceeding with the analysis of your first salt or mixture of salts.

AsO_4^{3-} _____

_____

PO_4^{3-} _____

_____

BO_2^- or BO_3^{3-} _____

_____

CrO_4^{2-} _____

_____

SO_4^{2-} _____

_____

SO_3^{2-} _____

_____

CO_3^{2-} _____

_____

S^{2-} _____

_____

I^- _____

_____

Br^- _____

_____

Cl^- _____

_____

NO_3^- _____

_____

$C_2H_3O_2^-$ _____

_____

ANALYSIS OF UNKNOWN SALT

The following summary gives the steps that will be followed and the tests that will be made in the analysis of salts and salt mixtures.

I. Examine the solid and note its distinctive and significant physical properties (see Procedure 29).

II. Put the solid into solution (see Procedure 30 and the accompanying notes). Be sure to note any observable reaction that takes place when the solid dissolves. If the solid is soluble in water and the resulting solution has a characteristic color, indicating the presence of certain cations or anions, make all possible eliminations of cations and anions (see solubility rules, Chapter 4).

III. Determine what cations are present. This may be done either before, with, or after the anion analysis (see Procedure 31).

    A. *Observe the color of the solution.* A colorless (water-clear) solution cannot contain Cu^{2+}, Ni^{2+}, Co^{2+}, Cr^{3+}, Fe^{3+}, Fe^{2+}, or Mn^{2+} ions.

    B. *Determine whether or not any cations can be eliminated on the basis of the solubility in water and the results of the anion analysis.* For example, if a salt mixture is soluble in water and has been shown to contain carbonate ion, the only cations that can possibly be present are Na^+, K^+, and NH_4^+. A water solution of a chromate can contain only Na^+, K^+, NH_4^+, and Mg^{2+} (see solubility rules, Chapter 4).

    C. *Flame test* (see Note 1, Procedure 24 in Chapter 12).

    D. *Blanket tests with Na_2CO_3, NaOH, and aqueous NH_3 may be made if the salt is soluble in water* (see Procedure 31).

    E. *Carry out a systematic cation analysis,* as outlined in Procedures 1 to 27, for the cations not eliminated by (A), (B), (C), and (D).

IV. Determine what anions are present. This may be done either before, with, or after the cation analysis (see Procedure 32).

    A. *Observe the color of the solution.* If the salt is soluble in water and the color of the solution indicates the presence of one or more of the cations that form colored solutions, certain anions, referred to in Rules 5 and 6 of the solubility rules (Chapter 4), can be eliminated. The colors of the CrO_4^{2-} and $Cr_2O_7^{2-}$ ions are very distinctive; therefore, a colorless solution cannot contain these anions.

    B. *Determine whether or not any anions can be eliminated on the basis of the results of the cation analysis.* If no arsenic or chromium was found in the cation analysis, arsenates and chromates cannot be present. If a salt is soluble in water and the cation analysis shows that it contains lead ion, the only anions that can be present are nitrate and acetate.

    C. *Treat the solid salt with concentrated H_2SO_4* (see Procedure 33, and the accompanying notes).

    D. *Treat a solution of the salt with silver nitrate* (see Procedure 34, and the accompanying notes).

    E. *Treat a solution of the salt with barium chloride* (see Procedure 35, and the accompanying notes).

F. *Perform specific tests for the anions that have not been definitely eliminated or verified as present by* (A), (B), (C), (D), *and* (E). The specific tests are given in the following procedures:

Arsenate	Procedure 36
Phosphate	Procedure 37
Borate	Procedure 38
Chromate	Procedure 39
Sulfate	Procedure 40
Sulfite	Procedure 41
Carbonate	Procedure 42
Sulfide	Procedure 43
Iodide	Procedure 44
Bromide	Procedure 45
Chloride	Procedure 46
Nitrate	Procedure 47
Acetate	Procedure 48

The results of the analysis of salts and salt mixtures will be reported on the blank on page 286 in the manner illustrated on page 287.

Although the steps in the analysis of a salt are discussed on the next few pages in the order in which they are listed in the summary above, it should be emphasized that this exact order need not always be followed when a particular salt or salt mixture is analyzed. For example, if a mixture is soluble in water, it usually is advantageous to run the anion analysis before the cation analysis. On the other hand, if the mixture is insoluble in water, it is usually best to do the cation analysis first. In many instances, it will be a good idea to run the anion analysis and cation analysis simultaneously. In most cases, it is desirable to make the treatment of the solid with concentrated sulfuric acid (Procedure 33) one of the first steps in the analysis. A flame test applied to a sample of the original solid in the manner directed in Note 1 of Procedure 24 in Chapter 12 is quickly and easily carried out and may give valuable information. If the salt is found to be soluble in water, the *blanket tests* outlined in Procedure 31 and the *preliminary tests* with $AgNO_3$ and $BaCl_2$ (Procedures 34 and 35) should be the next steps. The wise analyst will be guided from one step to another by the behavior of the particular material being analyzed.

EXAMINATION OF THE SOLID

Review the discussion on "the examination of the solid" section at the beginning of this chapter first.

PROCEDURE 29

Physical Examination of the Sample

Examine the unknown carefully (see Note 1), noting the color or colors (see Note 2) and whether the material is crystalline or noncrystalline, homogeneous or heterogeneous. Record your observations on a report blank on page 286 in the manner illustrated on page 287.

Notes

1. The information gained by the examination of the solid may not, by itself, be of much value in the analysis of a sample. When combined with the entire body of information, however, it may prove to be very useful. The fact that a solid is green is itself of some value; when combined with the fact that the solid is soluble in water, the fact of its green color becomes, as pointed out later, extremely valuable.

2. The fact that the sample is white does not prove the absence of all ions that usually form colored salts. Anhydrous copper sulfate is white, and so is ferric phosphate. The salts with characteristic colors, moreover, may be present in such small amounts that they are covered up by an excess of other salts.

DISSOLUTION OF THE SOLID

Before a solid can be analyzed, it must be dissolved. The choice of solvent is made by testing the solubility of a portion of the solid in various solvents. Water is the most desirable solvent; the water solution can be used for all procedures except those that specifically call for the use of some of the solid. If the solid is insoluble in water, two separate solutions should be prepared: one for cation analysis, the other for anion analysis. The solution for cation analysis should be prepared by dissolving the solid in hydrochloric acid. The solution for anion analysis should be prepared by boiling with sodium carbonate.

The reactions that take place when a solid is dissolved in acid may give definite information about the presence or absence of certain ions. Carbonates, sulfites, and some sulfides give off CO_2, SO_2, and H_2S gas, respectively. Chromates change in color from yellow to orange-red. When effervescence occurs, the odor of the gas evolved often gives valuable clues about the anions present. Review the information on the "examination of the solid" section at the beginning of this chapter before you attempt to dissolve your solid sample.

PROCEDURE 30

Dissolving the Solid

(A) Test a bit of the solid with 20 drops of water (first cold, then hot) to see if it will dissolve (see Note 1). If the solid completely dissolves, prepare a stock solution by dissolving a quantity, the volume of 2 drops of water, in 5 mL of water, and use this solution for cation analysis, preliminary tests for anions, and specific tests for anions. Note the color of the solution.

(B) If the solid is not soluble in water, prepare a solution for cation analysis by dissolving in HCl (see Note 2) or HNO_3 (see Note 3), or, if necessary, in aqua regia (see Procedure 28 in Chapter 13). First try 6 M HCl and then 12 M HCl, followed by dilution with an equal volume of hot water; then try 16 M HNO_3 (see Note 4), followed by dilution with water (see Note 5); finally try aqua regia (see Note 6).

(C) If the solid is not soluble in water, prepare a solution for anion analysis as follows (see Note 7). Place a quantity of the solid, the volume of 2 drops of water, in a casserole; then add 5 mL of a saturated solution of Na_2CO_3 and boil gently for 2 min (see Note 8). Transfer to test tubes, centrifuge, and decant, discarding the precipitate (carbonates of metals) (see Note 6). A part of this decantate can be saved for making specific tests for all anions except carbonates. Transfer the remainder of the decantate to a casserole, making acidic with 3 M HNO_3, and boil gently until all CO_2 has been driven off. This solution can be used for the group tests with $AgNO_3$ and $BaCl_2$ and for the specific tests for all anions except nitrate, sulfide, sulfite, sulfate, and carbonate.

Notes

1. Bismuth and antimony salts hydrolyze strongly in water to give the insoluble white basic salts:

$$Bi^{3+} + Cl^- + H_2O \rightleftharpoons BiOCl + 2H^+$$
$$Sb^{3+} + Cl^- + H_2O \rightleftharpoons SbOCl + 2H^+$$

The basic salts are readily dissolved by an excess of acids.

2. The evolution of chlorine gas when a solid is dissolved in concentrated HCl shows the presence of some oxidizing agent such as a nitrate, a chromate, a dioxide, or a peroxide.

$$2CrO_4^{2-} + 6Cl^- + 16H^+ \rightleftharpoons 2Cr^{3+} + 3Cl_2 + 8H_2O$$
$$MnO_2 + 2Cl^- + 4H^+ \rightleftharpoons Mn^{2+} + Cl_2 + 2H_2O$$

3. The evolution of I_2, Br_2, or Cl_2 when a substance is dissolved in concentrated HNO_3 shows the presence of iodides, bromides, or chlorides (see Note 2).

$$6I^- + 2NO_3^- + 8H^+ \rightleftharpoons 2NO + 4H_2O + 3I_2 \text{ (violet gas)}$$

The evolution of brown fumes of NO_2 when a substance is dissolved in HNO_3 shows the presence of reducing agents such as sulfides, sulfites, iodides, or bromides.

$$ZnSO_3 + 2NO_3^- + 2H^+ \rightleftharpoons Zn^{2+} + SO_4^{2-} + 2NO_2 + H_2O$$

If a solid mixture containing a nitrate and a chloride, bromide, iodide, sulfide, or sulfite is treated with concentrated H_2SO_4, brown fumes of NO_2 will be noted for the reason given above. Likewise, if a solid mixture of a halide and a chromate, dioxide, or peroxide is treated with concentrated H_2SO_4, free halogen will be formed, as discussed in Note 2.

4. Concentrated HNO_3, because of its powerful oxidizing property, dissolves many metallic sulfides that are unattacked by HCl. Sulfur is usually liberated.

Sulfides will yield a precipitate of sulfur when dissolved in nitric acid or aqua regia. This sulfur should be discarded.

$$CuS + 2NO_3^- + 4H^+ \rightleftharpoons Cu^{2+} + S(s) + 2NO_2 + 2H_2O$$

A solution for use in anion analysis can be prepared by dissolving the solid mixture in nitric acid. Its use will involve the following limitations:

a. When such a solution is treated with $AgNO_3$ in Procedure 34, the only insoluble silver salts that can precipitate are AgCl, AgBr, AgI, and Ag_2S. All other silver salts are soluble in HNO_3.

b. When such a solution is treated with $BaCl_2$ in Procedure 35, it must not first be made alkaline with aqueous NH_3. The only insoluble barium salt that can then precipitate is $BaSO_4$; all other barium salts are soluble in HNO_3.

Obviously, then, if such a nitric acid solution is used, SO_3^{2-}, CO_3^{2-}, PO_4^{3-}, AsO_4^{3-}, and BO_2^- must be eliminated or verified by cation analysis, by sulfuric acid treatment (Procedure 33), or by specific tests. Since SO_3^{2-} and CO_3^{2-} are either eliminated or indicated by sulfuric acid treatment and AsO_4^{3-} will be detected in the cation analysis, the only anions that will always require elimination or detection by specific tests are PO_4^{3-} and BO_2^-.

5. Many chlorides—such as $NaCl$ and $BaCl_2$—though soluble in water, are not dissolved by concentrated HCl; hence the directions to dilute the mixture with water after heating with hot concentrated acids.

6. If the solid is insoluble in all acids, it will be necessary to prepare a solution for cation analysis by boiling with sodium carbonate as directed in Procedure 30(C). The metals will be precipitated as carbonates. The precipitate of the metal carbonates must, therefore, be saved, washed three times with hot water, and then dissolved in 6 M HCl. This series of steps gives a solution of metal chlorides.

7. The reason a solution of a water-insoluble substance that is to be used in making the anion group tests is prepared by boiling with sodium carbonate rather than by dissolving in HNO_3 or HCl is that all cations except sodium, potassium, and ammonium must be removed before the group test with barium chloride (Procedure 35) can be made. To illustrate the necessity of removing these metals by precipitation with Na_2CO_3, suppose that the unknown contains $Ca_3(PO_4)_2$ and suppose it is dissolved in dilute HNO_3. The solution will then contain Ca^{2+} ions and PO_4^{3-} ions. In making the group test with $BaCl_2$, the acid solution must first be made alkaline with aqueous NH_3. As soon as this is done, $Ca_3(PO_4)_2$ precipitates; there will then be no phosphate ions left in solution, and the group test with $BaCl_2$ will be negative.

If, instead of being dissolved in dilute HNO_3, the unknown above is boiled with Na_2CO_3, the $Ca_3(PO_4)_2$ will react as indicated in the following equation:

$$Ca_3(PO_4)_2 + 3CO_3^{2-} \rightleftharpoons 3CaCO_3 + 2PO_4^{3-}$$

Most of the $Ca_3(PO_4)_2$ is converted into insoluble $CaCO_3$. On centrifuging and decanting, all the calcium will be found in the precipitate as $CaCO_3$ and $Ca_3(PO_4)_2$, whereas most of the phosphate will be present in the decantate as the phosphate ion. When the decantate is now made alkaline with aqueous NH_3, no precipitate will form since all the calcium ions have been removed. On addition of $BaCl_2$, a white precipitate of $Ba_3(PO_4)_2$ will form.

Insoluble salts of other metals will behave like $Ca_3(PO_4)_2$.

$$2AgCl + CO_3^{2-} \rightleftharpoons Ag_2CO_3 + 2Cl^-$$

See the discussion on the formation of weak electrolytes in Chapter 7 for the detailed reactions that take place in these solution processes.

The solution prepared by boiling a water-insoluble salt with sodium carbonate contains a large amount of carbonate ion added as Na_2CO_3 in preparing this solution. These

CO_3^{2-} ions must be destroyed; otherwise they will react with the Ba^{2+} ions when the group test with $BaCl_2$ is made.

$$Ba^{2+} + CO_3^{2-} \rightleftharpoons BaCO_3$$

The carbonate ions are removed by adding HNO_3 and boiling.

$$CO_3^{2-} + 2H^+ \rightleftharpoons H_2O + CO_2$$

This addition of HNO_3 results in the following complications when the test for certain of the anions is made.

a. Any carbonate present in the original unknown will be removed as CO_2 by the boiling with HNO_3.

b. Sulfites, if present, will to some extent be either removed as SO_2 or oxidized.

$$SO_3^{2-} + 2H^+ \rightleftharpoons H_2O + SO_2$$
$$3SO_3^{2-} + 2NO_3^- + 2H^+ \rightleftharpoons 3SO_4^{2-} + 2NO + H_2O$$

c. Sulfides will, to some extent, either be driven off as H_2S or oxidized to sulfur.

$$S^{2-} + 2H_+ \rightleftharpoons H_2S$$
$$3S^{2-} + 2NO_3^- + 8H^+ \rightleftharpoons 3S(s) + 2NO + 4H_2O$$

Therefore, the solution prepared by boiling with Na_2CO_3 followed by digestion with HNO_3 will not give dependable tests for carbonates, sulfites, sulfates, or sulfides. Furthermore, since HNO_3 was used in its preparation, this solution cannot be used in testing for nitrates.

It will not always be necessary or wise to prepare a solution of a water-insoluble mixture for anion analysis, as directed in Procedure 30(C). Suppose, for instance, that the treatment with concentrated sulfuric acid (Procedure 33) shows the absence of all anions except SO_4^{2-}, PO_4^{3-}, AsO_4^{3-}, and BO_2^-, whereas the cation analysis shows the absence of AsO_4^{3-}. Specific tests for SO_4^{2-}, PO_4^{3-}, and BO_2^- can then be carried out on the solid or on solutions prepared with dilute HNO_3 according to Procedure 30(B), and the group tests with $AgNO_3$ and $BaCl_2$ (Procedures 34 and 35) can be omitted entirely. The same reasoning will hold if the sulfuric acid treatment (Procedure 33) shows the absence of all anions except CO_3^{2-}, SO_4^{2-}, PO_4^{3-}, AsO_4^{3-}, and BO_2^-.

8. Some solids must be fused with sodium carbonate in order to prepare a solution for analysis. Directions for carrying out such a fusion can be obtained from your instructor.

DETERMINE WHAT CATIONS ARE PRESENT

After the sample to be analyzed has been dissolved in an appropriate solvent, the next step could be to go through a complete cation analysis according to Procedures 1 to 27 and follow it with a complete anion analysis according to Procedures 32 to 48. In many analyses, that is exactly what is done. In some instances, however, observations of specific behavior make it possible to eliminate certain cations or anions and, thereby, modify and shorten the overall procedure.

Elimination of Anions and Cations Based on Solubility

To illustrate the arguments used to eliminate species based on solubility rules, consider the following cases:

Suppose a salt mixture dissolves readily and completely in cold water to give a green solution. A green solution means the presence of copper, nickel, iron(II), or chromium. The mixture cannot, therefore, contain carbonates, sulfites, phosphates, borates, chromates, arsenates, or sulfides, since they all form insoluble salts with copper, nickel, iron(II), and chromium. The preliminary tests for anions involving treatment of the solution with $BaCl_2$ can be omitted since all but one of the anions detected by this test are already eliminated.

In another illustration, a salt mixture dissolves in water to give a yellow solution. This solution turns orange when HCl is added in the first step in Procedure 1 in Chapter 9. Such behavior clearly indicates the presence of a chromate. Since the chromates of all metals except magnesium, sodium, potassium, and ammonium are insoluble, there is no need to test for any cations except these four.

In another case, a salt mixture dissolves readily and completely in cold water to give a water-clear solution. When HCl is added in the first step in Procedure 1, vigorous effervescence takes place and a colorless, odorless gas is evolved. Such behavior clearly indicates the presence of carbonate; since sodium, potassium, and ammonium alone form soluble carbonates, they are the only cations that can be present.

In still another case, a student first does the anion analysis on a water-soluble mixture and finds nitrates, chlorides, and phosphates. Sodium, potassium, and ammonium are the only cations whose nitrates, chlorides, and phosphates are soluble. A complete cation analysis is, therefore, unnecessary in this case.

 Eliminations such as those cited above require a knowledge of the solubilities of the salts of the various acids. The rules governing the solubility of the common salts are given in Chapter 4 and should be reviewed at this time.

PROCEDURE 31

Analysis for Cations

Proceed with the blanket tests as follows:

1. Make a list of those cations that can be definitely eliminated or verified on the basis of the solubility of the solid, the color of the solution, and the anion analysis (if the anion analysis has been carried out).

2. Carry out a flame test as directed in Note 1, Procedure 24 in Chapter 12.

3. If the salt is soluble in water, carry out the following blanket tests:

    a. To a few drops of the water solutions, add Na_2CO_3 solution until it is basic. If no precipitate forms, all cations except Na^+, K^+, and NH_4^+ are absent. Na_3PO_4 may be substituted for Na_2CO_3.

    b. To a few drops of the water solution, add 8 M NaOH until distinctly alkaline and stir for 1 min. If no precipitate forms, all cations except Na^+, K^+, NH_4^+, Ba^{2+}, Ca^{2+}, Zn^{2+}, Cr^{3+}, Al^{3+}, Pb^{2+}, Sn^{2+}, Sn^{4+}, and Sb^{3+} are eliminated (see Note 1).

    c. To a few drops of the water solution, add 15 M NH_4OH until distinctly alkaline and stir for 1 min. If no precipitate forms, cations except Na^+, K^+, NH_4^+, Ag^+, Ba^{2+}, Ca^{2+}, Cu^{2+}, Cd^{2+}, Zn^{2+}, Ni^{2+}, and Co^{2+} are eliminated (see Note 1).

4. Carry out a cation analysis (Procedures 1 to 27) for those cations not eliminated in Steps 1, 2, and 3. Use a solution prepared according to Procedure 30(A) or (B) (see Note 2).

Notes

1. If no precipitate forms with either 8 M NaOH or 15 M aqueous NH$_3$, the cation(s) must be Na$^+$, K$^+$, NH$_4^+$, Ba^{2+}, Ca^{2+}, and/or Zn^{2+}.
2. If a solution is prepared in Procedure 30(B) by dissolving the solid in cold HCl, Ag$^+$, Hg$_2^{2+}$, and Pb^{2+} must be absent since they would form insoluble chlorides. The cation analysis will then start with Procedure 5 in Chapter 10.

PROCEDURE 32

Analysis for Anions

Using the solution prepared in Procedure 30(A) or (C) and samples of the solid unknown, make a complete anion analysis in the following order:

1. Make a list of the anions that can be eliminated on the basis of the solubility of the solid (see Chapter 4), the cation analysis, and the color of the solution (see page 246).
2. Carry out the three preliminary tests described in Procedure 33, Procedure 34, and Procedure 35. List the anions definitely eliminated by these tests. Also list the anions definitely proved present by these tests.
3. Carry out specific tests (see page 250) for those anions not definitely proved to be either absent or present by Steps 1 and 2.

Summarize the results of the cation and anion analysis on the report blank in Chapter 15.

TREATMENT OF THE UNKNOWN WITH CLASS REAGENTS

PROCEDURE 33

Treatment of the Solid with H$_2$SO$_4$

The reagents H$_2$SO$_4$ (cold and hot), AgNO$_3$, and BaCl$_2$ give characteristic reactions that provide useful information about the identity of the anions present. Sulfuric acid is used with the solid unknown, whereas the other two reagents are employed on solutions of the unknown.

Place as much of the solid as can be carried on $\frac{1}{4}$ in. of the tip of the spatula in a small test tube. Add 1 or 2 drops of 18 M H$_2$SO$_4$ (see Note 1). Notice everything that happens, particularly the color and odor of escaping gases (see Note 2; do not place your nose over the mouth of the test tube, but hold the tube a few inches away and fan any gas toward your nose). Then heat, but not so strongly that the H$_2$SO$_4$ is boiled, and note what happens. Finally heat the sides of the test tube over its entire length and note whether or not brown fumes (NO$_2$) are formed (see Note 3). **Do not look down into the test tube! Do not point the test tube at yourself or at your neighbor!**

List each anion whose presence or absence is indicated by this test.

1. The reactions observed when single salts are treated with concentrated H_2SO_4 are usually very definite and easy to interpret. When a mixture of salts is treated with H_2SO_4, however, the observations may be both indefinite and misleading because the reactions of one salt may completely mask other reactions. Thus, if iodides are present, their very violent reactions with H_2SO_4 may cover up the presence of all other anions. Likewise, bromides may cover up everything except iodides and chromates. The number of instances when this test yields precise information, however, are numerous enough to justify the short time required to carry it out.

   If only small quantities of certain salts, such as chlorides, nitrates, acetates, and sulfides, are present, treatment with sulfuric acid may show no definite reaction. Even iodide will not react with concentrated sulfuric acid if it is present as the highly insoluble compound HgI_2. For this reason, elimination of anions made on the basis of this test should not be considered as absolutely final. Rather, the information obtained should supplement information gained in other tests.

2. When concentrated sulfuric acid comes in contact with moisture, a strongly exothermic reaction takes place. This may result in the evolution of steam accompanied by sputtering if the sample is moist. Care should be taken not to mistake this effect for effervescence.

   Attempting to identify a gas by its odor generally is a dangerous practice because such substances may be toxic at worst, an irritant at best. Thus, the American Conference of Governmental Industrial Hygienists established the threshold values, given in parentheses in ppm, for the following gases: H_2S (10), HCl (<5) HBr (3), Br_2 (0.1), I_2 (<0.1), and SO_2 (5). For comparison, the corresponding threshold values for CO_2 and NO_2 are 5000 ppm and <5 ppm, respectively. Recall from Chapter 10 also that H_2S is toxic.

3. Nitric acid decomposes only at relatively high temperatures. Furthermore, it is most readily decomposed by heating its vapors. Consequently, the test tube is heated over its entire length.

PROCEDURE 34

Treatment of a Solution of the Anions with AgNO₃

If the salt or mixture of salts is completely soluble in water, follow (A). If the salt is not soluble in water, follow (C).

(A) *The salt or salt mixture is soluble in water* (see Figure 14.1 for a scheme of analysis). Place 10 drops of the water solution in a test tube; then add 3 to 4 drops of 0.2 M $AgNO_3$. If no precipitate forms, Cl^-, Br^-, I^-, S^{2-}, AsO_4^{3-}, PO_4^{3-}, CrO_4^{2-}, SO_3^{2-}, and BO_2^- are shown to be absent (see Note 1). If a precipitate forms, any or all of the 13 anions may be present. If the precipitate is white, Br^-, I^-, S^{2-}, AsO_4^{3-}, PO_4^{3-}, CrO_4^{2-} (see Figure 14.1 and Note 2), and possibly CO_3^{2-} are absent (see Note 3), but the others may be present (see Note 4); if the precipitate is black, all 13 anions may be present since black will cover up all other colors, but Ag_2S or possibly Ag_2CO_3 (see Note 3) must be present to account for the black color. Note the color of the precipitate and try, by means of the list given above, to decide what it is.

   Centrifuge and decant, discarding the decantate. Wash the precipitate in the test tube once with cold water; then treat it with a few drops of 3 M HNO_3 and stir. If the precipitate dissolves completely, Cl^-, Br^-, I^-, and S^{2-} are absent (see Note 2); if the precipitate is not completely dissolved, one or more of the four anions just enumerated is present (see Note 1). Note the appearance of the residue and try to decide what it is. Centrifuge and decant into a test tube. Save the decantate for (B).

(B) Make the decantate from (A) just alkaline with 15 M aqueous NH_3; then make it just acidic with 5 M acetic acid and add a few drops of 0.2 M $AgNO_3$. The following precipitates will form if the required anions are present: Ag_2CrO_4, brownish red; Ag_3AsO_4, chocolate brown; Ag_3PO_4, yellow; $AgBO_2$, white; Ag_2SO_3, white.

On the basis of the observations made in (A) and (B), what anions, if any, are definitely shown to be absent? What anions are shown to be present?

(C) *The salt or salt mixture is not soluble in water.* A solution for use in this procedure and also in Procedure 35 must first be prepared as directed in Procedure 30(C).

Place 10 drops of this solution in a test tube, acidify with 3 M HNO_3, and then add 3 to 4 drops of 0.2 M $AgNO_3$. If no precipitate forms, Cl^-, Br^-, I^-, and S^{2-} are absent (see Note 1). If a precipitate forms, it may be $AgCl$, $AgBr$, AgI, and Ag_2S. Note the color of the precipitate.

On the basis of the observations made in (C), what anions, if any, are definitely shown to be absent? What anions are shown to be present?

Notes

1. In the presence of HNO_3, Ag_2CrO_4 will not precipitate on addition of $AgNO_3$ unless the concentration of CrO_4^{2-} ions is very high. Once Ag_2CrO_4 has precipitated, however, it is only slowly dissolved by HNO_3.

   The silver compounds $AgBO_2$, Ag_2CO_3, Ag_2SO_3, Ag_2CrO_4, Ag_3PO_4, and Ag_3AsO_4 are salts of weak acids. Therefore, they dissolve in the strong acid HNO_3 in the manner discussed in Chapter 7 on the formation of weak electrolytes.

   a. $Ag_2SO_3 \leftrightarrows 2Ag^+ + SO_3^{2-}$
   b. $SO_3^{2-} + 2H^+ \leftrightarrows H_2SO_3 \leftrightarrows H_2O + SO_2$

   These compounds are not dissolved by $HC_2H_3O_2$ because it is such a weak acid that it does not provide a high enough concentration of H^+ ions to reduce the concentration of the anion in the first equation below the equilibrium solubility product value.

2. Ag_2S will not dissolve in HNO_3 even though it is the salt of a weak acid, as discussed in Chapter 7 on the formation of weak electrolytes.

3. Although Ag_2CO_3 is white, the precipitate formed when $AgNO_3$ is added to a solution of CO_3^{2-} ions darkens quickly, because Ag_2CO_3 is unstable and tends to decompose to form black Ag_2O.

$$Ag_2CO_3 \leftrightarrows Ag_2O + CO_2$$
$$\text{(white)} \quad \text{(black)}$$

   When Ag_2CO_3 is dissolved in HNO_3, CO_2 gas escapes. Consequently, a precipitate of Ag_2CO_3 may not be formed during the subsequent treatment with aqueous NH_3, $HC_2H_3O_2$, and $AgNO_3$.

4. Silver acetate, $AgC_2H_3O_2$ (white), being sparingly soluble, may precipitate when $AgNO_3$ is added if the acetate ion concentration is quite high.

PROCEDURE 35

Treatment of a Solution of the Anions with BaCl₂

(See Figure 14.2 for a scheme of analysis.) If the salt or salt mixture is soluble in water, use the water solution; if the salt is not soluble in water, use the solution prepared as directed in Procedure 30(C). Place 4 to 5 drops of this solution in a test tube, make the solution just alkaline with 5 M aqueous NH₃, and mix thoroughly (see Note 1). If any precipitate forms (hydroxides of metals), centrifuge and decant into a clean test tube, discarding the precipitate. Treat the decantate or solution with 2 to 3 drops of 0.2 M BaCl₂. Formation of a precipitate shows the presence of one or more of the following anions: SO_3^{2-}, SO_4^{2-}, CO_3^{2-}, CrO_4^{2-}, BO_2^-, PO_4^{3-}, or AsO_4^{3-} (see Note 2). A yellow precipitate proves the presence of chromate; a white precipitate proves the absence of chromate. Make the mixture acid with 2 M HCl and stir well. If the precipitate dissolves completely, the absence of sulfate is proved. If the precipitate is not completely dissolved by HCl, the presence of sulfate is proved.

On the basis of the observations made in this procedure, what anions, if any, are definitely shown to be absent? What anions are shown to be present?

Notes

1. Because all the water-insoluble barium salts except BaSO₄ are soluble in strong acids, it is imperative that the solution not be acidic when BaCl₂ is added. Accordingly, the solution is first made alkaline with aqueous NH₃. This will precipitate, as insoluble hydroxides, those cations whose hydroxides are insoluble in the presence of excess aqueous NH₃.

2. Since barium borate precipitates very slowly, absence of a precipitate in the group test for the Sulfuric Acid Group does not positively eliminate borates. A specific test for borates should always be made.

   Although boron may be present as either the metaborate (BO_2^-), the orthoborate (BO_3^{3-}), or the tetraborate ($B_4O_7^{2-}$), it is largely precipitated as the metaborate Ba(BO₂)₂.

SPECIFIC TESTS FOR ANIONS

The order in which the specific tests for the anions should be performed is governed by the following facts. Arsenates, if present, will interfere with the test for phosphates and must be removed before the test for carbonates. Sulfide, bromide, and iodide ions will interfere with the test for chloride. Iodide ions will interfere with the test for bromide. Consequently, the test for sulfite must be made before the test for the carbonate ion, the test for arsenate must be made before the test for the phosphate ion, and the halides and sulfides must be tested for in the order sulfide, iodide, bromide, and chloride. The order of performance of the tests for sulfate, borate, chromate, nitrate, and acetate is not important.

PROCEDURE 36

Test for the Arsenate Ion (AsO₄³⁻)

This test can be omitted if the cation analysis has proved the absence of arsenic (see Note 1). Place 6 to 8 drops of the solution in a small test tube and add Na₂CO₃ solution, drop by drop and with constant stirring, until no more precipitation of metal carbonates takes place and the solution is no longer acidic (see Note 2). If a precipitate has formed, centrifuge and decant into a clean test tube, discarding any precipitate. Make the decantate acid with 2 M HCl and boil to expel all CO_2. Then make just alkaline with 5 M aqueous NH_3. If any precipitate forms, centrifuge and decant into a clean test tube, discarding the precipitate. Add 4 to 5 drops of magnesia mixture and mix thoroughly (see Note 3). If no precipitate forms, arsenate is absent. If a white precipitate forms, it may be $MgNH_4AsO_4$, $MgNH_4PO_4$, or both (see Note 4). Centrifuge and decant; wash the precipitate three times with hot water. Add to the precipitate a drop of 0.2 M $AgNO_3$ solution. A chocolate brown precipitate (Ag_3AsO_4) proves the presence of arsenate. A yellow precipitate (Ag_3PO_4) proves the presence of phosphate and, at the same time, proves the absence of arsenate.

Notes

1. If an unknown substance contains arsenate, the cation analysis of that substance would give a positive test for arsenic in Procedure 13 in Chapter 10. If no positive test for arsenic is obtained in the cation analysis, arsenate cannot be present. Therefore, if arsenic has already been shown to be absent in the cation analysis, the test for the arsenate ion may be omitted. Performance of the test for arsenate will, however, serve as a check on the cation analysis.

2. The addition of Na_2CO_3 will precipitate out all cations except sodium, potassium, and ammonium as carbonates. These cations, if not removed as carbonates, will precipitate as insoluble arsenates when the solution is made alkaline with aqueous NH_3. The arsenate ion would, thus, be lost. Cations not completely removed as carbonates may precipitate on addition of aqueous NH_3.

3. Magnesia mixture is a solution containing Mg^{2+}, OH^-, and excess NH_4^+. Common sources of these ions are $MgCl_2$, aqueous NH_3, and NH_4Cl. The important reactions involved in the test for arsenate are

$$Mg^{2+} + NH_4^+ + AsO_4^{3-} \rightleftharpoons MgNH_4AsO_4$$
$$MgNH_4AsO_4 + 2Ag^+ \rightleftharpoons Ag_3AsO_4 + Mg^{2+} + NH_4^+$$

4. The precipitate formed by the magnesia mixture is not sufficient confirmation of the arsenate ion, since phosphates—when present—precipitate as $MgNH_4PO_4$, which looks exactly like the $MgNH_4AsO_4$. The addition of $AgNO_3$ makes it possible to distinguish the arsenate from the phosphate—$AgNO_3$ reacts with both $MgNH_4AsO_4$ and $MgNH_4PO_4$, but Ag_3AsO_4 is chocolate brown and Ag_3PO_4 is light yellow.

   Magnesium ammonium arsenite ($MgNH_4AsO_3$) is soluble and is not precipitated when the magnesia mixture is added to a solution containing an arsenite. The decantate left after the removal of the $MgNH_4AsO_4$ may contain AsO_3^{3-}. If this decantate is made acid with HCl, and H_2S is passed into the solution or thioacetamide is added as in Procedure 5(C) in Chapter 10, any AsO_3^{3-} present will be precipitated as yellow As_2S_3.

PROCEDURE 37

Test for the Phosphate Ion (PO_4^{3-})

(A) *If arsenates are absent.* Place 4 to 5 drops of the solution in a test tube, acidify with 3 M HNO_3, add 3 to 4 drops of ammonium molybdate solution [$(NH_4)_2MoO_4$], mix thoroughly, and heat almost to boiling for 2 min. Formation, sometimes very slowly, of a finely divided yellow precipitate [$(NH_4)_3PO_4 \cdot 12MoO_3$] confirms the presence of the phosphate ion.

(B) *If arsenates are present* (see Note 1). Place a small amount of the solid or a few drops of the solution in a test tube, add 7 to 8 drops of 12 M HCl, and heat just to boiling. Treat with H_2S for about 20 s or add 4 drops of 1 M thioacetamide and heat in the boiling water bath for 2 min. Add 10 drops of water, mix thoroughly, centrifuge, and decant into a casserole, discarding the precipitate (As_2S_5). Evaporate to dryness, but do not bake. Cool; then add 5 drops of water and 5 drops of 3 M HNO_3, stir thoroughly, and transfer to a test tube. If necessary, centrifuge and decant, discarding the precipitate. Treat the decantate with 3 to 4 drops of ammonium molybdate solution, mix thoroughly, and heat almost to boiling for 2 min. A yellow precipitate [$(NH_4)_3PO_4 \cdot 12MoO_3$], which may form slowly, proves the presence of phosphate (see Note 2).

(C) In the presence of iodide, the phosphate test gives a green solution, which sometimes precipitates a green solid upon standing. The green color is the result of a mixture of species, which arises from the oxidation of I^- by the 12-molybdophosphate ion. Heating the green solution in a boiling water bath (under a hood) causes iodine vapor to escape; the green color fades and the yellow ammonium 12-molybdophosphate precipitate becomes visible.

Alternatively, if the solution is known to contain iodide, it can be removed by acidification with HNO_3 and treatment with silver nitrate, the resulting AgI being removed by centrifugation and decantation. The decantate is then tested for phosphate according to the usual procedure.

Notes

1. Both phosphates and arsenates react with ammonium molybdate to form insoluble yellow precipitates: ammonium 12-molybdophosphate [$(NH_4)_3PO_4 \cdot 12MoO_3$] and ammonium arsenomolybdate [$(NH_4)_3AsO_4 \cdot 12MoO_3$]. As a consequence, arsenates must be removed before the confirmatory test for phosphates can be made.

   Arsenates, if present, are removed as As_2S_5 by precipitation with H_2S.

$$2AsO_4^{3-} + 5H_2S + 6H^+ \leftrightarrows As_2S_5 + 8H_2O$$

2. The formation of the yellow precipitate, $(NH_4)_3PO_4 \cdot 12MoO_3$, takes place as follows:

$$PO_4^{3-} + 12MoO_4^{2-} + 24H_+ + 3NH_4^+ \leftrightarrows (NH_4)_3PO_4 \cdot 12MoO_3 + 12H_2O$$

PROCEDURE 38

Test for the Borate Ion (BO_2^- or BO_3^{3-})

Place a small quantity of the solid material in a casserole, add 3 to 4 drops of 18 M H_2SO_4, and stir thoroughly. Add 10 to 12 drops of methyl alcohol (CH_3OH, methanol) and again mix thoroughly; then set fire to the mixture. *Do not stir after the mixture has taken fire.* If it burns with a green flame the instant that it takes fire, the borate ion (see Note 1) is present (see Note 2). A green flame that does not appear until 20 or 30 s after the mixture has taken fire, or unless the mixture is stirred while on fire, is due to copper or barium and should be ignored.

Notes

1. There are three boric acids: orthoboric (H_3BO_3, the boric acid of commerce), metaboric (HBO_2), and tetraboric ($H_2B_4O_7$). All three of these boric acids form stable salts.
   All borates react with H_2SO_4 to form orthoboric acid, H_3BO_3.

$$BO_2^- + H^+ + H_2O \leftrightharpoons H_3BO_3$$

$$B_4O_7^{2-} + 2H^+ + 5H_2O \leftrightharpoons 4H_3BO_3$$

$$BO_3^{3-} + 3H^+ \leftrightharpoons H_3BO_3$$

In the presence of sulfuric acid, H_3BO_3 reacts with alcohol to form the volatile ester, methyl borate, $(CH_3)_3BO_3$.

$$3CH_3OH + H_3BO_3 \leftrightharpoons (CH_3)_3BO_3 + 3H_2O$$

The methyl borate burns with a characteristic green flame, forming solid B_2O_3 (boric acid anhydride):

$$2(CH_3)_3BO_3 + 9O_2 \leftrightharpoons 6CO_2 + 9H_2O + B_2O_3$$

2. Copper and barium do not interfere with the borate test because their salts are not volatilized when the mixture first takes fire.

PROCEDURE 39

Test for the Chromate(VI) Ion (CrO_4^{2-})

(This procedure is to be omitted if no chromium was found in the complete cation analysis.) Place 2 drops of the solution in a test tube, add 10 drops of water, and make just acid with 3 M HNO_3. Add 5 to 6 drops of ether and 1 drop of 3 percent H_2O_2, stir well, and then allow to settle. A blue coloration of the ether layer (see Note 1) confirms the presence of the chromate ion (see Note 2).

Notes

1. The blue coloration in the ether layer is due to the presence of chromium peroxide, CrO_5.

$$2CrO_4^{2-} + 2H^+ \leftrightharpoons Cr_2O_7^{2-} + H_2O$$

$$Cr_2O_7^{2-} + 4H_2O_2 + 2H^+ \leftrightharpoons 2CrO_5 + 5H_2O$$

See Note 1, Procedure 22 in Chapter 11, for further details concerning this reaction.

2. If a chromate is carried through the complete cation analysis, it is reduced to Cr^{3+} when the solution is evaporated down with 6 M HCl in Procedure 5 in Chapter 10.

$$2CrO_4^{2-} + 2H^+ \leftrightharpoons Cr_2O_7^{2-} + H_2O$$

$$Cr_2O_7^{2-} + 6Cl^- + 14H^+ \leftrightharpoons 2Cr^{3+} + 7H_2O + 3Cl_2$$

The resulting Cr^{3+} ions remain in the decantate from Procedure 5 and are subsequently precipitated as $Cr(OH)_3$ in Procedure 15 in Chapter 11. A positive test for chromium is then obtained in Procedure 22 in Chapter 11.

The fact that a positive test for chromium is obtained in Procedure 22 does not mean, however, that the particular salt mixture contains chromate. The chromium may be present as Cr^{3+}.

PROCEDURE 40

Test for the Sulfate Ion (SO_4^{2-})

Place a few drops of the solution in a test tube, acidify with 6 M HCl, and then add a drop of 0.2 M $BaCl_2$. A white precipitate ($BaSO_4$) proves the presence of sulfate (see Note 1).

Note

1. Sulfites are slowly oxidized to sulfates by atmospheric oxygen. Consequently, sulfites commonly show a positive test for sulfates.

$$2SO_3^{2-} + O_2 \leftrightharpoons 2SO_4^{2-}$$
$$Ba^{2+} + SO_4^{2-} \leftrightharpoons BaSO_4$$
$$(white)$$

PROCEDURE 41

Test for the Sulfite Ion (SO_3^{2-})

Place 7 to 8 drops of the solution in a test tube, acidify with 6 M HCl, add 2 to 3 drops of 0.2 M $BaCl_2$, and mix thoroughly. If a precipitate ($BaSO_4$) forms, remove it by centrifuging and decanting. To the clear decantate, add a drop of 3 percent H_2O_2 (see Note 1). The formation of a white precipitate ($BaSO_4$) proves the presence of sulfite.

Note

1. H_2O_2 oxidizes sulfite to sulfate.

$$SO_3^{2-} + H_2O_2 \leftrightharpoons SO_4^{2-} + H_2O$$

The barium ions in solution react with these sulfate ions to form insoluble $BaSO_4$.

Any sulfate ions in the original solution are removed, as barium sulfate, when the $BaCl_2$ is first added.

PROCEDURE 42

Test for the Carbonate Ion (CO_3^{2-})

(A) *When sulfites are absent.* Place a small amount of the solid in a test tube. Then add a few drops of 2 *M* HCl. Test the escaping gas for CO_2 by holding a drop of barium hydroxide solution, suspended from the tip of a medicine dropper, a short distance down into the mouth of the test tube. The "clouding" of the drop, due to the formation of a white precipitate of barium carbonate ($BaCO_3$), proves the presence of carbonate (see Note 1).

(B) *When sulfites are present* (see Note 2). Place a small amount of the solid in a test tube and add an equal amount of solid Na_2O_2. Then add 3 to 4 drops of water and mix thoroughly. Proceed as directed in the second and succeeding sentences of (A).

Notes

1. Carbonates react with acids to evolve CO_2 gas. This CO_2 reacts with $Ba(OH)_2$ solution to form a white precipitate of $BaCO_3$.

$$\text{carbonate of a metal} + nH^+ \rightleftharpoons \text{metal cation}^{+n} + H_2O + CO_2$$

$$CO_2 + Ba^{2+} + 2OH^- \rightleftharpoons BaCO_3 + H_2O$$
$$\text{(white)}$$

2. Sulfites evolve SO_2 gas when treated with acids. The SO_2 gas will react with $Ba(OH)_2$ solution to form a white precipitate of $BaSO_3$. Sulfites must, therefore, be removed or destroyed before the test for carbonates is carried out. Sodium peroxide oxidizes sulfites to sulfates.

$$SO_3^{2-} + Na_2O_2 + H_2O \rightleftharpoons SO_4^{2-} + 2Na^+ + 2OH^-$$

PROCEDURE 43

Test for the Sulfide Ion (S^{2-})

Place a small quantity of the solid in a test tube and (under the hood) add 10 drops of 6 *M* HCl. Hold a strip of filter paper moistened with 0.2 *M* $Pb(C_2H_3O_2)_2$ solution over the mouth of the test tube so that any gas that is being evolved will come in contact with the lead acetate. A brownish-black or silvery-black stain (PbS) on the paper confirms the presence of sulfides (see Note 1).

If no blackening of the lead acetate occurs after 1 min, heat the tube gently; if still no reaction occurs, add a small amount of granulated zinc to the contents of the tube (see Note 2). If the lead acetate is not darkened, the sulfide ion is absent; if it is darkened, sulfide is present.

Notes

1. The reactions that take place in the test sulfides are:

$$\text{sulfide of metal} + nH^+ \rightleftharpoons \text{metal cation}^{+n} + H_2S$$
$$Pb^{2+} + H_2S \rightleftharpoons PbS + 2H^+$$

2. The sulfides of nickel and cobalt and the metals of the silver and copper-arsenic groups are not soluble in dilute HCl. When zinc is added, the sulfides are reduced and H_2S is liberated.

$$Zn + HgS + 2H^+ \leftrightharpoons Zn^{2+} + Hg + H_2S$$

PROCEDURE 44

Test for the Iodide Ion (I^-)

Place 5 drops of the solution in a test tube and acidify with 5 M $HC_2H_3O_2$. Then add 2 drops of 0.2 M KNO_2. A reddish-brown coloration, due to liberation of iodine, proves the presence of iodide. If the brown color is very faint, add about 10 drops of carbon tetrachloride (CCl_4) or chloroform ($CHCl_3$), shake, and then allow to settle. A violet coloration in the CCl_4 layer shows the presence of iodine (see Note 1; see Note 2 for an alternative test for iodide).

Notes

1. The used CCl_4 or $CHCl_3$ solution should *not* be poured down the sink; a waste bottle with a screw cap should be provided in the hood to receive the CCl_4 or $CHCl_3$ solutions.
2. *Alternative test for iodide ion.* Place 5 drops of the solution in a test tube, acidify with 3 M HNO_3, and add a drop of 0.2 M $Fe(NO_3)_3$. A reddish-brown coloration, due to free iodine, proves the presence of iodide. If the solution is shaken with CCl_4 or $CHCl_3$, the iodine will concentrate in the CCl_4 or $CHCl_3$ layer and will give it a violet coloration.

     Nitrite ions and ferric ions will both oxidize iodides, but will not oxidize either bromides or chlorides (see Table 6.1). For that reason, both the regular test and the alternative test can be carried out in the presence of bromides and chlorides. The reactions that take place in the two tests are

$$2HNO_2 + 2I^- + 2H^+ \leftrightharpoons 2NO + I_2 + 2H_2O$$
$$2Fe^{3+} + 2I^- \leftrightharpoons 2Fe^{2+} + I_2$$

PROCEDURE 45

Test for the Bromide Ion (Br^-)

Iodides will interfere with this test (see Note 1). Place 5 drops of the solution in a test tube; add 5 drops of chlorine water. A yellow-to-brown coloration, due to liberated bromine, shows the presence of bromide. If the solution is shaken with about 10 drops of CCl_4 or $CHCl_3$, the brown color will concentrate in the lower CCl_4 or $CHCl_3$ layer (see Note 2 for an alternative test for bromides).

Notes

1. Iodide ions, if present, must be removed before the test for bromide ions is carried out. The iodide ions can be removed by either of the following methods. *Method A:* Acidify the solution with 3 M HNO_3 and add 0.2 M KNO_2, dropwise, with constant stirring, until there

is no further increase in the depth of the brown color. Extract once, with 5 drops of CCl_4 or $CHCl_3$, discarding the CCl_4 or $CHCl_3$ layer. Boil the water layer carefully until the iodine has been largely driven off. Test the colorless, or near-colorless, water solution for bromides as directed above. *Method B:* Acidify the solution with 3 M HNO_3 and add 0.2 M $Fe(NO_3)_3$ dropwise, with constant stirring, until there is no further increase in the depth of the brown color. Extract once with 5 drops of CCl_4 or $CHCl_3$. Boil the water layer until the iodine has been largely driven off. Test the near-colorless water solution for bromides as directed above (see Note 1, Procedure 44).

2. *Alternative test for bromide ion* (iodides will interfere with this test). Place 5 drops of the solution in a test tube, acidify with 3 M HNO_3, add a drop of 0.02 M $KMnO_4$, and mix thoroughly. Decolorization of the permanganate and formation of a yellow-to-brown coloration, due to free bromine, shows the presence of bromide. If the solution is shaken with CCl_4 or $CHCl_3$, the bromine will concentrate in the CCl_4 or $CHCl_3$ layer.

Bromide ions are oxidized by chlorine and by permanganate ions according to the following equations (see Table 6.1).

$$Cl_2 + 2Br^- \leftrightarrows 2Cl^- + Br_2$$
$$2MnO_4^- + 10Br^- + 16H^+ \leftrightarrows 2Mn^{2+} + 5Br_2 + 8H_2O$$

The violet color of MnO_4^- will disappear, due to formation of colorless Mn^{2+}. Persistence of the violet MnO_4^- color is a sign that all bromide ions have been oxidized. Iodide ions will also be oxidized by both chlorine and MnO_4^-.

$$Cl_2 + 2I^- \leftrightarrows 2Cl^- + I_2$$
$$2MnO_4^- + 10Br^- + 16H^+ \leftrightarrows 2Mn^{2+} + 5Br_2 + 8H_2O$$

PROCEDURE 46

Test for the Chloride Ion (Cl^-)

Sulfides, bromides, and iodides will interfere with this test (see Note 1). Place 5 to 6 drops of the solution in a test tube, acidify with 3 M HNO_3, and add a drop of 0.2 M $AgNO_3$. A white, curdy precipitate (AgCl) proves the presence of chlorides.

Note

1. Since Ag_2S, AgBr, and AgI also are insoluble in acid solution, this test is not conclusive unless S^{2-}, Br^-, and I^- are definitely shown to be absent. Chromate ions, if present in high concentrations, may also interfere.

Interference from chromate ions can be eliminated by dilution with 3 M HNO_3.

Sulfide ions can be removed by boiling the solution with 2 M H_2SO_4 until the escaping vapors give no test for H_2S gas with filter paper moistened with lead acetate solution.

Iodide and bromide ions can be removed as described below.

Separation and detection of chlorides, bromides, and iodides in the presence of one another. Place 15 drops of the solution in a casserole; add 15 drops of 5 M $HC_2H_3O_2$ and 30 drops of water; mix thoroughly. Then add a small quantity of solid potassium peroxydisulfate ($K_2S_2O_8$) and heat. A brown coloration, due to the liberation of iodine, proves the presence of iodide.

Boil the solution until all the iodine is removed, replenishing the liquid with water. Test for complete oxidation of iodide by adding a small quantity of $K_2S_2O_8$ and 2 drops of 5 M $HC_2H_3O_2$. When iodide is completely removed, add 15 drops of 2 M H_2SO_4 and another small quantity of $K_2S_2O_8$ and heat to boiling. A brown coloration, due to bromine, proves the presence of bromides.

Boil the solution until all the bromine is driven off, replenishing the liquid with water. Test for complete oxidation of bromide by adding a small quantity of $K_2S_2O_8$ and 2 drops of 2 M H_2SO_4. When removal of bromide is complete, cool the solution, acidify with 3 M HNO_3, and add 2 to 3 drops of 0.2 M $AgNO_3$. A white precipitate (AgCl) proves the presence of chlorides.

This separation depends on the fact that, in acetic acid solution, $K_2S_2O_8$ will oxidize iodide but will not oxidize bromide or chloride.

$$2I^- + S_2O_8^{2-} \leftrightharpoons 2SO_4^{2-} + I_2$$

In a more strongly acid solution it will oxidize iodide and bromide but will not oxidize chloride.

$$2Br^- + S_2O_8^{2-} \leftrightharpoons 2SO_4^{2-} + Br_2$$

PROCEDURE 47

Test for the Nitrate Ion (NO_3^-)

Iodides, bromides, and chromates interfere with this test and must be removed (see Note 1) if they are present. Place 2 drops of the water solution, or the supernatant liquid obtained by treating the solid with hot water (see Note 2), in a test tube and carefully add 10 drops of 18 M H_2SO_4. Mix thoroughly and cool. Carefully add 3 to 4 drops of 0.2 M $FeSO_4$ solution, allowing the latter to float on top of the sulfuric acid solution. Allow to stand for 1 or 2 min. A brown coloration at the junction of the two layers due to the presence of the complex nitrosyliron(II) ion, $Fe(NO)^{2+}$, proves the presence of nitrate (see Note 3).

Notes

1. Iodides and bromides react with concentrated H_2SO_4 to liberate I_2 and Br_2.

$$SO_4^{2-} + 8I^- + 10H^+ \leftrightharpoons H_2S + 4I_2 + 4H_2O$$
$$SO_4^{2-} + 2I^- + 4H^+ \leftrightharpoons SO_2 + I_2 + 2H_2O$$
$$SO_4^{2-} + 2Br^- + 4H^+ \leftrightharpoons SO_2 + Br_2 + 2H_2O$$

Chromate(VI) ions, if present, will be reduced by Fe^{2+} to green Cr^{3+}.

$$2CrO_4^{2-} + 2H^+ \leftrightharpoons Cr_2O_7^{2-} + H_2O$$
$$Cr_2O_7^{2-} + 6Fe^{2+} + 14H^+ \leftrightharpoons 2Cr^{3+} + 6Fe^{3+} + 7H_2O$$

The colors of I_2, Br_2, and Cr^{3+} will interfere with detection of the brown color of $Fe(NO)^{2+}$. Consequently, I^-, Br^-, and CrO_4^{2-} must be removed as follows: Place 4 drops of the water solution in a test tube and add 0.2 M $Pb(C_2H_3O_2)_2$ until precipitation is complete. Centrifuge and decant, discarding the precipitate ($PbCrO_4$). Treat the decantate with silver sulfate solution until precipitation is complete. Centrifuge and decant into a casserole, discarding the precipitate (AgI, $AgBr$, $PbSO_4$). Evaporate the solution down to a volume of about 2 drops, add 2 drops of water, and transfer to a test tube. Test this solution for nitrate according to Procedure 47.

2. Since all nitrates are soluble in water, the nitrate ions will be contained in the supernatant liquid obtained by digesting the solid material with water, even though the solid may not be completely soluble.

3. The test for nitrate ion is based upon the fact that in the presence of concentrated H_2SO_4, Fe^{2+} reduces NO_3^- to NO.

$$3Fe^{2+} + NO_3^- + 4H^+ \rightleftharpoons 3Fe^{3+} + NO + 2H_2O$$
$$NO + Fe^{2+} \rightleftharpoons Fe(NO)^{2+}$$
$$\text{(excess)} \quad \text{(brown)}$$

The NO formed combines with excess Fe^{2+} ions to form the characteristic brown complex ion, $Fe(NO)^{2+}$. This complex ion is unstable at higher temperatures; hence, the test must be made at room temperature.

PROCEDURE 48

Test for the Acetate Ion ($C_2H_3O_2^-$)

Place a small amount of the solid in a test tube, add 2 to 3 drops of 18 M H_2SO_4, and mix thoroughly (see Note 1). Add 4 drops of ethyl alcohol (C_2H_5OH, ethanol) and again mix thoroughly. Heat the tube in the boiling water bath for about 1 min. Carefully smell the odor of the escaping fumes. A fruity odor like airplane glue or nail polish, due to ethyl acetate ($C_2H_5O_2CCH_3$), proves the presence of acetate (see Note 2). If the test is doubtful, place a pinch of the solid in a test tube, heat strongly, and note whether there is any charring of the material. Charring is indicated by the escape of fumes that have the sharp penetrating odor of burning hair or singed feathers; also, the solid darkens in color. Charring shows the presence of acetate (see Note 3).

Notes

1. If sulfites, carbonates, iodides, bromides, chlorides, or sulfides are present, the mixture should be warmed gently for some time after addition of the sulfuric acid and before addition of the alcohol. This heating will drive off all SO_2, CO_2, I_2, Br_2, HCl, and H_2S. If they are not driven off, their odors will mask the characteristic odor of the ethyl acetate. Care should be taken that the heating is not so intense that $HC_2H_3O_2$ is distilled off.

2. When an acetate is treated with concentrated H_2SO_4, acetic acid (ethanoic acid) is liberated. The acetic acid reacts with ethyl alcohol to form the volatile ester $C_2H_5O_2CCH_3$; the characteristic fruity odor of the latter compound makes possible its identification.

$$H^+ + C_2H_3O_2^- \rightleftharpoons HC_2H_3O_2$$
$$C_2H_5OH + HC_2H_3O_2 \rightleftharpoons C_2H_5O_2CCH_3 + H_2O$$

3. The presence of acetate is shown in the preliminary treatment with sulfuric acid (Procedure 33), in which it gives the characteristic odor of vinegar (acetic acid).

   Charring on heating is characteristic of the salts of organic acids. Since acetic acid is the only organic acid among the 13 anions, the incidence of charring when the solid is heated is proof of the presence of acetate.

PROBLEMS

Simple Solubility

14.1 From the following list of soluble salts, select (a) 10 pairs that, when brought together in solution, will not react to form a precipitate; (b) 10 pairs that, when brought together in solution, will form a precipitate. Give the formula of the precipitate in each case and write the net equation for its formation.

$Pb(NO_3)_2$, Ag_2SO_4, $MgCrO_4$, BaS, $FeCl_3$, NiI_2, $ZnBr_2$, Na_3PO_4, K_2CO_3, $(NH_4)_3AsO_4$, $NaBO_2$, $Al_2(SO_4)_3$, $Bi(NO_3)_3$, $MnCl_2$, $(NH_4)_2SO_3$, CaS

14.2 Give the formula of each acid described.

   a. Its barium salt is insoluble in water, but its copper salt is soluble.
   b. Its copper salt is insoluble in water, but its calcium salt is soluble.
   c. Its lead salt is insoluble in water, but its zinc salt is soluble.
   d. Its silver salt is insoluble in water, but its copper salt is soluble.
   e. Its mercury(I) (mercurous) salt is insoluble in water, but its mercury(II) (mercuric) salt is soluble.
   f. Its manganese salt is insoluble in water, but its potassium salt is soluble.
   g. Its nickel salt is insoluble in water, but its magnesium salt is soluble.

14.3 A mixture of barium and silver salts was readily and completely soluble in cold water. What anions are not present in this mixture?

14.4 An unknown salt or mixture of salts is completely soluble in water and contains carbonate and nitrate as the only anions. What cations may be present?

14.5 An unknown salt or mixture of salts is completely soluble in water and contains Cu^{2+}, Ag^+, and Na^+ as the only cations. What anions may be present?

Solubility Plus Other Characteristics

14.6 An unknown salt or mixture of salts is completely and readily soluble in cold water. On treating this solution with dilute HCl, effervescence takes place; a colorless, odorless gas is evolved. What cations may be present?

14.7 An unknown salt or mixture of salts is completely soluble in cold water. On being treated with dilute HCl the water solution changes in color from yellow to orange-red. What cations may be present?

14.8 A homogeneous powder that is known to contain only one metallic ion dissolves completely in cold water to give a pale blue solution, which turns a very deep blue when treated with NH_4OH. What anions may be present?

14.9 The chloride of a metal is soluble in cold water. Its hydroxide is white and is insoluble in water and in NH_4OH, but dissolves readily in HCl and in KOH. What is the metal?

14.10 An unknown salt or salt mixture is readily and completely soluble in cold water. The resulting solution is green in color. List the anions that can be eliminated on the basis of solubility.

14.11 The sulfide of a metal is insoluble in water, but soluble in dilute HCl. Its hydroxide is insoluble in water but soluble in both NaOH and NH_4OH. What is the metal?

14.12 A mixture of salts is completely soluble in water. When a solution of sodium carbonate is added to the water solution, there is no sign of any reaction. What cations can be eliminated on the basis of this observation?

14.13 An unknown salt or mixture of salts is readily and completely soluble in cold water. On being treated with dilute HCl, effervescence takes place, a colorless gas with a very sharp, penetrating odor being evolved. What cations may be present?

14.14 A solid mixture of salts was known to contain sulfate and bromide as the only anions.

a. The solid dissolved readily and completely in cold water to give a clear, colorless solution showing no sign of milkiness or opalescence.

b. When a sample of the water solution prepared in part (a) was treated with an excess of NaOH, a clear solution with no precipitate was obtained.

c. When a sample of the water solution prepared in part (a) was treated with an excess of NH_4OH, a clear solution with no precipitate was obtained.

What cations may be present?

14.15 A mixture consists of equivalent amounts of two salts, A and B. Each of the two salts taken alone is soluble in water.

When a sample of the mixture is treated with cold water, half of the ions are present in solution and half are present as a precipitate.

When a second sample of the mixture is treated with cold 3 M HCl, a clear colorless solution with no precipitate is formed.

When a third sample of the mixture is treated with cold 3 M HNO_3, a precipitate is present in the solution that is formed.

When a fourth sample is treated with cold 3 M H_2SO_4, a precipitate is present in the solution that is formed.

What can salt A be? (If there is more than one possibility, give all of them.)

What can salt B be? (If there is more than one possibility, list all of them.)

14.16 A mixture consists of equivalent amounts of two salts, A and B.

a. Each salt is readily and completely soluble in cold water. However, when an adequate amount of cold water is added to the equimolar mixture of the two single salts, followed by thorough stirring, none of the four ions that constitute salts A and B remains in solution. Instead, a mixture of two white compounds is formed as a solid residue.

b. When 3 M HCl is added to the solid residue of two white compounds, one of the compounds dissolves, and the other does not dissolve.

c. On the basis of the evidence given in parts (a) and (b), give the possible formulas of the white residue compound that did not dissolve in 3 M HCl and of the white residue compound that did dissolve in 3 M HCl.

Solubility in Various Reagents

14.17 A yellow solid that is known to be a single salt is completely insoluble in hot water, but dissolves in hot dilute HCl to give an orange-red solution. When this solution is cooled, a white crystalline precipitate forms. This precipitate redissolves when the solution is heated, but does not redissolve when cold water is added. What is the salt?

14.18 Give the symbol for each metal described.

   a. Its sulfide is insoluble in water, but soluble in dilute HCl.

   b. Its sulfide is insoluble in dilute HCl, but soluble in concentrated HCl.

   c. Its sulfide is insoluble in dilute HCl, but soluble in dilute HNO_3.

   d. Its hydroxide is insoluble in water, but soluble in NaOH.

   e. Its hydroxide is insoluble in water, but soluble in NH_4OH.

   f. Its hydroxide is insoluble in NH_4OH, but soluble in KOH.

   g. Its hydroxide is insoluble in NaOH, but soluble in NH_4OH.

   h. Its hydroxide is insoluble in water, but soluble in NH_4OH and also in NaOH.

   i. Its hydroxide is amphoteric.

   j. Its oxide is insoluble in HNO_3, but soluble in HCl.

   k. Its chloride hydrolyzes in water to give an insoluble basic chloride.

   l. Its sulfide is completely hydrolyzed in water.

14.19 Give the simplest and best solvent that could be used to dissolve each of the following: ZnS, $MgCrO_4$, $CuSO_4$, $PbCO_3$, Na_3AsO_4, HgS, AgCl, $BiCl_3$, $NiSO_3$, $BaSO_4$.

14.20 A solid salt mixture contains $PbSO_4$, HgS, AgCl, $FePO_4$, and $Mg(NO_3)_2$. Describe in detail how you would prepare solutions of the solid mixture for both cation and anion analysis. Show schematically how you would proceed and what would happen. Write detailed equations and net equations for all reactions that take place.

Treatment with H_2SO_4

14.21 How would treatment with concentrated H_2SO_4 enable you to distinguish between the following pairs of dry salts?

   (a) $BaSO_4$ and $Pb(NO_3)_2$ **(b)** KCl and Na_3AsO_4

   (c) $MnSO_4$ and Na_2CO_3 **(d)** $NaC_2H_3O_2$ and NaCl

   (e) $MgBr_2$ and K_3PO_4 **(f)** $CaCrO_4$ and KI

   (g) ZnS and $Ba(BO_2)_2$ **(h)** KBr and KNO_3

14.22 Five different solid, dry, anhydrous sodium salts were treated, independently, with cold, concentrated H_2SO_4. The results listed below for each salt were obtained.

   a. Effervescence. The evolved gas had a light reddish-brown color and a sharp odor and fumed strongly in moist air.

   b. No effervescence. Color of solid changed from yellow to orange.

   c. Effervescence. The evolved gas was colorless and odorless and did not fume in moist air.

   d. Effervescence. The evolved gas was colorless, had a sharp odor, and fumed strongly in moist air.

   e. Effervescence. The evolved gas was colorless and had a very sharp odor, but did not fume in moist air and did not discolor piece of filter paper that had been moistened with a solution of lead nitrate.

   Identify each salt.

14.23 A silver salt was but slightly soluble in water. Upon heating the dry salt with concentrated H_2SO_4, there was no evidence of any reaction. Name the anions possibly present.

14.24 A pure single salt is white, crystalline, and completely soluble in water. When the water solution is acidified with HCl and then treated with H_2S, a black precipitate forms. When H_2SO_4 is added to the water solution, a white precipitate is formed. When cold concentrated H_2SO_4 is added to the dry salt, nothing happens; when the H_2SO_4 is heated, however, a brown gas is given off. What is the salt?

Preliminary Tests with $AgNO_3$ and $BaCl_2$

14.25 A mixture of salts is readily and completely soluble in cold water. Separate samples of this water solution give no precipitate with $AgNO_3$ or with NH_4OH and $BaCl_2$. What anions could be present in this salt mixture?

14.26 An unknown salt or mixture of salts is completely soluble in water. The colorless solution gives a red flame test. When the water solution is treated with $AgNO_3$, nothing happens. What anions may be present?

14.27 An unknown was found to be completely soluble in water. Addition of solutions of Na_2HPO_4, $BaCl_2$, and $AgNO_3$ to separate portions of an aqueous solution of the unknown failed to yield any precipitate. Name the cations and anions possibly present.

14.28 An unknown salt or salt mixture was readily and completely soluble in cold water. Separate samples of this water solution showed the following behavior.

  a. No precipitate when treated with $AgNO_3$ solution.
  b. No precipitate when treated with HNO_3 and $AgNO_3$ solution.
  c. No precipitate when treated with slight excess of NH_4OH and then with $BaCl_2$ solution.
  d. No precipitate when acidified with HCl and then treated with $BaCl_2$ solution.

  What anions and what cations are eliminated in each test?

14.29 A white solid known to be a mixture of sodium salts showed the following behavior.

  a. There was no sign of any reaction whatsoever when a sample of the solid was warmed with concentrated H_2SO_4.
  b. A water solution of the solid gave a white precipitate when made acid with HCl and then treated with $BaCl_2$.
  c. A water solution of the solid gave no precipitate when acidified with HNO_3 and then treated with $AgNO_3$.
  d. A water solution of the solid gave a white precipitate when treated with a few drops of $AgNO_3$.

  What anions are shown to be absent?

14.30 A white solid that is known to be a single salt is completely soluble in cold water. This solution is acid to litmus. When the solid is treated with HCl, nothing happens; when treated with H_2SO_4, effervescence takes place, the solid remaining colorless. On treating the water solution with NH_4OH, a white precipitate is formed that redissolves when more NH_4OH is added. When $BaCl_2$ is added to the water solution, nothing happens; but when $AgNO_3$ is added, a white precipitate is formed. What is the solid?

14.31 A white crystalline, homogeneous solid known to be a single salt dissolves in cold water to give a water-clear solution that reacts acid to litmus paper. Separate samples of this water solution give the following results when treated separately with the given reagents.

  a. Excess NaOH, no precipitate.
  b. Excess NH_4OH, no precipitate.
  c. Na_2CO_3, white precipitate.

    d. H_2S, white precipitate.

    e. Excess NH_4OH + $BaCl_2$, white precipitate.

    f. HNO_3 + $AgNO_3$, no precipitate.

    g. HCl, no sign of any reaction.

What is the salt?

14.32 A green, crystalline substance that is known to be a single salt is completely soluble in cold water. When an excess of NH_4OH is added to the water solution, no precipitate is formed but a deep blue solution results. When H_2S is added to a water solution that has been acidified with HCl, a black precipitate forms. When $BaCl_2$ is added to the water solution, nothing happens. When cold, concentrated H_2SO_4 is added to the water solution, nothing happens. When cold concentrated H_2SO_4 is added to the solid, effervescence takes place; a colorless, strongly fuming gas with a sharp odor is evolved.

What could the single salt be?

14.33 A single salt is found to contain Zn^{2+} as the only cation. The salt is insoluble in water. No visible reaction takes place in the H_2SO_4 treatment, no gas being evolved at either high or low temperatures. A white precipitate is obtained in the $BaCl_2$ preliminary test made on the solution prepared according to Procedure 30(C).

What can the salt be?

Specific Tests for Anions

14.34 Describe the method by which you would make a specific test for each.

    a. NO_3^- in an unknown that contains I^-.

    b. Borate in an unknown that contains Cu^{2+}.

    c. CO_3^{2-} in an unknown that contains SO_3^{2-}.

    d. PO_4^{3-} in an unknown that contains AsO_4^{3-}.

    e. Br^- in an unknown that contains I^-.

    f. Cl^- in an unknown that contains S^{2-}.

    g. SO_3^{2-} in an unknown that contains SO_4^{2-}.

    h. $C_2H_3O_2^-$ in an unknown that contains SO_3^{2-}.

    i. NO_3^- in an unknown that contains CrO_4^{2-}.

    j. AsO_4^{3-} in an unknown that contains PO_4^{3-}.

14.35 Describe the method by which you would test for each.

    a. Boric acid in an eyewash.

    b. Nitrate in a solid fertilizer.

    c. Phosphate in bath salts.

    d. Carbonate in limestone.

    e. Arsenate in an insecticide.

    f. Sulfite in a bleaching solution.

    g. Iodide in table salt.

14.36 A white solid is completely soluble in cold water. Its water solution is neutral to litmus. No change takes place when the solid is warmed with concentrated H_2SO_4. There is no change when the water solution is treated with H_2SO_4 and $FeSO_4$. Which one of the following substances may the solid be: KNO_3, $AlCl_3$, Na_2CO_3, Na_2SO_4, $NaBr$, $BaSO_4$, K_2CrO_4, KCl, NiS, $Cr_2(SO_4)_3$?

14.37 A salt mixture that was readily and completely soluble in cold water showed the following behavior.

    a. One sample of the water solution gave a yellow precipitate when acidified with HNO_3 and treated with ammonium molybdate.

    b. Another sample of the water solution gave a dark brown coloration when treated with a mixture of concentrated H_2SO_4 and 0.2 M $FeSO_4$.

    c. A sample of the solid effervesced when treated with cold, concentrated H_2SO_4. The evolved gas did not fume in moist air, but had a very sharp, penetrating odor.

    What anions are definitely shown to be present? What cations may be present?

14.38 A single salt is known to contain magnesium as the only cation. Effervescence takes place when the solid is treated with cold, concentrated H_2SO_4, a brown gas being evolved. The $AgNO_3$ test on the water solution yields a cream-colored precipitate. When H_2SO_4 and $FeSO_4$ are added to the water solution, a brown ring forms at the junction of the two liquids. What is the salt?

GENERAL QUESTIONS

14.39 How would you distinguish, by means of simple tests, between the following solids?

    (a) $BaSO_4$ and $Pb(NO_3)_2$ **(b)** KCl and ZnS

    (c) $MnSO_3$ and Na_2CO_3 **(d)** $AgBr$ and $FeCl_3$

    (e) $HgCO_3$ and $(NH_4)_3PO_4$ **(f)** $Bi_2(CrO_4)_3$ and $CuSO_4$

    (g) $Cr(NO_3)_3$ and $Cd(NO_3)_2$ **(h)** $CoCl_2$ and $AlCl_3$

    (i) $NiCO_3$ and KBr **(j)** $BaCl_2$ and $PbCO_3$

    (k) Mn and Zn **(l)** $BaSO_4$ and $PbSO_4$

14.40 By means of what simple test would you be able to distinguish between the following solutions? State what happens to each solution that enables you to make the distinction.

    (a) K_3AsO_4 and $CdSO_4$ **(b)** Na_3PO_4 and Na_3AsO_4

    (c) $Cu(NO_3)_2$ and $Ni(NO_3)_2$ **(d)** KNO_3 and KBr

    (e) $Ca(OH)_2$ and $Ca(HCO_3)_2$ **(f)** Fe^{2+} and Fe^{3+}

    (g) $CdCl_2$ and $ZnCl_2$ **(h)** 1 M Na_2SO_3 and 1 M Na_2CO_3

14.41 A solid is known to be made up of a mixture of equivalent amounts of two or more of the solids listed below.

$$CuSO_4,\ AgNO_3,\ (NH_4)_2SO_4,\ PbCl_2,\ Ba(NO_3)_2,$$
$$K_2CrO_4,\ Cr(NO_3)_3,\ Zn(NO_3)_2,\ Na_2CO_3$$

When water is added to the mixture, a white precipitate, B, and a solution, C, are formed; solution C shows an acidic reaction toward litmus paper. The precipitate, B, is insoluble in 2 M H_2SO_4, but is dissolved by 6 M $NaOH$; there was no evidence of any reaction when B was treated with 2 M H_2SO_4. When the solution C is treated with an excess of 6 M $NaOH$, a colorless solution that smells strongly of ammonia is formed. On the basis of this information, indicate whether each solid is absent, present, or impossible to determine.

14.42 Treatment of a solid unknown with 0.3 M HCl gives a black precipitate, A, and a solution, B. The precipitate, A, after washing, dissolves in hot dilute HNO_3, leaving a small amount of brownish-yellow residue. Solution B, treated with excess $NaOH$, gives a green solution and no precipitate. Indicate which of the following substances are definitely present, which are definitely absent, and which are undetermined.

$$ZnS,\ PbCl_2,\ FeCl_3,\ CrCl_3,\ KHSO_4,\ HgS$$

14.43 A solution known to contain only Na^+ and K^+ as cations was treated in the following manner. To one portion of the solution, $AgNO_3$ was added; a yellow precipitate resulted. When $BaCl_2$ was added to a second part of the solution, the result was a white precipitate that dissolved in HCl. A third part of the solution decolorized a $KMnO_4$ solution. A fourth portion of the solution gave a brown coloration when treated with a mixture of concentrated H_2SO_4 and $FeSO_4$ solution. Indicate which of the following anions are definitely present in the solution, which are definitely absent, and which are undetermined.

$$Cl^-, I^-, SO_4^{2-}, PO_4^{3-}, NO_3^-, AsO_4^{3-}$$

14.44 A solid is known to be a mixture of two or more salts. It dissolves readily and completely in cold water to give a yellow solution. Separate samples of this yellow solution show the following behavior.

a. Addition of 0.2 M $AgNO_3$ forms a red precipitate.

b. Addition of 0.2 M $BaCl_2$ forms a yellow precipitate.

c. When acidified with 2 M HNO_3 the solution becomes orange-red in color. Nothing happens when 0.2 M $AgNO_3$ is added to this orange-red solution.

d. When acidified with 2 M HCl the solution becomes orange-red, but there is no sign of any other change. Nothing happens when 0.2 M $BaCl_2$ is added to this orange-red solution.

e. Addition of an excess of 5 M NH_4OH forms a white precipitate.

What anions are definitely shown to be present? What cations are definitely shown to be present? What additional anions, if any, could be present? What additional cations, if any, could be present?

14.45 A mixture of equivalent amounts of solid salts shows the following behavior.

a. When a sample of the mixture is treated with hot, concentrated H_2SO_4, a vapor (gas) is given off; this gas shows an acidic reaction toward litmus paper.

b. A sample of the mixture dissolves readily and completely in cold water to give a colorless, water-clear solution; this solution shows a basic reaction toward litmus paper.

c. When a sample of the solution prepared in (b) is treated with 0.2 M $AgNO_3$, nothing happens.

d. When a sample of the solution prepared in (b) is treated with 0.2 M $Ba(NO_3)_2$, a white precipitate forms.

What anions are definitely shown to be present? What other anions could be present? What cations are definitely shown to be present? What cations are definitely shown to be absent?

14.46 An unknown mixture of dry salts was known to consist of equivalent amounts of two or more of the substances listed below.

$$AgNO_3, ZnCl_2, K_2CO_3, MgSO_4,$$
$$Ba(C_2H_3O_2)_2, NH_4NO_3$$

When the solid was treated with an amount of water sufficient to yield a 0.4 M solution of any salt that dissolved, a precipitate, A, and a solution, B, were formed.

The precipitate, A, was washed with water. One sample of this washed precipitate was found to be completely soluble in dilute HNO_3. Another sample of the precipitate yielded a precipitate and a clear supernatant liquid when treated with dilute HCl.

Solution B yielded a precipitate when treated with excess NH_4OH. On the basis of the information above, indicate whether each substance is present, absent, or impossible to determine.

14.47 The following formulas represent salts, each of which may, or may not, be present in an unknown salt mixture. Those salts that are in the mixture are present in approximately equivalent amounts.

$$CuSO_4, FeSO_4, AgNO_3, (NH_4)_2SO_4, ZnSO_4,$$
$$Ba(NO_3)_2, Na_2CO_3, Na_2CrO_4, PbCl_2, Cr_2(SO_4)_3$$

When the unknown is treated with hot water, a white solid is left, which is separated and found to be insoluble in 3 M H_2SO_4, but soluble in 6 M NaOH. The hot water decantate is tested with litmus and found to be faintly acid. Addition of 6 M NaOH to the decantate gives a colorless solution with an odor of ammonia and a green precipitate. On the basis of the observations given, indicate whether a certain salt is present, absent, or undetermined.

14.48 An unknown mixture of dry salts was known to consist of equivalent amounts of two or more of the following substances.

$$Fe_2(SO_4)_3, ZnCO_3, Ag_3BO_3, NaI,$$
$$MgCO_3, Ba(NO_3)_2, PbBr_2$$

A sample of the solid unknown dissolved completely in cold, dilute HCl.

When cold water was added to the solid unknown, a dark, yellow-brown solution and a solid residue formed. The solid residue was separated from the solution; this solution was then extracted with CCl_4 and the CCl_4 layer became deeply colored.

A sample of the solid dissolved completely in dilute HNO_3. This HNO_3 solution was boiled for 1 min and was then cooled and treated with an *excess* of NH_4OH; a dark precipitate formed. This dark precipitate was removed and the solution that remained was treated with *excess* NaOH; a white precipitate formed.

On the basis of the information above, indicate whether a particular substance is present, absent, or impossible to determine.

14.49 A solid unknown is known to contain equivalent amounts of two or more of the following substances.

$$Pb(NO_3)_2, NH_4Cl, MgBr_2, Na_3PO_4, FeCl_3, Bi_2(SO_4)_3$$

The solid was readily and completely soluble in cold, dilute HCl.

When a sample of the solid was treated with cold water, a clear, colorless solution was formed.

When a sample of the solid was treated with cold, concentrated H_2SO_4, a brown gas that fumed in moist air was evolved.

When the solution formed by dissolving the solid in dilute HCl was boiled for 1 min, cooled, and then treated with NH_4Cl and excess NH_4OH, there was no precipitate; however, on further treatment with excess NaOH a white precipitate formed.

Indicate whether each of the substances is present, absent, or indeterminate.

14.50 A solid unknown is known to contain one or more of the following substances.

$$MgI_2, K_2CrO_4, ZnCO_3, CaCl_2, Fe_2(SO_4)_3,$$
$$AgNO_3, NiBr_2, Na_3AsO_4$$

The solid was completely soluble in water.

When a sample of the water solution was treated with excess NH_4OH, a clear, colorless solution, with no precipitate, formed.

When a sample of the water solution was treated with excess NaOH, a dark precipitate formed.

Indicate whether each of the substances is present, absent, or indeterminate.

14.51 You are given the following facts.

a. A cation, M^{2+}, forms the water-insoluble compounds MS (black), $MCrO_4$ (yellow), MSO_3 (white), and MSe (green).

b. All four of the compounds above are soluble in the halogen acid HX.

c. None of the compounds above are soluble in the acid HY.

d. MSO_3 and $MCrO_4$ are soluble in the acid HZ; MS and MSe are insoluble in the acid HZ.

e. When a mixture of MS and MSe is treated with NH_4OH, the MS dissolves, but the MSe is not affected.

f. When 1 M NaI is added to solid MSO_3, the precipitate changes in color from white to red. When 1 M NaI is added to solid $MCrO_4$, nothing is observed to take place.

On the basis of these facts, arrange the four insoluble compounds in the order of increasing solubility products.

14.52 A student prepared the following five solutions and placed each solution in a separate unlabeled beaker.

a. 0.1 M $ZnCl_2$

b. A solution 0.1 M in $Al(OH)_4^-$ and also containing Na^+ and OH^-.

c. A solution 0.1 M in SnS_3^{2-} and also containing NH_4^+.

d. A solution of 0.3 M in $Ag(NH_3)_2^+$ and 0.1 M in AsO_4^{3-} and also containing NH_3.

e. A solution 0.1 M in Sn^{2+}, 0.1 M in Bi^{3+}, 1.0 M in Cl^-, and 0.5 M in H^+.

While the student was gone from the laboratory, the five unlabeled beakers were moved and were placed in a disordered group at the end of the desk. In order to identify the solution in each beaker the student proceeded as follows:

She took the five disordered beakers and labeled them A, B, C, D, and E. She then tested the solution in each beaker, with these results.

Solution A, when treated drop by drop with 3 M HNO_3, gave a brown precipitate that dissolved in excess HNO_3 to form a clear solution.

Solution B, when treated drop by drop with 12 M HCl, gave a yellow precipitate that dissolved in excess HCl to form a clear colorless solution.

Solution C, when treated drop by drop with 5 M NH_4OH, gave a white precipitate that dissolved in excess NH_4OH to form a clear colorless solution.

Solution D, when treated with 6 M NaOH, gave a white precipitate that quickly turned black.

Solution E, when treated drop by drop with 6 M HCl, gave a white precipitate that dissolved in excess HCl to form a clear solution.

What is the composition of each of the solutions A, B, C, D, and E?

14.53 A solid unknown that is readily soluble in dilute HCl but insoluble in water is known to contain CrO_4^{2-}. How would you determine whether or not it also contains Cr^{3+}?

14.54 A solid unknown that is insoluble in water but soluble in dilute HCl is known to contain Cr^{3+}. How would you determine whether or not it also contains CrO_4^{2-}?

14.55 Using reactions employed in qualitative analysis, how could you form the following in the laboratory?

a. Metallic bismuth, starting with Bi_2S_3.

b. $FeCl_3$, starting with $Fe_3(SO_4)_3$.

c. Metallic mercury, starting with $Hg(NO_3)_2$.

d. Metallic antimony, starting with Sb_2S_3.

e. $KMnO_4$, starting with $MnCl_2$.

f. CrO_5, starting with $Cr(NO_3)_3$.

g. ZnS, starting with $ZnCO_3$.

h. Hg_2Cl_2, starting with HgS.

i. Na_2SnO_3, starting with metallic tin.

j. $CuCO_3$, starting with $CuSO_4$.

k. $SnCl_2$, starting with $Sn_3(PO_4)_4$.

Recording and Reporting Analysis

An acceptable suggested format, in the sense of clarity and ease of reading, used in recording and reporting the analyses of salts and salt mixtures follows and is illustrated by example. All solids analyzed should be reported in this manner, preferably in a bound notebook, or as specified by your instructor. A version of this format can be used for unknowns that are in solution.

REPORT OF THE ANALYSIS OF SAMPLE: _____ DATE: _____

I   **Physical Examination** _____

_____

II **Solubility** _____

_____

III **Analysis for Cations**

(a) Cations eliminated by the color of the solution obtained in **II**: _____

_____

(b) Cations eliminated by anion analysis: _____

_____

(c) Cations eliminated by solubility: _____

_____

(d) Results of blanket tests: _____

_____

(e) Results of cation analysis: _____

_____

IV **Analysis for Anions**

(a) Anions eliminated by color of solution obtained in **II**: _____

_____

(b) Anions eliminated by cation analysis: _____

_____

(c) Anions eliminated by solubility: _____

_____

(d) Results of H_2SO_4 treatment: _____

_____

(e) Results of $AgNO_3$ treatment: _____

_____

(f) Results of $BaCl_2$ treatment: _____

_____

(g) Results of specific tests: _____

_____

V   **Summary**

Cations present: _____

Anions present: _____

REPORT OF THE ANALYSIS OF SAMPLE: Sample 1   **DATE:** 2/14/05

I   **Physical Examination** White, crystalline, heterogeneous.

II   **Solubility** Soluble in cold water. Colorless solution. No gases evolved.

III   **Analysis for Cations**

(a) Cations eliminated by the color of the solution obtained in **II:** Cu^{2+}, Fe^{3+}, Co^{2+}, Ni^{2+}, Cr^{3+}; CrO_4^{2-} also absent.

(b) Cations eliminated by anion analysis: None; cation analysis performed before anion analysis.

(c) Cations eliminated by solubility: None.

(d) Results of blanket tests: Tests with NaOH and with NH_4OH showed absence of all cations except Na^+, K^+, NH_4^+, Ba^{2+}, Ca^{2+}, Mg^{2+}, and Zn^{2+}.

(e) Results of cation analysis: Zn^{2+} and Na^+ present.

IV   **Analysis for Anions**

(a) Anions eliminated by color of solution obtained in **II:** CrO_4^{2-}.

(b) Anions eliminated by cation analysis: AsO_4^{3-}, CrO_4^{2-}.

(c) Anions eliminated by solubility: SO_3^{2-}, CO_3^{2-}, PO_4^{3-}, BO_3^{3-}, S^{2-}, CrO_4^{2-}, AsO_4^{3-}.

(d) Results of H_2SO_4 treatment: Effervescence; gas was colorless, had sharp odor, and fumed in moist air; Cl^- indicated; Br^- and I^- absent, NO_3^- probably absent.

(e) Results of $AgNO_3$ treatment: White, curdy precipitate, insoluble in HNO_3. Cl^- present.

(f) Results of $BaCl_2$ treatment: White precipitate, insoluble in HCl. SO_4^{2-} present.

(g) Results of specific tests: Specific tests made for NO_3^- and $C_2H_3O_2^-$. NO_3^- and $C_2H_3O_2^-$ absent.

V   **Summary**

Cations present: Zn^{2+}, Na^+.

Anions present: SO_4^{2-}, Cl^-.

A Brief Review of Oxidation Numbers

The oxidation number is a numerical representation of the oxidation state of an atom in a chemical species. Oxidation is a process in which electrons are lost (Eq. I.1).

$$M \rightarrow M^{+n} + ne^-$$ (I.1)

Thus, the oxidation state of an atom is an indication of its electronic environment in a chemical species.

In assigning oxidation states to the various elements in a species or compound, we use the following conventions (or rules) that are applied in the order given:

1. All elements in the elementary state are assigned an oxidation state of zero.
2. The alkali metals (Li, Na, K, Rb, and Cs) in any compound are assigned an oxidation state of $+1$; the alkaline earth metals (Be, Mg, Ca, Sr, Ba, and Ra), and also Zn and Cd in any compound, are assigned an oxidation state of $+2$.
3. Oxygen is assigned an oxidation state of -2 in all of its compounds except the peroxides and superoxides (e.g., Na_2O_2 and KO_2).
4. Hydrogen is assigned an oxidation state of $+1$ in all of its compounds except the hydrides (e.g., NaH).
5. Oxidation states of the other elements in a species usually can be determined by the requirement that the algebraic sum of the oxidation state values for all of the elements in the species must equal the net charge on the species.

These rules have their origin in the atomic structures and electronegativities of the elements. However, it is important to realize that the oxidation state does not, in general, correspond to the actual charge on an element in a chemical species. Oxidation states are primarily a convenient bookkeeping device that is useful in balancing chemical reactions.

If a conflict arises, then the rule with the lower number takes precedence. In other words, apply the rules in the order given above until there is only one element left, which is then assigned an oxidation state consistent with the net-charge condition (rule 5).

The rules, as given, sometimes are not sufficient to cover *all* chemical species and, in such cases, it is best to work by analogy, using the periodic chart, with other compounds covered by the rules. In any case, rule 5 must always be satisfied.

Solution of Quadratic Equations

Equilibrium problems often involve the solution of quadratic equations as in example problems 3.4, 3.6, and 4.7 in this text. Since the solution to a quadratic equation can be generalized (see II.4 below), the numeric answer to such problems can be obtained readily using the following process. For example, in Chapter 4, the analysis of the solubility of TlCl in a solution containing 0.10 M KCl comes to the point in Eq. (4.20) where the solubility s appears in the quadratic equation.

$$0.10s + s^2 = 1.7 \times 10^{-4} \tag{II.1}$$

Rearranging Eq. (II.1) in the usual quadratic form,

$$ax^2 + bx + c = 0 \qquad (\text{for } a \neq 0) \tag{II.2}$$

we obtain

$$s^2 + 0.1s - (1.7 \times 10^{-4}) = 0 \tag{II.3}$$

The standard solution to the quadratic equation in (II.2) is of the form

$$x = \frac{-b \pm \sqrt{b^2 - 4ac}}{2a} \tag{II.4}$$

where a, b, and c are the coefficients of the quadratic equation when it is expressed in the form of Eq. (II.2). Thus, if we identify a with 1, b with 0.10, and c with -1.7×10^{-4}, which are coefficients in Eq. (II.3), the solution to Eq. (II.3) is obtained from Eq. (II.4).

$$s = \frac{-0.1 \pm \sqrt{(0.10)^2 - (4)(1)(-1.7 \times 10^{-4})}}{(2)(1)}$$

$$s = \frac{-0.1 \pm \sqrt{(0.010) + (6.8 \times 10^{-4})}}{2}$$

$$s = \frac{-0.1 \pm \sqrt{0.01068}}{2}$$

$$s = \frac{-0.1 \pm 0.1033}{2}$$

$$s = -0.1017 \quad \text{or} \quad 0.00165$$

Note that two solutions to a quadratic equation exist. Both are mathematically acceptable, but only one has physical meaning in the context of the problem at hand. Thus, the negative solution (-0.1017) has no meaning when we realize that s represents the solubility of TlCl; there is no physical meaning to a negative solubility. The physically acceptable solution is, thus, the positive one, 0.00165.

In Chapter 4 we made the assumption that $s \ll 0.10$ [Eq. (4.21)] in order to make the solution to the quadratic easier mathematically. Under this assumption, the solubility was estimated to be 1.7×10^{-3} molar [Eq. (4.22)]. The exact solution (1.65×10^{-3}) of the problem is, of course, obtained from the quadratic equation; however, the approximation gives a value, which is in error by about 3 percent. Often solutions to equilibrium problems are not required to a high degree of accuracy, and it becomes a value judgment whether a solution with a 3 percent error is sufficient, or if it is necessary to engage in the extra mathematical manipulations to obtain a more exact solution.

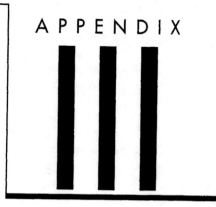

Reagents and Equipment Used

All liquid reagents and known solutions should be kept on the shelves in small bottles (about 100 mL) provided with stoppers fitted with medicine droppers. Solid reagents (except the aluminum and zinc) should all be finely powdered or pulverized and should be kept in 250-mL, wide-mouth bottles provided with one-hole No. 8 rubber stoppers fitted with spatulas for dispensing the solid. The handle of the spatula is thrust through the hole of the stopper in such a way that, when the bottle is stoppered, the metal tip of the spatula will be inside the bottle and partly buried in the powdered solid.

ACIDS

Acetic dilute, 5 M. Dilute 287 mL of glacial acetic acid with water to 1 liter.
Hydrochloric special concentrated, sp. gr. 1.18–1.20, 12 M.
Hydrochloric concentrated, 6 M. Add 1 volume of 12 M HCl to 1 volume of water.
Hydrochloric dilute, 2 M. Add 1 volume of 12 M acid to 5 volumes of water.
Nitric concentrated, sp. gr. 1.42, 16 M.
Nitric dilute, 3 M. Dilute 188 mL of 16 M acid with water to 1 L.
Sulfuric concentrated, sp. gr. 1.84, 18 M.
Sulfuric dilute, 2 M. Add 1 volume of 18 M acid to 8 volumes of water.

BASES

Aqueous ammonia concentrated, sp. gr. 0.90, 15 M.
Aqueous ammonia dilute, 5 M. Add 1 volume of 15 M NH$_4$OH to 2 volumes of water.
Barium hydroxide saturated solution.
Sodium hydroxide 8 M. Dissolve 356 g of solid NaOH in water and dilute to 1 L.

SALTS AND OTHER REAGENTS

Aluminon 1 g of the ammonium salt of aurintricarboxylic acid in 1 L of water.

Ammonium acetate $NH_4C_2H_3O_2$, 1 M. 77 g per liter of solution.

Ammonium chloride NH_4Cl, 2 M. 107 g per liter of solution.

Ammonium molybdate $(NH_4)_2MoO_4$. Dissolve 20 g of MoO_3 in a mixture of 60 mL of distilled water and 30 mL of 15 M NH_4OH. Add this solution slowly and with constant stirring to a mixture of 230 mL of water and 100 mL of 16 M HNO_3.

Ammonium oxalate $(NH_4)_2C_2O_4$, 0.2 M. 25 g per liter of solution.

Ammonium sulfate $(NH_4)_2SO_4$, 0.2 M. 26 g per liter of solution.

Ammonium sulfide $(NH_4)_2S$. Add 1 volume of the reagent-grade ammonium sulfide liquid (manufactured by the leading chemical companies) to two volumes of water *or* saturate 5 M NH_4OH with H_2S gas.

Ammonium thiocyanate NH_4SCN. Saturated solution in ethyl alcohol.

Barium chloride $BaCl_2 \cdot 2H_2O$, 0.2 M. 49 g per liter of solution.

Chlorine water Saturated solution.

Dimethylglyoxime $(CH_3)_2C_2(NOH)_2$. Dissolve 10 g in 1000 mL of 95 percent alcohol.

Diphenylthiocarbazone paper Soak filter paper with a concentrated solution of diphenylthiocarbazone in acetone or ethyl alcohol. Allow to dry; then cut into small rectangular pieces.

Disodium hydrogen phosphate Na_2HPO_4, 0.2 M. 28 g per liter of solution.

Hydrogen peroxide H_2O_2, 3 percent solution.

Iron(III) nitrate $Fe(NO_3)_3$, 0.2 M. 48 g per liter of solution.

Iron(II) sulfate $FeSO_4 \cdot 7H_2O$, 1 M. 278 g per liter of solution. Place clean scraps of iron in the solution and acidify with a few milliliters of 2 M H_2SO_4.

Lead acetate $Pb(C_2H_3O_2)_2 \cdot 3H_2O$, 0.2 M. 78 g per liter of solution. Add 10 mL of 5 M acetic acid per liter of solution.

Magnesia mixture Dissolve 55 g of $MgCl_2 \cdot 6H_2O$ and 140 g of NH_4Cl in 500 mL of water. Add 131 mL of 15 M NH_4OH and dilute with water to 1000 mL.

Magnesium reagent Dissolve 0.1 g of *p*-nitrobenzeneazoresorcinol [called also 4-(*p*-nitrophenyl-azo)resorcinol or 2,4-dihydroxy-4'-nitroazobenzene] in 1000 mL of 0.025 M $NaOH$.

Mercury(II) chloride $HgCl_2$, 0.1 M. 27 g per liter of solution.

Potassium chromate K_2CrO_4, 0.2 M. 38 g per liter of solution.

Potassium cyanide KCN, 0.2 M. 13 g per liter of solution.

Potassium hexacyanoferrate(II) $K_4Fe(CN)_6 \cdot 3H_2O$, 0.2 M. 84 g per liter of solution.

Potassium hexacyanoferrate(III) $K_3Fe(CN)_6$, 0.2 M. 66 g per liter of solution.

Potassium nitrite KNO_2, 0.2 M. 17 g per liter of solution.

Potassium permanganate $KMnO_4$, 0.02 M. 3.2 g per liter of solution.

Potassium thiocyanate $KSCN$, 0.2 M. 19 g per liter of solution.

Silver nitrate $AgNO_3$, 0.2 M. 34 g per liter of solution.

Sodium acetate $NaC_2H_3O_2 \cdot 3H_2O$, 0.2 M. 27 g per liter of solution.

Sodium carbonate Na_2CO_3. Saturated solution.

Sodium hexanitritocobaltate(III) $Na_3Co(NO_2)_6$. Dissolve 10 g of $Co(NO_3)_2 \cdot 6H_2O$ in a mixture of 200 mL of distilled water and 52 mL of 5 M $HC_2H_3O_2$. Then, add 100 g of $NaNO_2$, mix well, and allow to stand 24 h before using.

Sodium thiosulfate $Na_2S_2O_3 \cdot 5H_2O$, 2.0 M. 496 g per liter of solution.

Thioacetamide CH_3CSNH_2, 1 M. 75 g per liter of water solution. Prepare in small quantities as needed. Solid reagent and solutions should be refrigerated.

Tin(II) chloride $SnCl_2 \cdot 3H_2O$, 0.2 M. Dissolve 45 g of the salt in 500 mL of 6 M HCl and dilute to 1 L. Keep in a well-stoppered bottle containing a few pieces of granulated tin.

Zinc uranyl acetate Add 10 g of uranyl acetate, $UO_2(C_2H_3O_2)_2 \cdot 2H_2O$, to 5 mL of 5 M acetic acid. Heat just short of boiling for 5 min or until solid is dissolved. Then, dilute to 50 mL with water and, if necessary, heat until the solid is completely dissolved. Add 30 g of zinc acetate, $Zn(C_2H_3O_2)_2 \cdot 2H_2O$, to 5 mL of 5 M acetic acid and heat just short of boiling for 5 min. Dilute with water to 50 mL and, if necessary, heat until the salt is completely dissolved. The two solutions are mixed to give a clear solution. Add about 0.2 g of NaCl and let stand 24 h. Decant the clear solution for use.

ORGANIC LIQUIDS

Carbon tetrachloride CCl_4.
Ether, diethyl $(C_2H_5)_2O$.
Alcohol, ethyl C_2H_5OH.
Alcohol, methyl CH_3OH.

SOLIDS

Aluminum wire 26 gauge, cut to 25-mm lengths.
H_2S source material Bulk mixture of 5 parts by weight of asbestos shreds, 25 parts by weight of finely shredded paraffin, and 15 parts by weight of finely ground sulfur. Sold commercially under the trade name Aitch-Tu-Ess.
Potassium peroxydisulfate $K_2S_2O_8$.
Sodium bismuthate $NaBiO_3$.
Sodium fluoride NaF.
Sodium peroxide Na_2O_2.
Sodium thiosulfate $Na_2S_2O_3 \cdot 5H_2O$.
Zinc granulated.

KNOWN SOLUTIONS

The silver group known solution is approximately 0.1 M with respect to silver, mercury, and lead. The other three known solutions are approximately 0.2 M with respect to each metal in the group except lead, which, because of the limited solubility of $PbCl_2$, is only 0.035 M, and NH_4^+, which—to increase the strength of the test—is 0.6 M.

1. **Silver group** 28 g of $HgNO_3 \cdot H_2O$, 17 g of $AgNO_3$, and 33 g of $Pb(NO_3)_2$ per liter of solution. Place the solid $HgNO_3 \cdot H_2O$ in a 2-L beaker, add 50 mL of 16 M HNO_3; then, add water in 200-mL portions with frequent stirring, to give a volume of about 950 mL. Allow to stand, with frequent stirring, until the $HgNO_3 \cdot H_2O$ is dissolved. Then, add the $AgNO_3$ and $Pb(NO_3)_2$ and dilute to 1 L with water. Stir until the solids are completely dissolved.
2. **Copper-arsenic group** 10 g of $PbCl_2$, 62 g of $Na_2AsO_4 \cdot 7H_2O$, 54 g of $HgCl_2$, 46 g of $CdCl_2 \cdot 2\frac{1}{2}H_2O$, 34 g of $CuCl_2 \cdot 2H_2O$, 63 g of $BiCl_3$, 46 g of $SbCl_3$, and 70 g of

$SnCl_4 \cdot 5H_2O$. Place the mixture of solids in a 2-L beaker and add 492 mL of 12 M HCl and 330 mL of water, to give a total volume of about 1 L. Stir until all solids are dissolved, warming if necessary.

3. **Aluminum-nickel group** 27 g of $ZnCl_2$, 40 g of $MnCl_2 \cdot 4H_2O$, 48 g of $NiCl_2 \cdot 6H_2O$, 54 g of $FeCl_3 \cdot 6H_2O$, 48 g of $CoCl_2 \cdot 6H_2O$, 53 g of $CrCl_3 \cdot 6H_2O$, and 48 g of $AlCl_3 \cdot 6H_2O$. Place the mixture of solids in a 2-L beaker and add 50 mL of 12 M HCl; then, add 740 mL of water, to give a total volume of about 1 L. Stir until the solids are completely dissolved.

4. **Barium-magnesium group** 12 g of NaCl, 15 g of KCl, 32 g of NH_4Cl, 29 g of $CaCl_2 \cdot 2H_2O$, 49 g of $BaCl_2 \cdot 2H_2O$, and 41 g of $MgCl_2 \cdot 6H_2O$. Place the mixture of solids in a 2-L beaker; add 10 mL of 6 M HCl; then, add 850 mL of water, to give a total volume of about 1 L. Stir until solids are completely dissolved.

UNKNOWN SOLUTIONS

The unknown solutions should be prepared in the same concentrations as the known solutions and the same salts should be used.

SOLIDS FOR EXPERIMENTS 1 AND 4

Zinc sulfide; borax or sodium metaborate; and the nitrate, sulfate, sulfite, carbonate, chromate, chloride, bromide, iodide, acetate, phosphate, and arsenate of sodium.

SOLUTIONS FOR EXPERIMENTS 2, 3, AND 4

0.2 M solutions of the sodium salts of each of the 13 anions ($C_2H_3O_2^-$, NO_3^-, Cl^-, Br^-, I^-, S^{2-}, SO_4^{2-}, SO_3^{2-}, CO_3^{2-}, CrO_4^{2-}, BO_2^-, PO_4^{3-}, AsO_4^{3-}).

SOLUTIONS OF CATIONS AS "KNOWNS"

It is desirable to have 0.2 M solutions of each of the cations available.

LIST OF EQUIPMENT NEEDED

(This equipment can be purchased in standard semimicro size at all laboratory supply houses.)

Beaker cover, aluminum, for 100-mL beaker, punched to hold 10-mm test tubes.
Beakers, 50, 100, 150 mL.
Burners, standard size, with rubber tubing.
Casserole, 20 mL, 45 mm porcelain.
Centrifuge, four-tube, to carry 10 × 75 mm test tubes. One centrifuge for each three to six students working in a laboratory at one time.
Cork stoppers, tapered.

File, three-cornered, small.

Flasks, 125 mL, Erlenmeyer.

Forceps, metal.

Glass tubing, 6 mm.

Graduated cylinder, 10 mL.

Hydrogen sulfide bubbling tubes, 125 mm.

Hydrogen sulfide generator (central dispensing system or desk-style generators). (See Figure 10.2 and the description thereof.)

Matches.

Medicine droppers and medicine-dropper bulbs.

Ring stand, rings, and clamps.

Spatula, semimicro style, nickel or stainless steel.

Sponge.

Stirring rods, 3×125 mm.

Test tube brush, semimicro.

Test tube holder, wire.

Test tube rack, wood, plastic, or metal, for 10-mm tubes.

Test tubes, 10×75 mm and 16×150 mm.

Thermometer, 110°C.

Towels.

Watch glass, 50 mm.

Wing top for burner.

Wire gauze, 100-mm square.

Index to Procedures

Index

Index